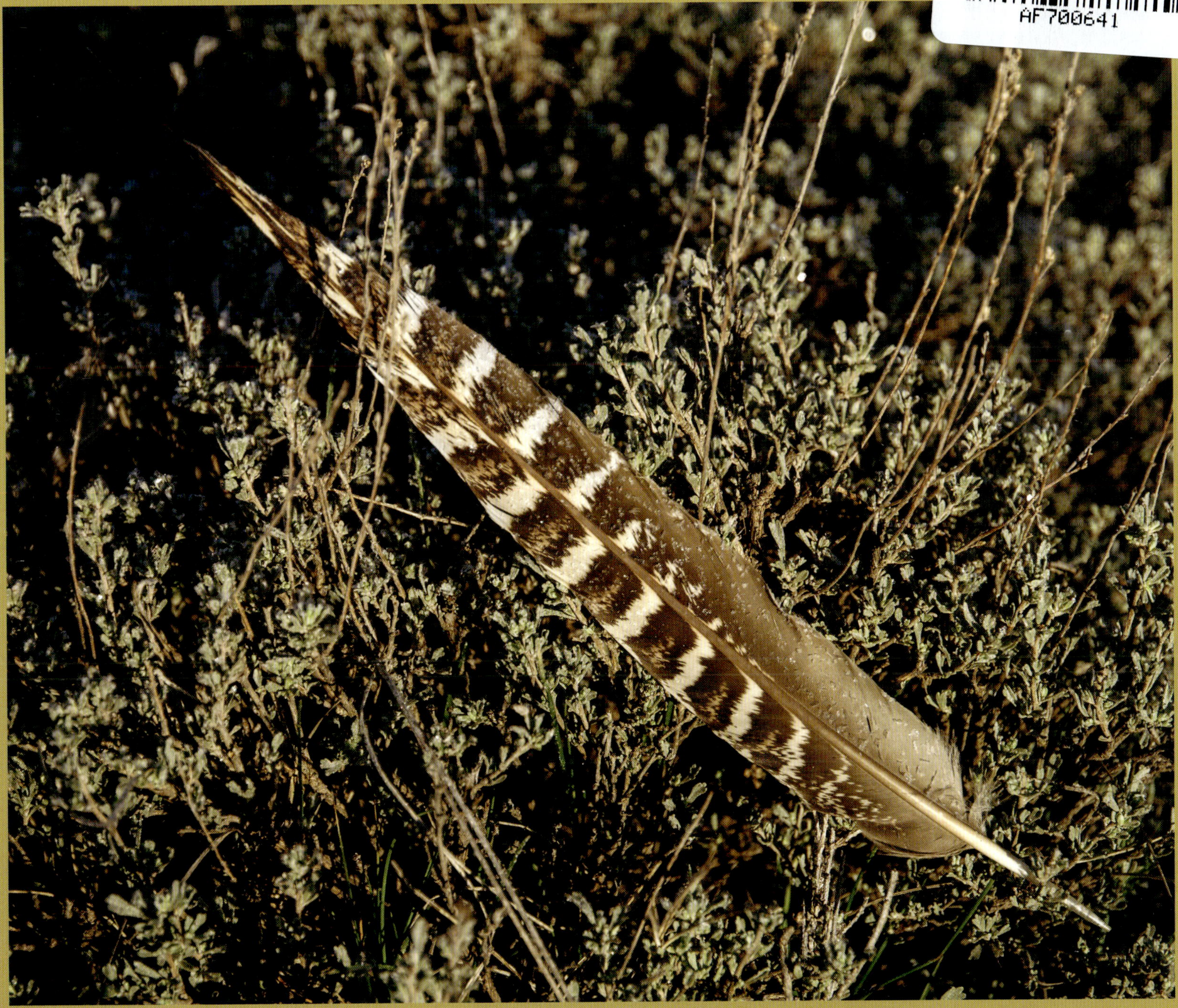

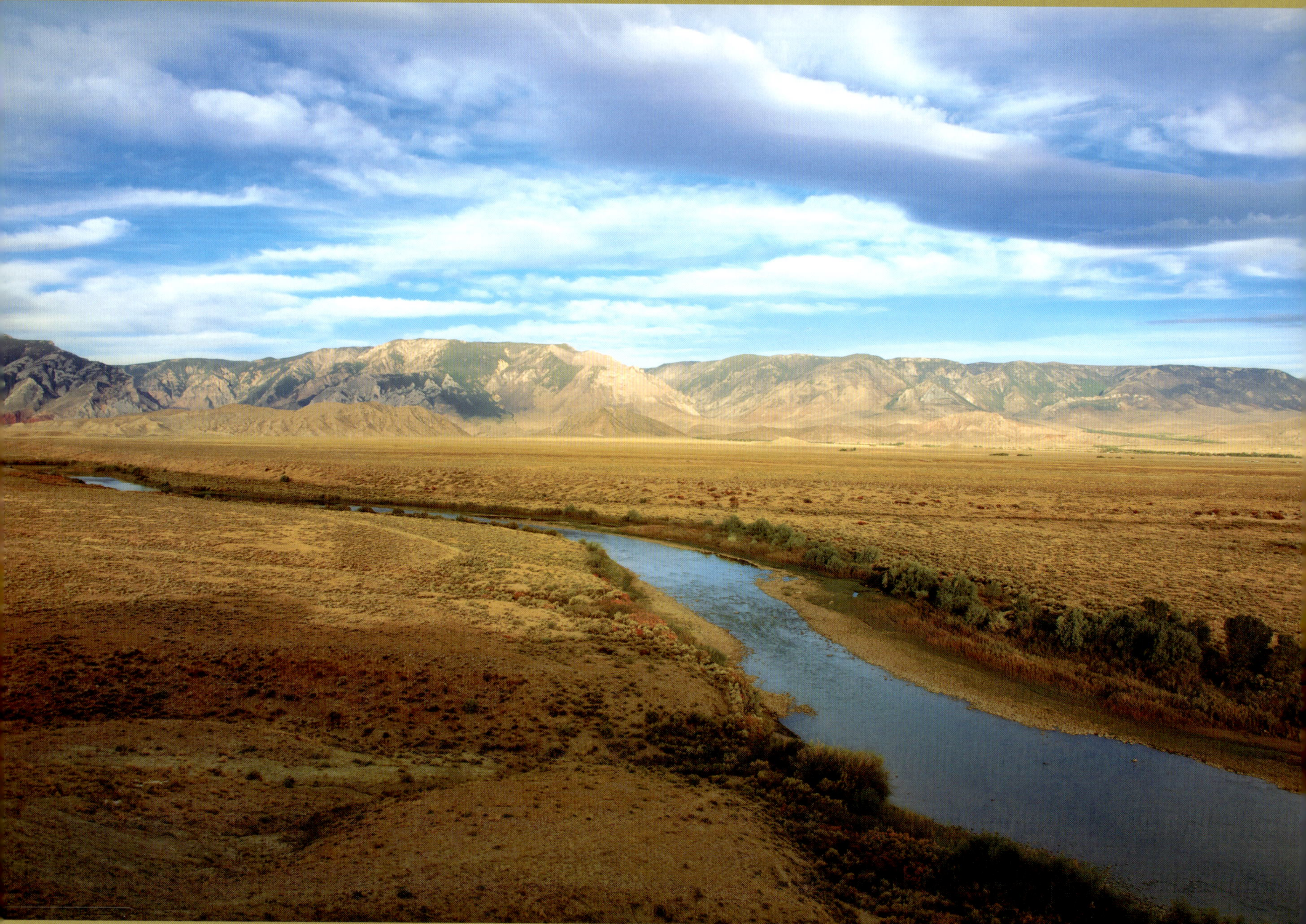

The American West
at a Crossroads

sage spirit

Dave Showalter
Foreword by David Allen Sibley

BRAIDED RIVER

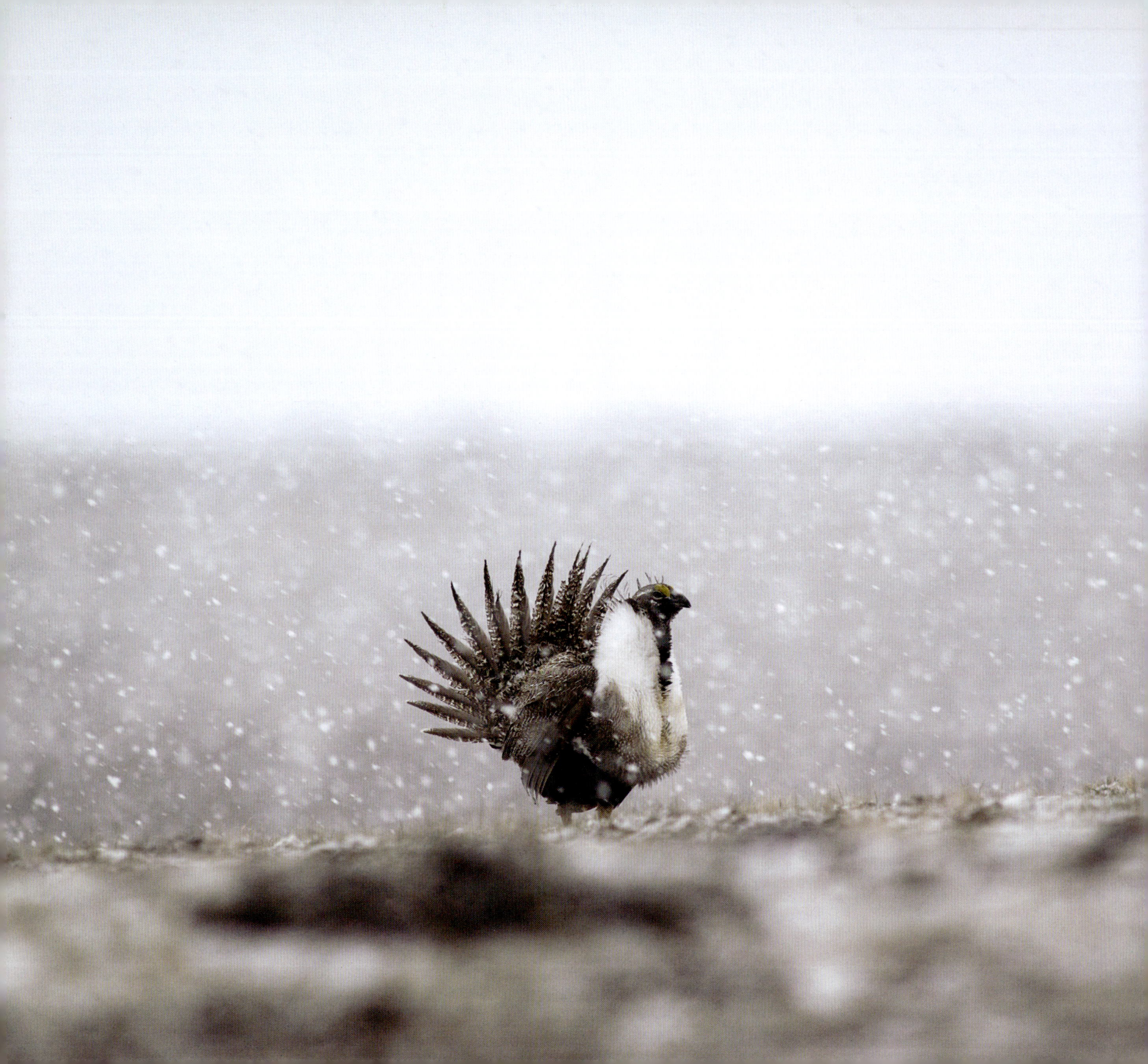

CONTENTS

MAPS

The Sagebrush Sea and the Sage Spirit Project

The sagebrush sea reaches across eleven states, as well as into Canada and Mexico. The Sage Spirit project focuses on critical habitat in the eastern portion of the ecosystem.

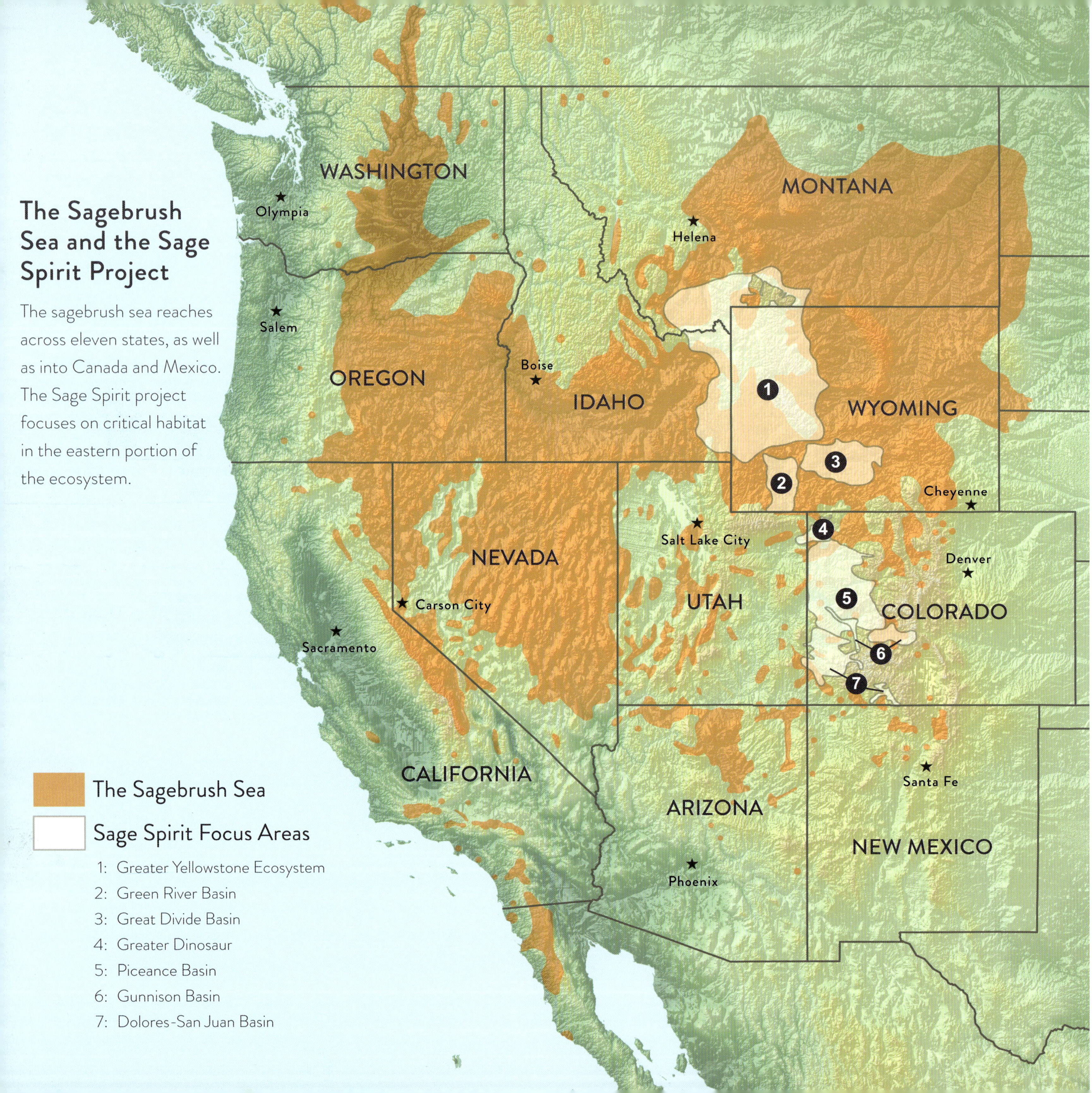

The Nexus of Habitat and Energy Development

Together, Montana, Wyoming, and Colorado harbor 60 percent of the sage-grouse population. These same states possess massive coal and gas reserves.

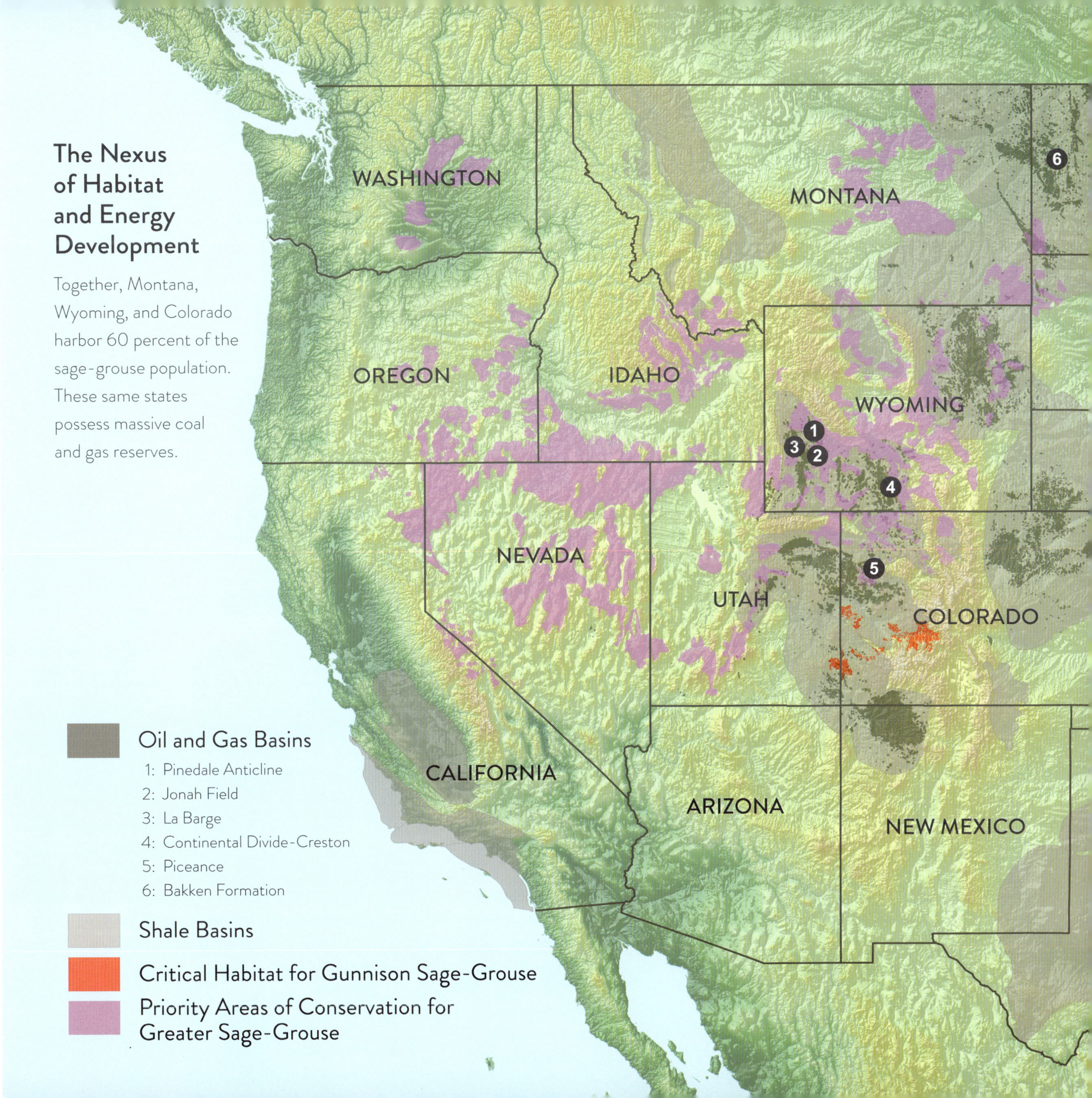

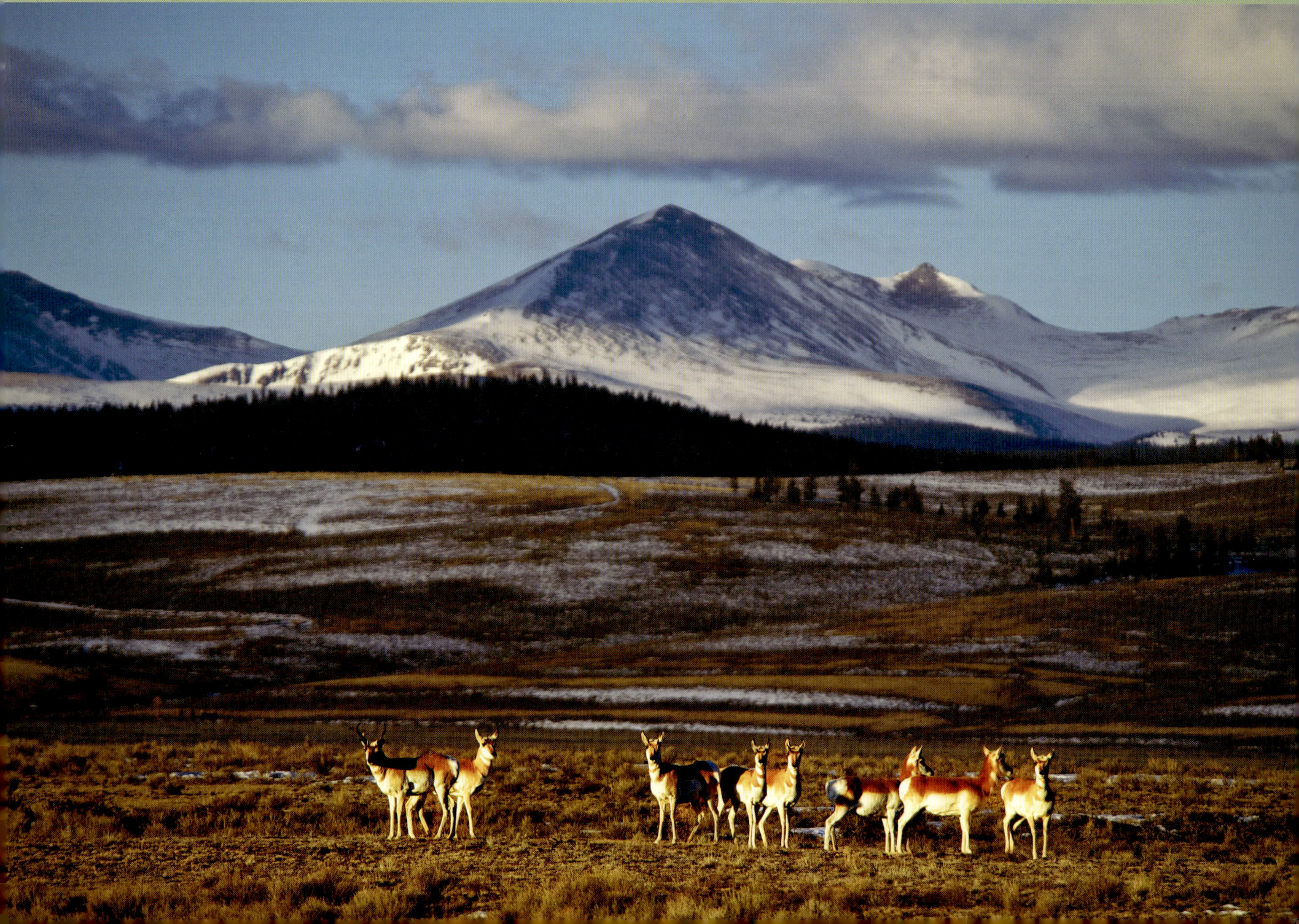

PUBLISHER'S NOTE
INCLUSIVE CONSERVATION
Helen Cherullo

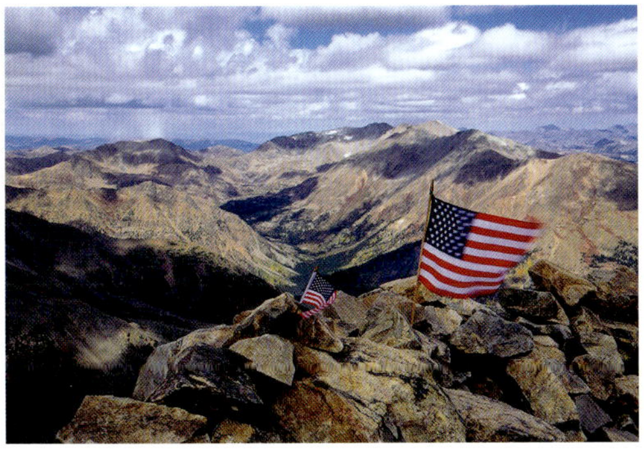

In the American West, there are still places where pronghorn migrate hundreds of miles as they have for untold generations, since the time of the Pleistocene.

WITH CAMERA IN HAND and an open mind, photographer Dave Showalter has explored the semi-arid sagebrush terrain of Montana on through Wyoming and Colorado to New Mexico for the last twenty-five years. In places, he photographed a natural world still rich with many of the same species encountered in the early nineteenth century by Lewis and Clark. But around the next bend, he captured a different picture: land blighted by careless development, overtaken by invasive species, or compromised by industrial-scale resource extraction.

As Dave met with people along the way, he came to know a country and way of life that is quickly waning. Yet he encountered those who are willing to wrestle with these problems without resorting to divisiveness and unworkable dogma. Colorado rancher Heather Stirling told Dave, "It's really important for my peace of mind to have harmony with people and animals. This land, this place is very healing and spiritual."

Dave also spent hours waiting for and watching the greater sage-grouse and even rarer Gunnison sage-grouse, species that became emblematic to him of current conditions in this vast sage country. Due to the fragmentation and degradation of its home ground, the sage-grouse's numbers have declined from conceivably as many as 16 million birds

ABOVE American flags on Mount Elbert five days after the 9/11 attacks. Elbert, at 14,431 feet (4398 meters), is the highest peak in the Rocky Mountains, the crown of the Sawatch Range. The west side of the Sawatch feeds the Roaring Fork, Eagle, and Gunnison Rivers—all tributaries of the Colorado River.
OPPOSITE A small pronghorn herd in the Cochetopa Hills are backed by 13,983-foot (4262-meter) Stewart Peak. Pronghorn in Gunnison Basin, Colorado, were decimated by deep snow in the winter of 2007–2008, then augmented by Colorado Parks and Wildlife; they have recovered well since.

Barrel-racing horse and rider dig for the finish line at the Cody Night Rodeo in Cody, Wyoming.

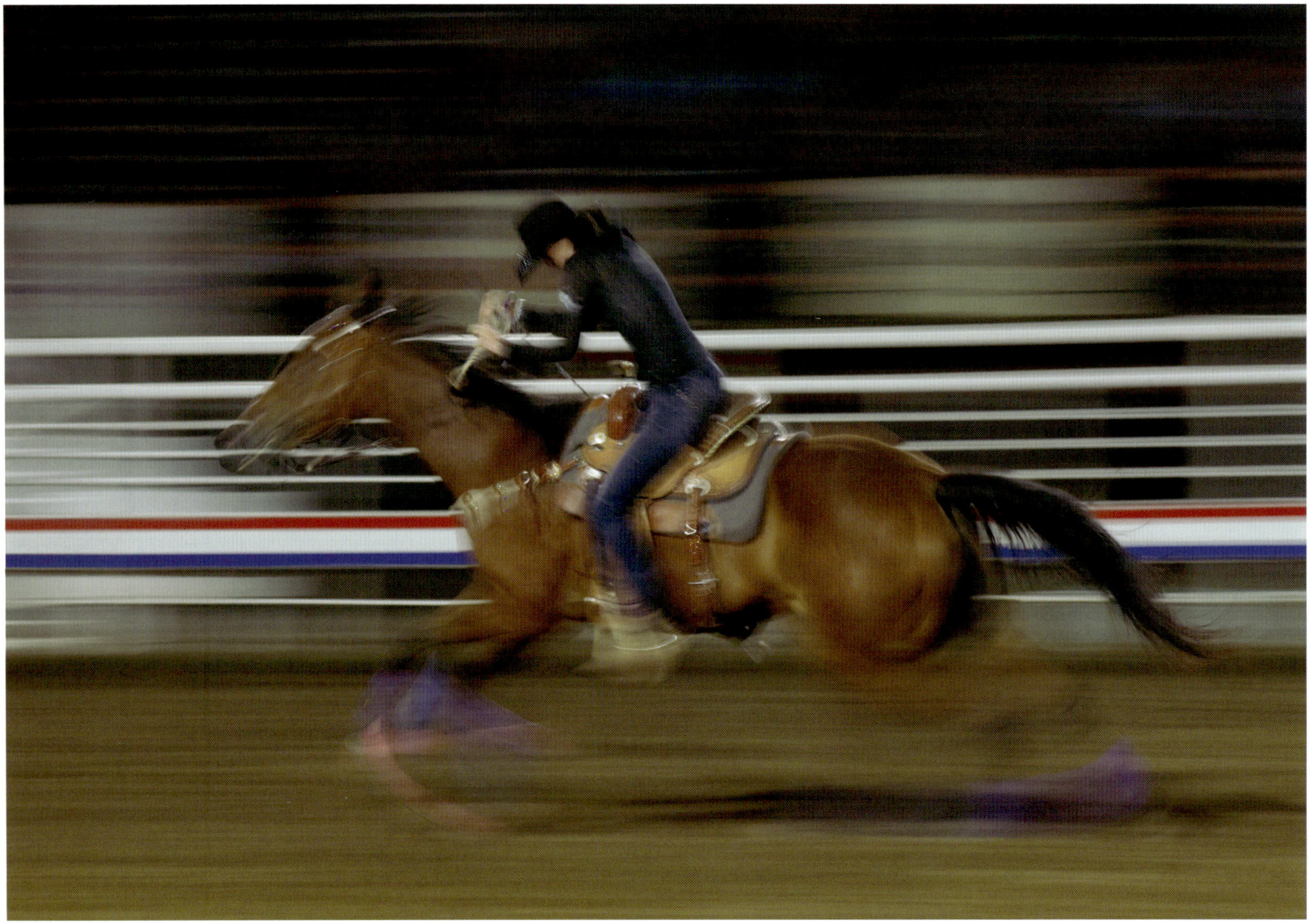

to between 200,000 and 500,000 for the greater sage-grouse—all in the span of a little more than 200 years. Gunnison sage-grouse numbers are even more alarming: fewer than 5000 birds.

The ever-diminishing numbers of sage-grouse and the array of pressures put on their habitat have led to serious consideration of their being federally listed as an endangered species. The possibility has met with emotions ranging from shock to defiance—and no clear-cut consensus along the usual partisan lines. The possibility of a federal listing startled people not only who live in the area but throughout the country. The underlying reasons include 150 years of unsustainable land management practices and the growing need for energy and acquiring it by any means. The introduction of "fracking"—a new way of extracting oil from shale deposits via hydraulic fracturing—places America at a crossroads in determining what kind of country we want to be and what condition we want to leave it in for future generations. The judgment that befalls the sage-grouse and many other western species, and conclusions reached about the extent of resource extraction, will ripple through the West—and beyond.

In addition, original arguments for "energy independence" have escalated. No longer trying just to satisfy America's thirst for energy, corporations now want to find a way to deal with the current surplus. Plans to expand infrastructure to export even more resources—including resources on public lands—will compromise the health and integrity of this land for short-term financial gains.

These are complicated, thorny, and contentious issues—not easy or simple to resolve. Climate change, carbon pollution, the

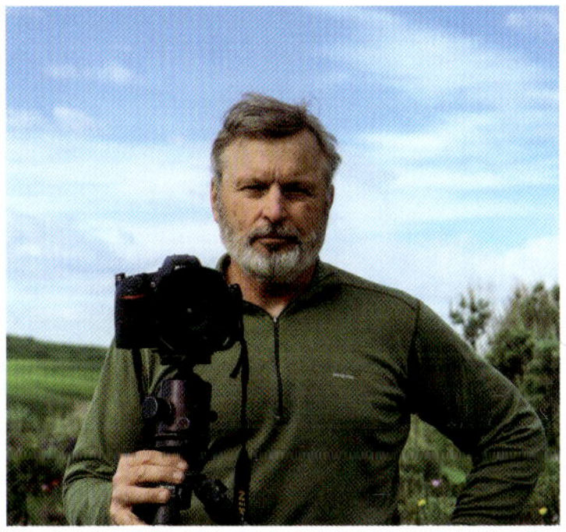

What do I want the West to be?
Through my photography, I am working for a sustainable West, one where future generations can also enjoy vistas free of drill pads and powerlines, hillsides replete with wildlife and birdsong, and the sweet aroma of sage.
—Dave Showalter

A brilliant autumn sunrise over McCullough Peaks Wilderness Study Area in Park County, Wyoming. McCullough Peaks is situated in the Bighorn Basin, home to a diverse range of endemic sagebrush-ecosystem wildlife and the highest concentration of golden eagles in North America.

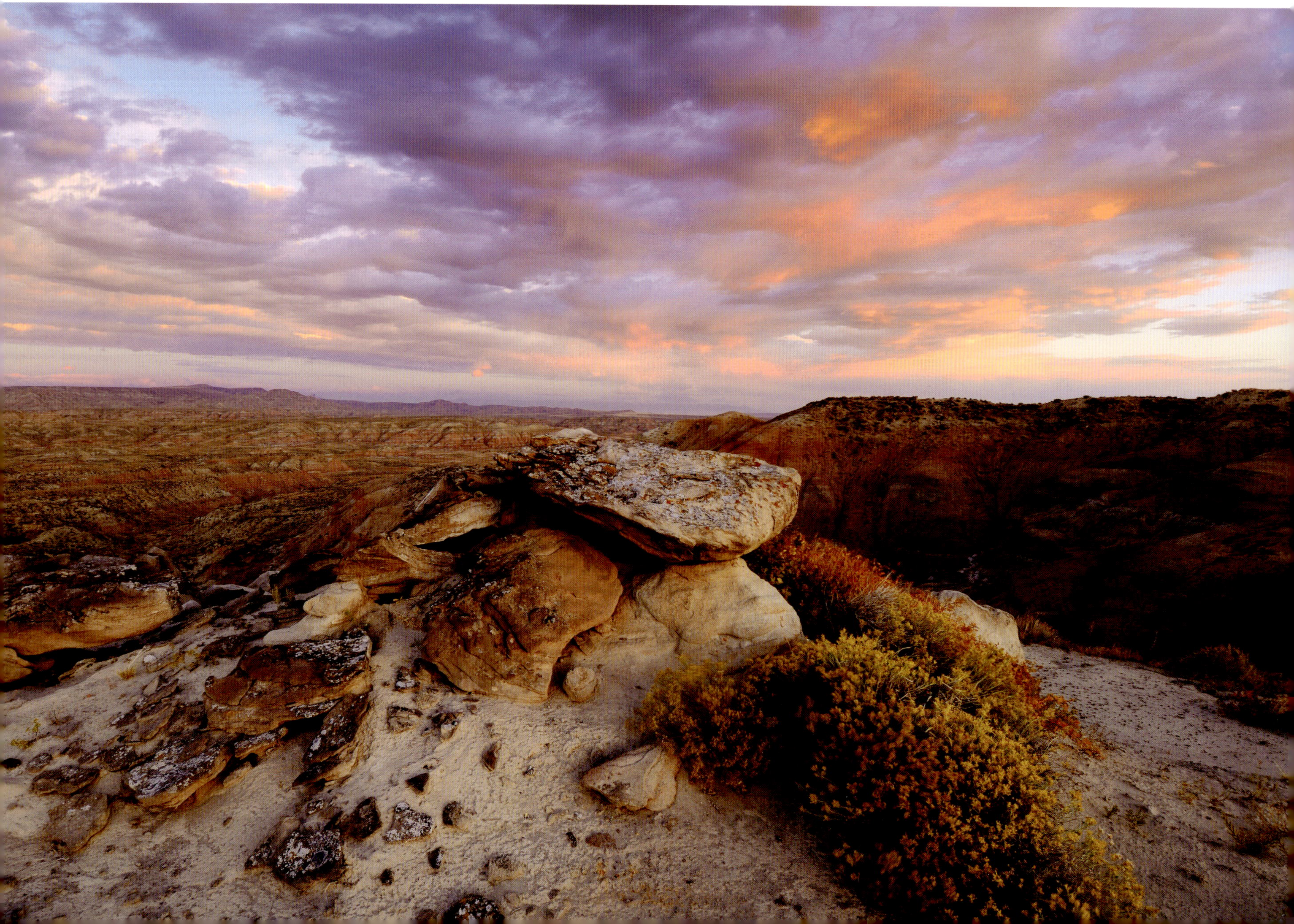

With this book, we invite you to envision a future where the western way of life can flourish and American innovation trumps short-term exploitation.

biodiversity crisis, global population, and energy issues all must be addressed. In a newfound inclusive approach to conservation, the people and partnerships involved at first seemed unlikely, especially given the contentious political landscape in America today. But in his travels, Dave found people working on solutions built on small steps taken close to home.

He talked with and photographed ranchers and environmentalists, business owners and hunters, land managers and politicians, energy developers and birders. Many of these people established common ground in a shared love of the land and an appreciation that the well-being of every species (including humans) depends on healthy expanses of sage in order to thrive.

Dave's desire to foster civil conversation brought him to Braided River. This book lays out the foundation for a much larger "Sage Spirit" campaign and commitment to ensure that the balance of the natural world that comprises the sagebrush sea is not destroyed in a quest for short-lived profits that ignore a healthy environment—not only for the sage-grouse but for everyone and every living creature that calls this place home.

The tactics from decades ago that protected the Pacific Northwest's northern spotted owl are out of place here and now—and in many ways, the resolution of the listing of the sage-grouse will provide the next chapter in land preservation in America today. Working with groups such as Audubon—which has identified the eastern sagebrush sea as a key flyway for migrating birds—will establish the groundwork for greater public dialogue, media exposure, and engagement of elected representatives. Innovative solutions are sought that go beyond the convenience of industry and address permanent protected habitat and unbroken migration corridors.

This campaign is about even more than the sage-grouse. The well-being of this bird reflects the overall health of the entire sagebrush environment. This matchless landscape is defined by generations of ranching families, epic wildlife, and the wide-open spaces emblematic of the great American West. The recovery of the sage-grouse will directly benefit the ecology—as well as the economy—of a uniquely western way of life. Braided River and Dave Showalter, along with steadfast grassroots partners, are committed to these principles and the work that lies ahead.

Photographer Dave Showalter extensively explored territory critical for the survival of the sage-grouse—the eastern segment of an eleven-state western ecosystem known as the sagebrush sea. This vast expanse of land includes parts of North Dakota, South Dakota, Colorado, Montana, Wyoming, Utah, Idaho, Nevada, Washington, Oregon, and California. It is divided by the Rocky Mountains, encompassing grasslands, deserts, mountains, rivers, valleys, and forests—held together by life that is connected and interdependent. An American Serengeti. §

FOLLOWING PAGES A male greater sage-grouse displays for the attention of females on a *lek* (mating ground) south of Pinedale, Wyoming.

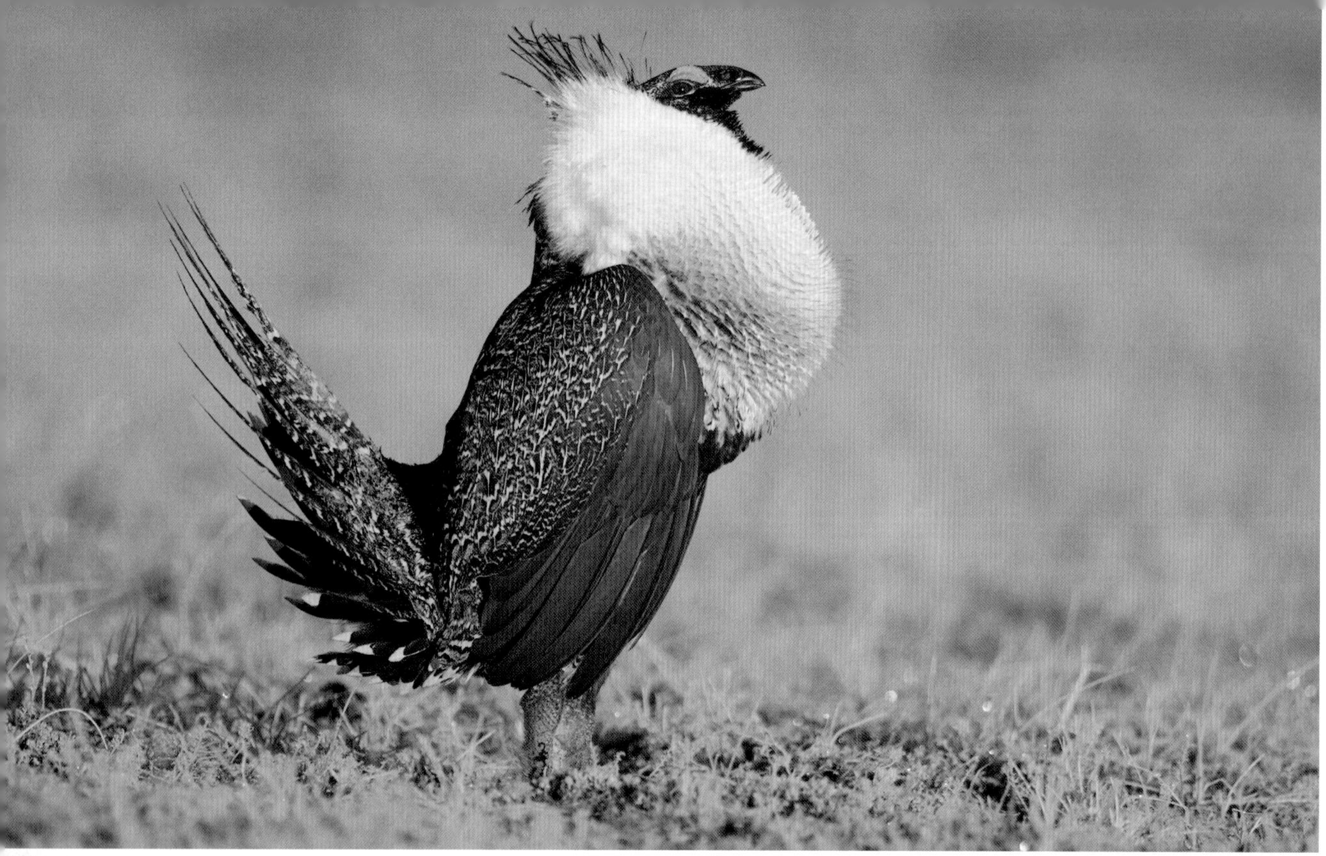

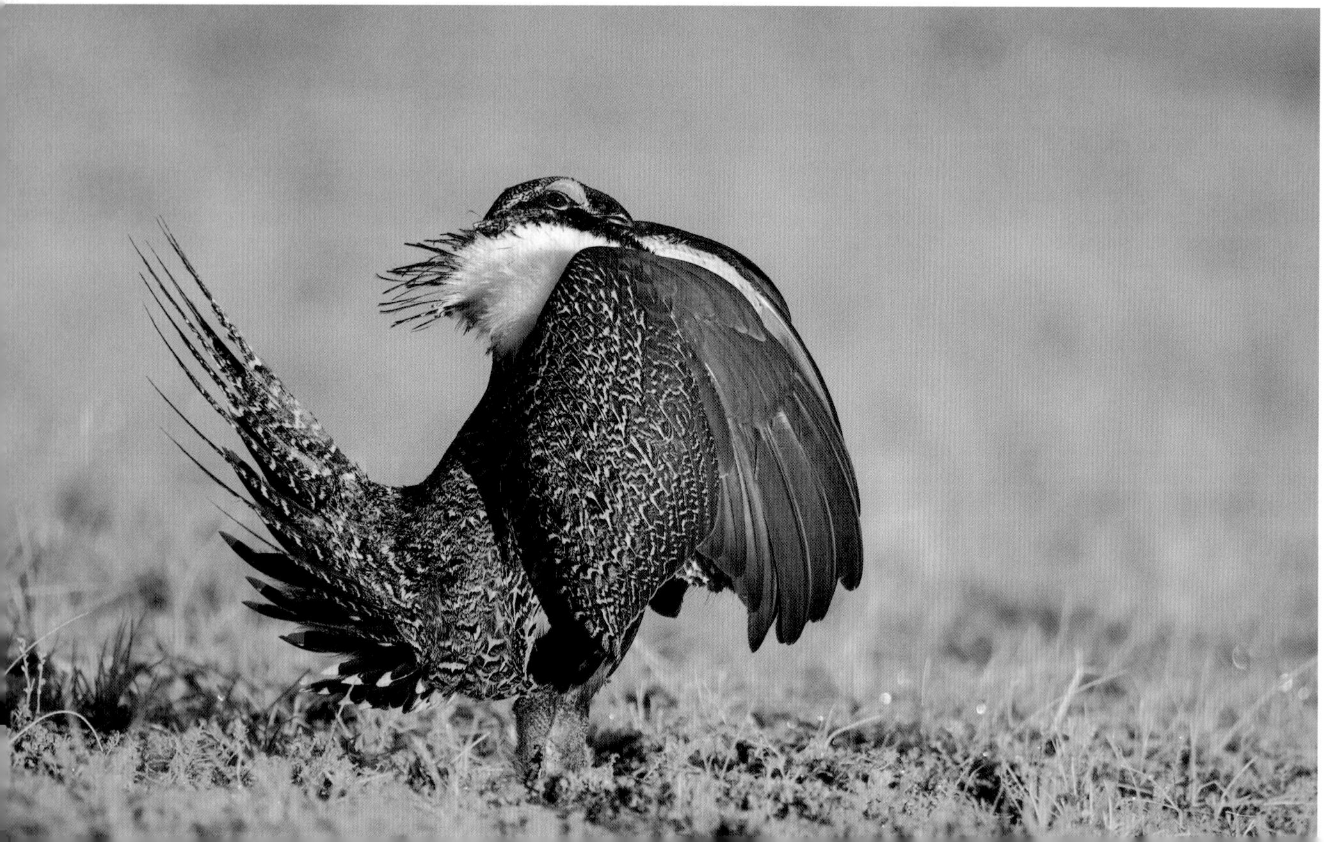

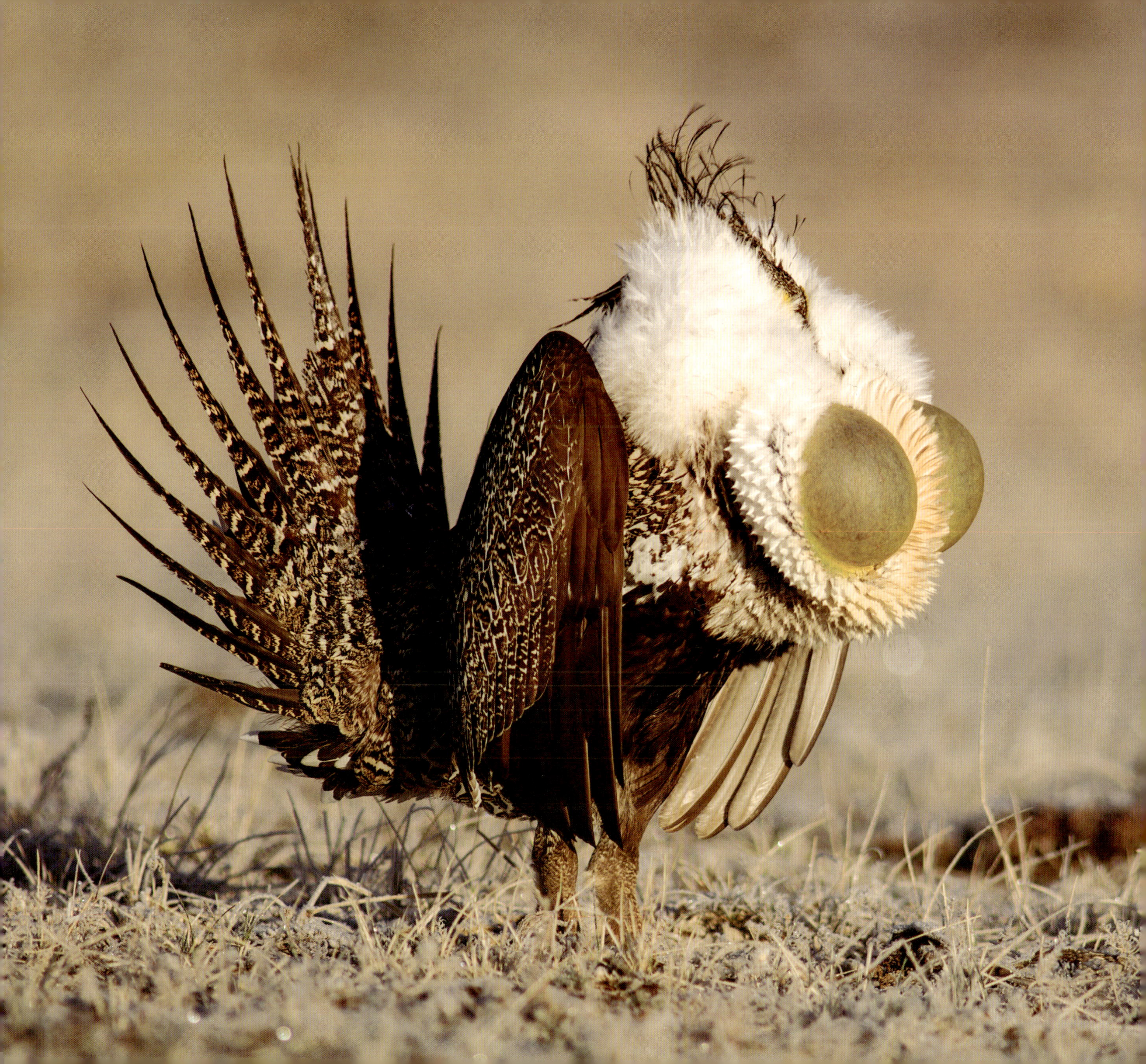

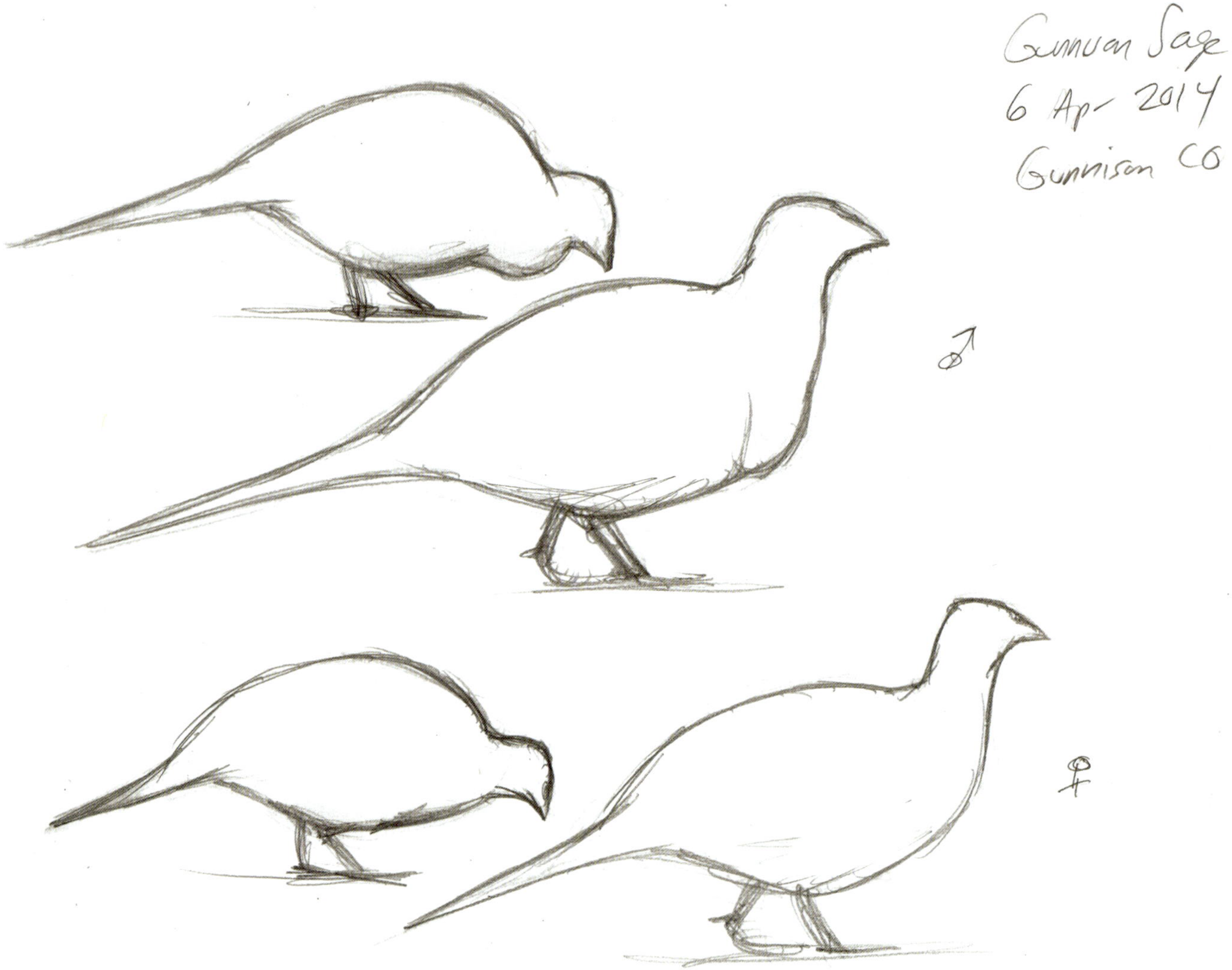
Gunnison Sage-Grouse
6 Apr 2014
Gunnison CO
♂
♀

FOREWORD

David Allen Sibley

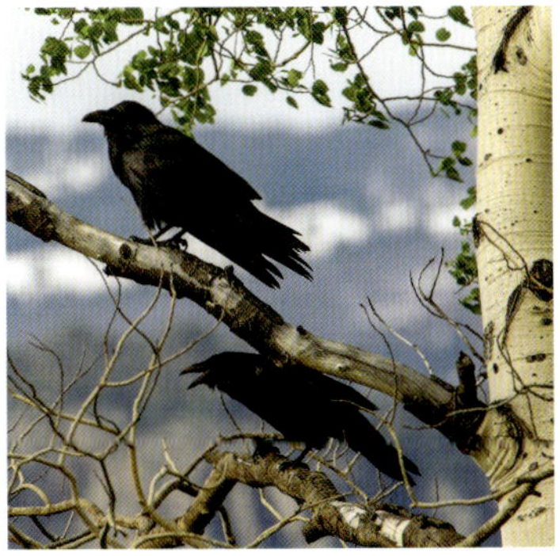

I HAVE BEEN BIRDING for nearly half a century. Birding became my single focus at a young age, and I have spent most of my life traveling back and forth across North America with the goal of seeing every species of bird known from the continent. Keeping a "life list" was a good way to track my progress, and as the list filled in, finding the missing species became a quest that guided much of my travel.

No bird lives in isolation, of course. They are all connected, so even as I focused on finding one species, I knew that my quest would always take me to rich areas filled with other birds. A trip in search of sage-grouse, for example, would lead me deep into sage-brush country, where some new experiences with mountain bluebird, Brewer's sparrow, Swainson's hawk, or any number of other species would surely happen.

Seeing a new bird and marking that space on my life list was just the beginning for me. I wanted to get to know each species, to understand their habitats, learn their voices, meet them at different seasons—and I tried to record all of that in my notes and sketches.

By the time my field guide to birds was published in 2000, I had seen all but three of the bird species that breed in North America. By 2010 I had seen two of those three: the island scrub-jay, found only on remote Santa Cruz Island off Southern California, and

ABOVE A pair of ravens in a Lamar Valley aspen tree in Yellowstone National Park, Wyoming. Ravens are opportunistic hunters and sage-grouse predators that take advantage of artificial perches and open dumps; thus humans have contributed to widespread increases of raven populations across the West.
OPPOSITE Sketches from David Allen Sibley's first sighting of the Gunnison sage-grouse

the whiskered auklet, found only in the central Aleutian Islands in Alaska and reached by flying about 1500 miles (2400 kilometers) west from Anchorage and then spending two days on a boat in the North Pacific.

It may seem strange that the quest to complete my list of North American breeding birds brought me, in April 2014, not to some remote outpost at the edge of the continent but to the sage-filled Gunnison Basin in southwestern Colorado.

I had spent a few weeks birding around Gunnison, Colorado, in the late 1980s, but back then everyone thought there was just one species of sage-grouse across the western United States. The isolated population of a few thousand grouse around Gunnison was an outlier geographically, but the biological differences that make this a distinct species were not then known. During the 1990s, researchers started taking a closer look at the Gunnison birds and found differences in body size, coloration, displays, and DNA. The new species—Gunnison sage-grouse—became official in 2000. During the entire twentieth century, only three new species of birds had been described from continental North America, the most recent in 1939, and then in 2000 there was Gunnison sage-grouse, a new species found in the heart of the United States!

Among many things the Gunnison sage-grouse has in common with its close relative, the more widespread greater sage-grouse, is losing ground in the face of human alterations to the landscape. About a century ago, sage-grouse were so numerous around Gunnison that hunters boasted of shooting eight hundred in a day. Intensive surveys of courtship *leks* (mating grounds) each spring now find a *total* of about eight hundred mature male Gunnison sage-grouse and the entire population is estimated at under five thousand individuals, making it one of the rarest bird species in the world.

The population of greater sage-grouse is about thirty times larger than that of Gunnison sage-grouse, spread out across eleven states and two Canadian provinces, but the scale of its decline in the last century is similar.

And yet, despite the fact that sage-grouse are reduced to a tiny fraction of their former numbers, the moments when we see them—a winter flock rocketing across a deep blue sky, a hen with tiny chicks picking insects from the new grass, a lek at dawn in early spring—are still fulfilling experiences. We don't see the overall plight of sage-grouse across the West; we see one bird, in one place, and in that moment our view is the same as that of a person one hundred or five hundred years earlier.

There is something magical about it. The lighting, the smells, the weather, and the birds can be a perfect moment and a link to the past all at once.

On my recent trip to Gunnison, biologist Pat Magee met me at 4:00 a.m. and we drove out near the lek site where male grouse were expected to display. We sat and talked quietly in the truck, in full darkness, with the moon and stars overhead, patches of snow on the ground, and the air temperature 19 degrees Fahrenheit (-7 degrees Celsius). When a faint glow was visible on the eastern horizon, we started to see the dark shapes of Gunnison sage-grouse walking onto the open lek from the surrounding sagebrush, and then we could hear the weird popping sounds of their display, like dripping water, so different from the sounds of greater sage-grouse. For about twenty minutes I was engrossed, watching and listening, as about seventy grouse gathered and performed their courtship rituals. And then, just as the sun appeared over the horizon, something startled them (probably one of the coyotes patrolling the area), and they all flew and melted away into the sagebrush.

I wish it had lasted longer, but that twenty minutes was idyllic—and I will go back. There is something primeval and timeless about

Autumn snow covers Antelope Flats and the Teton Range in Grand Teton National Park, Wyoming.

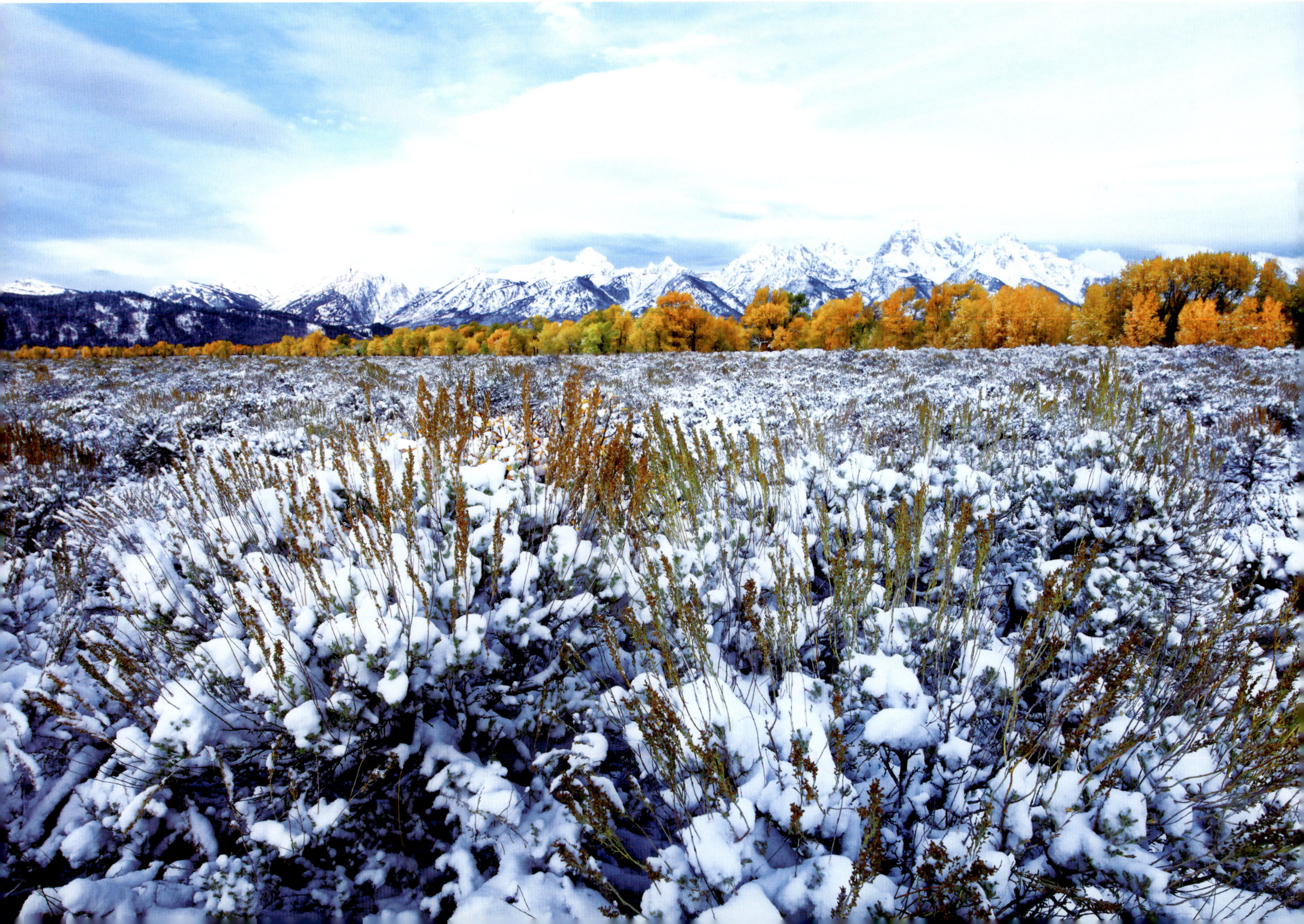

Displaying for a nearby female, this Gunnison sage-grouse male fans tail feathers that have broad white bands, which help distinguish the species from greater sage-grouse (Gunnison Basin, Colorado).

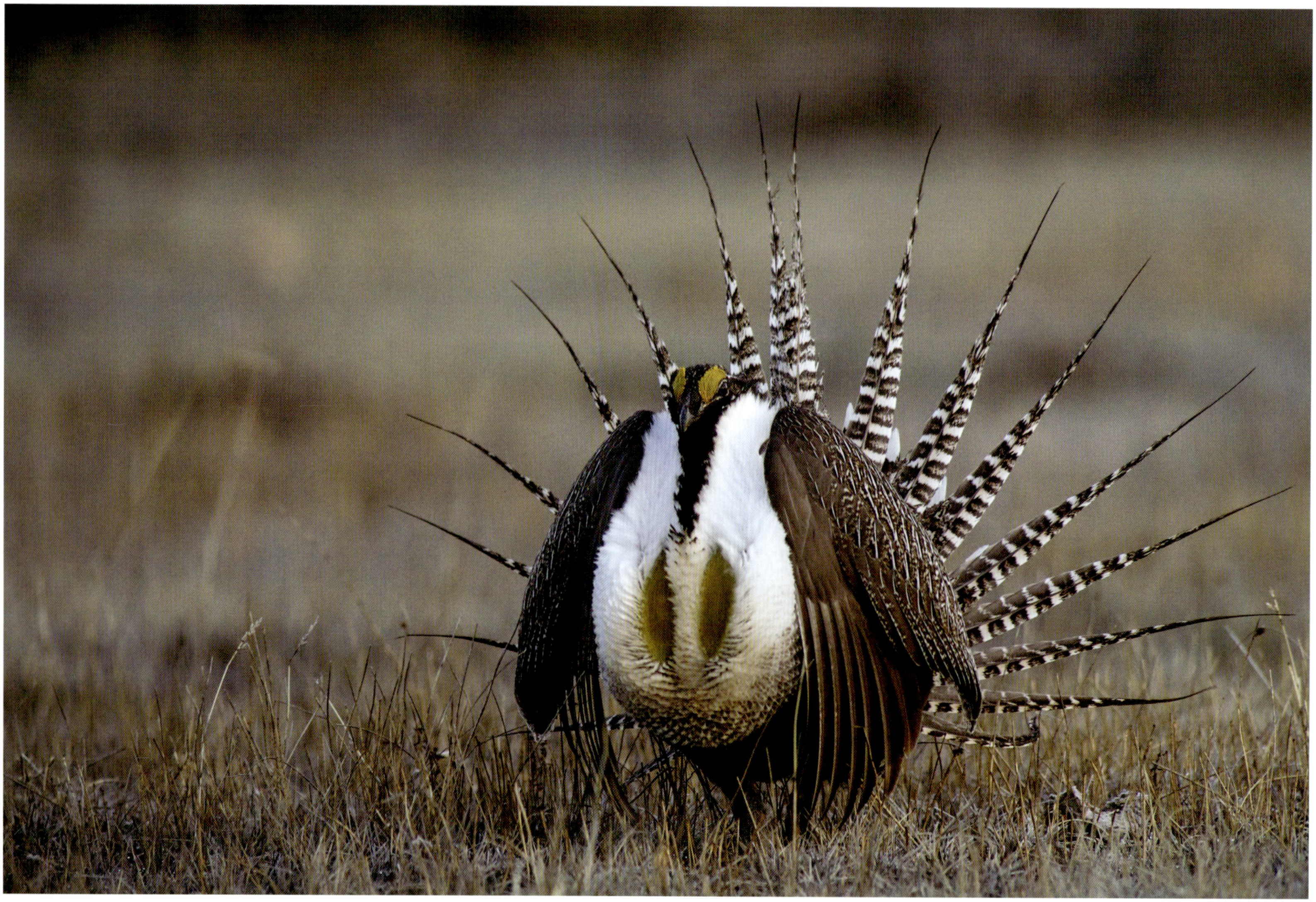

the courtship displays of sage-grouse and the first glow of dawn. It's easy to imagine people a thousand years ago being inspired by the same scene.

My personal quest to see Gunnison sage-grouse is trivial in the grand scheme of things, but it was deeply meaningful to me. The views, the smell and feel of the air, all reminded me of other birding experiences in the sagebrush when I was much younger. I remember looking out at sunrise across miles of blue-green sage on the edge of the Great Basin from a low ridge near Mono Lake, California, as the dawn chorus of Brewer's sparrows, sage thrashers, and gray flycatchers filled the cold, dry air like an auditory snowstorm. Or the family of long-billed curlews—two adults and two young chicks of these very large sandpipers—eating grasshoppers as they strolled, incongruously, across a grassy sagebrush plain at the Nature Conservancy's Matador Ranch in Montana.

The sagebrush plains of the western United States offer some of the wildest and most unaltered landscape views in the Lower 48. But more and more of it is being fragmented by powerlines, roads, and other development. Now much of the sagebrush is cattle-ranching and energy-extraction country, and the competitive American economic system affords little time or space for birds. But it is essential to find a balance. Economic booms will come and go, but the towns that dot the region will still be defined by the landscape and the sage.

The concept of an "umbrella species" is certainly apt here, and it goes beyond the purely ecological meaning—that conservation efforts focused on a particular species like sage-grouse will provide an "umbrella" that protects the entire ecosystem.

Sage-grouse can and do represent the habitat needs of many other birds, animals, and plants that also depend on the sagebrush. I see the umbrella extending beyond that, as all of those living things combine with the landscape and climate to create a sense of place—and the place supports a human culture and tradition and leads back to the whole span of history.

The book you hold in your hands is a celebration of the wild, intact sagebrush habitat. It is about the plants and animals that thrive there, because they are the foundation of what makes the region unique and special, and they are interwoven with the lives of every person who lives or visits there.

The sage-grouse is our link to all of these things. §

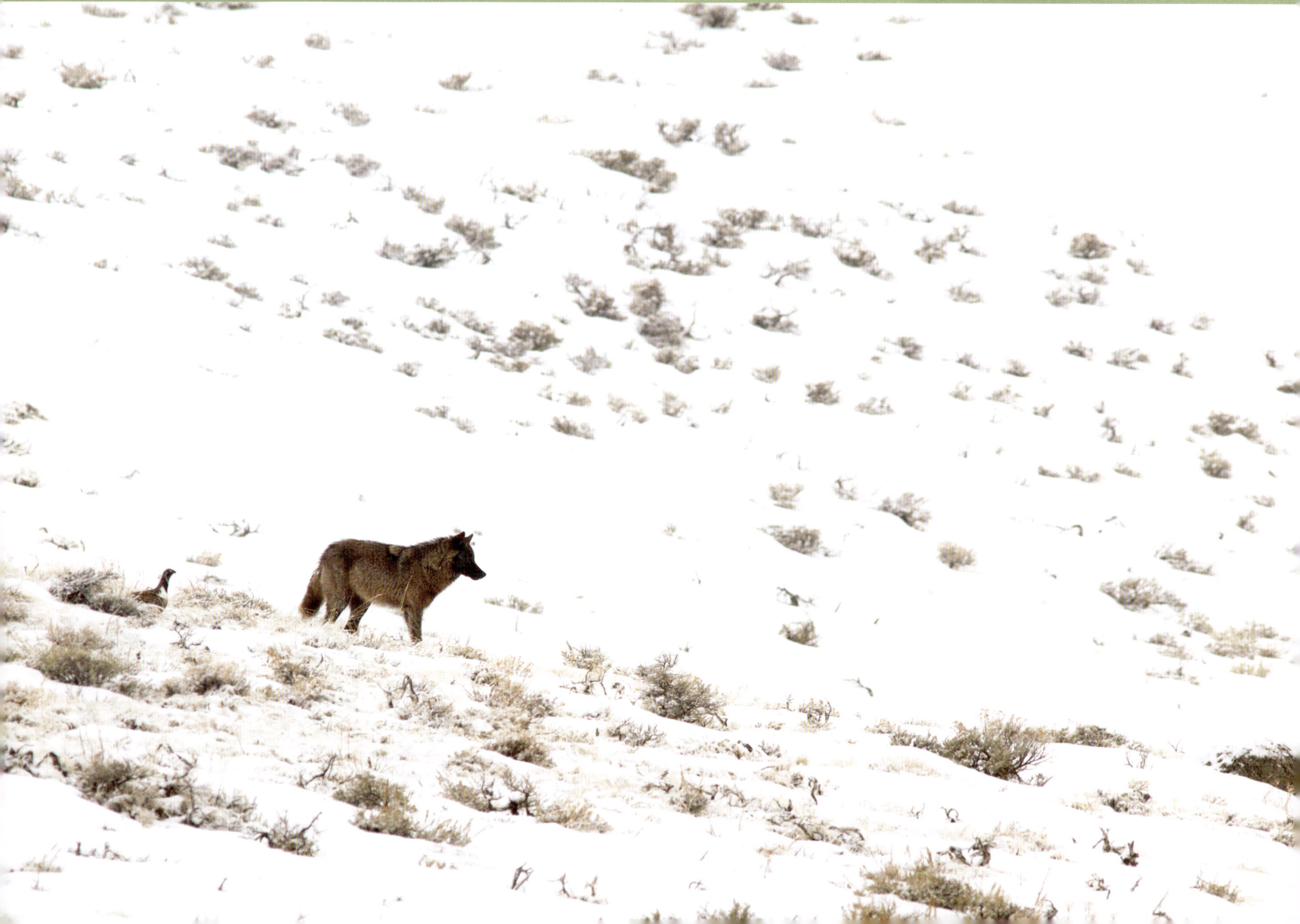

INTRODUCTION

SAGE SPIRIT: INTO THE HEART OF THE WEST

Dave Showalter

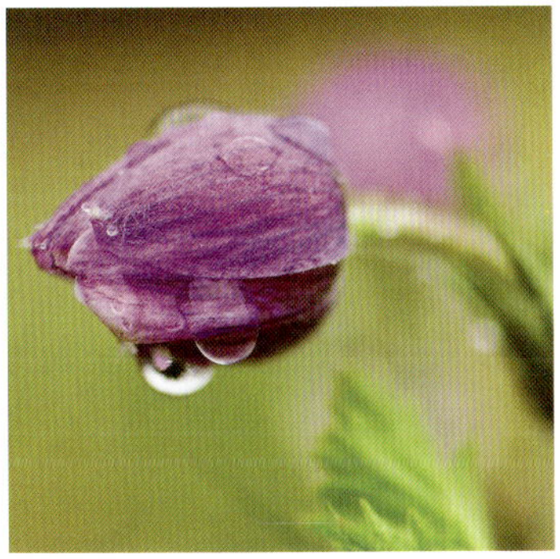

"One cannot be pessimistic about the West. This is the native home of hope. When it fully learns that cooperation, not rugged individualism, is the quality that most characterizes and preserves it, then it will have achieved itself and outlived its origins. Then it has a chance to create a society to match its scenery."

—Wallace Stegner, *The Sound of Mountain Water*

ONE MORNING IN JUNE 2004, my wife, Marla, and I were driving to the Front Range mountains from our home in Arvada, Colorado, and we saw a wolf hung up on the left-side guardrail, head turned toward us, eyes wide open. Stunned to see this beautiful wild creature, thought to be extinct in Colorado, dead by the side of the freeway, we wondered how it had gotten there—what was its story?

As we later found out, earlier that year a two-year-old female gray wolf designated Wolf 293F had traveled from Swan Lake Flats in Yellowstone National Park near Gardiner, Montana, in search of a mate. She shed her telemetry collar on her trek, so researchers had no idea what had happened to Wolf 293F until she was struck by a vehicle on Interstate 70 in Idaho Springs, Colorado, on that June morning.

If she had traveled in a straight line from Swan Lake Flats to Idaho Springs, Wolf 293F would have walked 450 miles (725 kilometers)—but she surely wandered much farther than that because wolves don't travel direct routes. Her meandering path would have linked some of the wilder sagebrush lands in this swath, traversing a large part of the eastern side of the American West's sagebrush ecosystem.

ABOVE A pasqueflower blooms in spring in Yellowstone National Park, Wyoming. OPPOSITE A gray wolf unknowingly flushes a greater sage-grouse on a wintry spring morning in Wyoming's Upper Green River Basin; these two imperiled species sharing the same space in a massive landscape.

A grizzly bear sow with two subadult cubs pauses while feeding on roots in spring sagebrush (Yellowstone National Park, Wyoming).

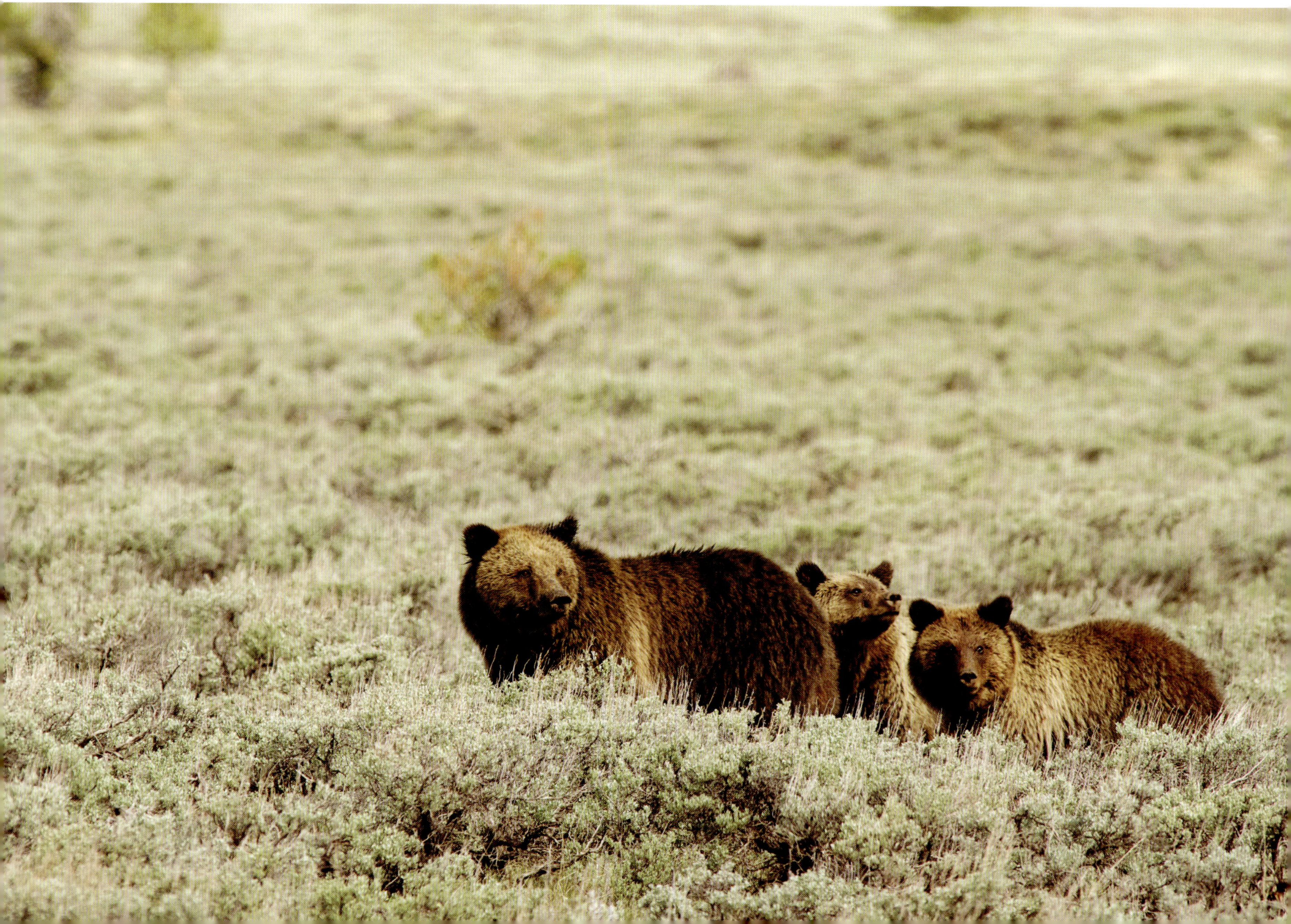

The impact of that moment, seeing the fate of a wolf that had navigated hundreds of miles of sagebrush terrain seeking a mate, is to me a fitting example of the potential of the sagebrush ecosystem to connect national parks, wilderness, and public lands that have unique ecological and cultural values. A wild creature sees no boundaries, only promise. The untamed spirit of the wolf and mystery of her journey, coupled with growing concern about a natural gas–drilling free-for-all in the sage, inspired the Sage Spirit advocacy project and the book you hold in your hands—built on the desire to conserve the heart of the American West.

The vast sagebrush ecosystem of the western United States is a thriving landscape so vibrant that it sustains virtually all western wildlife, including gray wolves like Wolf 293F, for at least part of each year. This ecosystem, often called the sagebrush sea, covers 166 million acres (67 million hectares), an expansive landmass of semi-arid lands divided in half by the north-south-running Rocky Mountains. The eleven western states that compose the ecosystem are Colorado, Wyoming, South Dakota, North Dakota, Montana, Idaho, Washington, Oregon, California, Nevada, and Utah.

The term "sagebrush sea" well describes a rolling gray-green landscape but suggests a featureless monoculture, which the sagebrush ecosystem is anything but. Anyone who's ever admired the Teton Range from Antelope Flats in Wyoming's Grand Teton National Park knows something about the sagebrush ecosystem. Antelope Flats is classic sagebrush habitat rimmed by glorious peaks, complete with roaming wolves and grizzly bears, a resident bison herd, moose traversing the Gros Ventre River corridor, greater sage-grouse, and pronghorn that migrate between summer and winter range on their ancient path. Tourists from around the world come to view all this wildlife that is free to roam a fully functioning sagebrush ecosystem in a relatively small section of a beloved national park.

The great migrations of pronghorn, mule deer, and elk flow through valleys covered with sagebrush east of the Rockies. Greater Yellowstone's grizzly bears, the highest concentration in the Lower 48, rely on the sagebrush ecosystem while roaming a large territory. In spring and fall, hundreds of bird species, a number of them biologically bound to sagebrush, migrate the Pacific Flyway and the Central Flyway on the eastern flank of the Rocky Mountains. To stand in open sagebrush on a late spring morning is to hear an unbelievably rich birdsong chorus beyond description.

But the sagebrush landscape that epitomizes the American West is rapidly disappearing. Songbirds and sage-grouse that breed and nest in sagebrush are proverbial canaries in the coal mine, indicators that this ecosystem is collapsing under the weight of development.

This book focuses primarily on the eastern sagebrush terrain that stretches from western Montana through Wyoming to southern Colorado: a land of cold rivers winding through deep canyons, rolling sagebrush-covered hills beneath serrated mountain ranges, high-desert buttes and bluffs, and plateaus that converge with aspen and conifer forest lapping at their edges. It's a complex, interconnected natural system. The east side of the ecosystem differs in a number of ways from the more western Great Basin of Nevada, Utah, southern Idaho, southeastern Oregon, and northeastern California, with its heart in Nevada's Basin and Range terrain. Washington State's sagebrush-steppe is found on the Columbia Plateau east of the Cascades.

Both east and west sides of the sagebrush sea are presented with unique challenges. Broadly, the east side of the Rockies faces heavier pressure to be developed, particularly by oil and natural gas drilling using hydraulic fracturing technology ("fracking") and coal-bed methane drilling for oil and natural gas. The more arid Great Basin is fighting an invasive cheatgrass incursion resulting from historically bad grazing practices, drought, and difficulty managing wild horses and burros. Conservation opportunity still exists in the eastern half of the sagebrush ecosystem, and some of the largest remaining populations of sage-grouse are concentrated in the sagebrush basins of Wyoming and northwest Colorado.

Growing up in Rockford, Illinois, I watched the Rock River flow while I fished for catfish and walleye along the riverbanks every chance I got. In the summer my folks hooked a pop-up camper to our car and took us somewhere; the trips out West are the most memorable, my sister and I pressed against backseat windows for each new view around the next turn. On one trip, my dad pulled into a primitive Montana campsite next to a burbling stream and we caught one trout after another with fly and bobber on spinning tackle. Back then you could just pull over, set up camp, and catch a mess of fish.

Later, while I attended the University of Wyoming in Laramie, where mountains, sagebrush, and shortgrass prairie intersect on the edge of the Snowy Range, sage began to imprint itself on me as I wandered my western surroundings. Over the last twenty-five years, with camera in hand, I have been launching western adventures from my Colorado home with Marla, exploring the prairie, sagebrush valleys, and Rocky Mountains that rim western valleys. Gradually, I began to think about ways to keep these places wild.

My original thought, in 2008, was to build a story that would advocate for the creation of a project to link the remaining wild sagebrush lands east of the Rockies, and central to the story would be the impact of habitat fragmentation and development on greater sage-grouse. It seems nearly possible, with a little imagination, that we could connect national parks, wilderness, and public and private lands with unique ecological and cultural values to conserve the heart of the West.

The project needed a starting point, so I visited a greater sage-grouse *lek* (mating ground). Here, on private lands north of Laramie, male sage-grouse strutted and made *bah-looping* sounds from puffed-up yellow air sacs, for hours, to impress a few smallish gray females. I had a long-distance view of thirty or so grouse across a pale gold landscape beneath a pastel blue sky, the April sun offering no warmth. I was with an Audubon Rockies group, most of us seeing the grand sage-grouse spectacle for the unforgettable first time. I was hooked, and the book that you hold in your hands was begun. That introductory grouse expedition has become part of my April ritual, to experience sage-grouse up close from a small portable hunting blind on a lek in the sage.

Perfectly named, sage-grouse spend their entire life in sagebrush, which makes them extremely vulnerable to development. The umbrella-species concept suggests that by protecting the core habitat of the sage-grouse (the umbrella), where the bird's population density is greatest, the 350 or so other bird, mammal, reptile, amphibian, and plant species that depend on healthy expanses of sage will thrive as well. It's a compelling concept that must be applied with some caution, because wildlife have varying habitat preferences; they roam, migrate, and use lands outside of the sagebrush.

The grizzly bear that you see in Yellowstone's Lamar Valley in May will be above timberline in the Absaroka Range east of the park in August, devouring army cutworm moths. It will use the forest, riverine corridors, and alpine, subalpine, and sagebrush ecosystems in a year of ranging several hundred square miles.

Sandhill cranes migrate from Bosque del Apache National Wildlife Refuge in central New Mexico to Colorado's Yampa River Valley and throughout the Greater Yellowstone Ecosystem in spring, then make the return trip in autumn, linking the entire eastern sagebrush ecosystem via their ancient migration. Like many species, Rocky Mountain sandhill cranes use a variety of seasonal habitats from their northern Chihuahuan Desert winter range to Greater Yellowstone; sagebrush is the thread that ties the Intermountain West together.

Hooves pound countless migration pathways in the sage. Mule deer, pronghorn, elk, bighorn sheep, and moose are among the migrating ungulates that must move between winter and summer range to survive. Remarkably, biologist Hall Sawyer recently documented a 150-mile (240-kilometer) mule deer migration between Wyoming's Red Desert and Green River Basin, now the longest known land mammal migration in the Lower 48. It's our obligation to protect these core habitats and migration corridors, to view the entire West for its full wild potential, unbounded by lines on maps—the way that Wolf 293F once roamed the sagebrush sea searching for a mate.

Grizzly bears, pronghorn, mule deer, wolves, sage-grouse, and hundreds of other species challenge us to think bigger, outside of boundaries drawn by humans, to see the West holistically, as inextricably linked natural systems on a landscape scale.

The energy boom that started with new hydraulic fracturing, or "fracking," technology in the mid-1990s has completely transformed vast areas in Wyoming and Colorado; habitat loss and wildlife declines go hand in hand. Greater sage-grouse and sagebrush-dependent songbirds that breed in sagebrush, as well as mule deer, have been hardest hit.

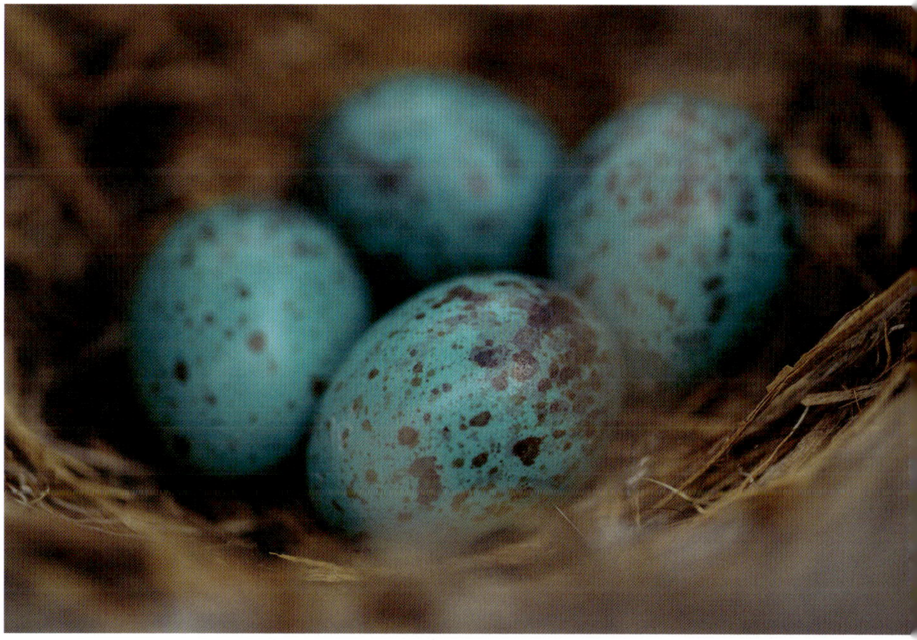

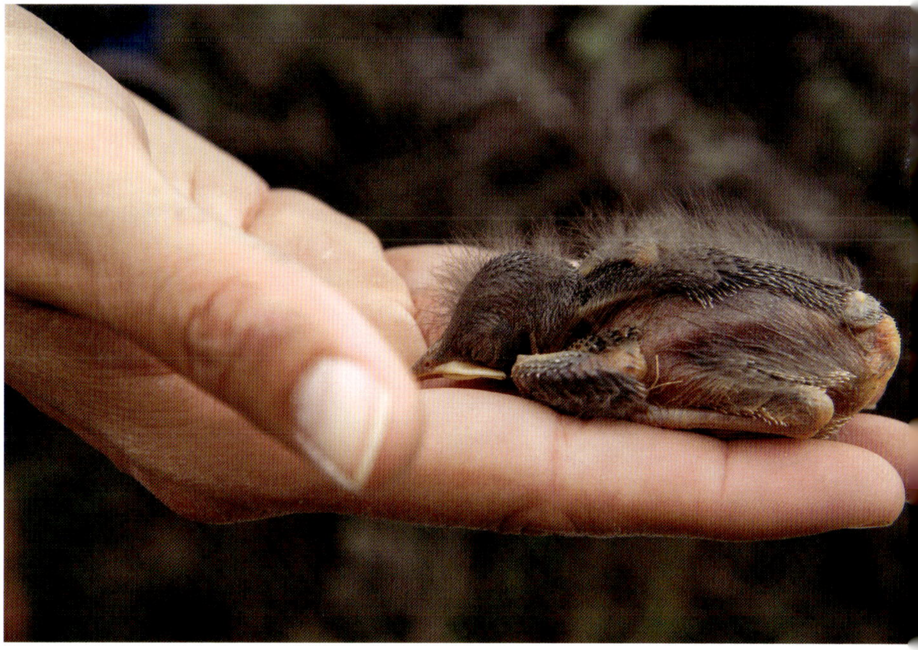

TOP RIGHT Sage thrasher eggs on a nest in dense sagebrush on the Pinedale Anticline, Sublette County, Wyoming.

BOTTOM RIGHT Biologist Tracey Johnson holds a sage thrasher chick in her hand while studying sagebrush songbirds; the bird was quickly returned to its nest unharmed (Pinedale Anticline, Wyoming).

In spring a male Brewer's sparrow sings from the highest perch of Wyoming big sagebrush, announcing his nesting territory. The Brewer's sparrow, a sagebrush obligate species, is declining at a rate of just over 2 percent per year (Black Canyon of the Gunnison National Park, Colorado).

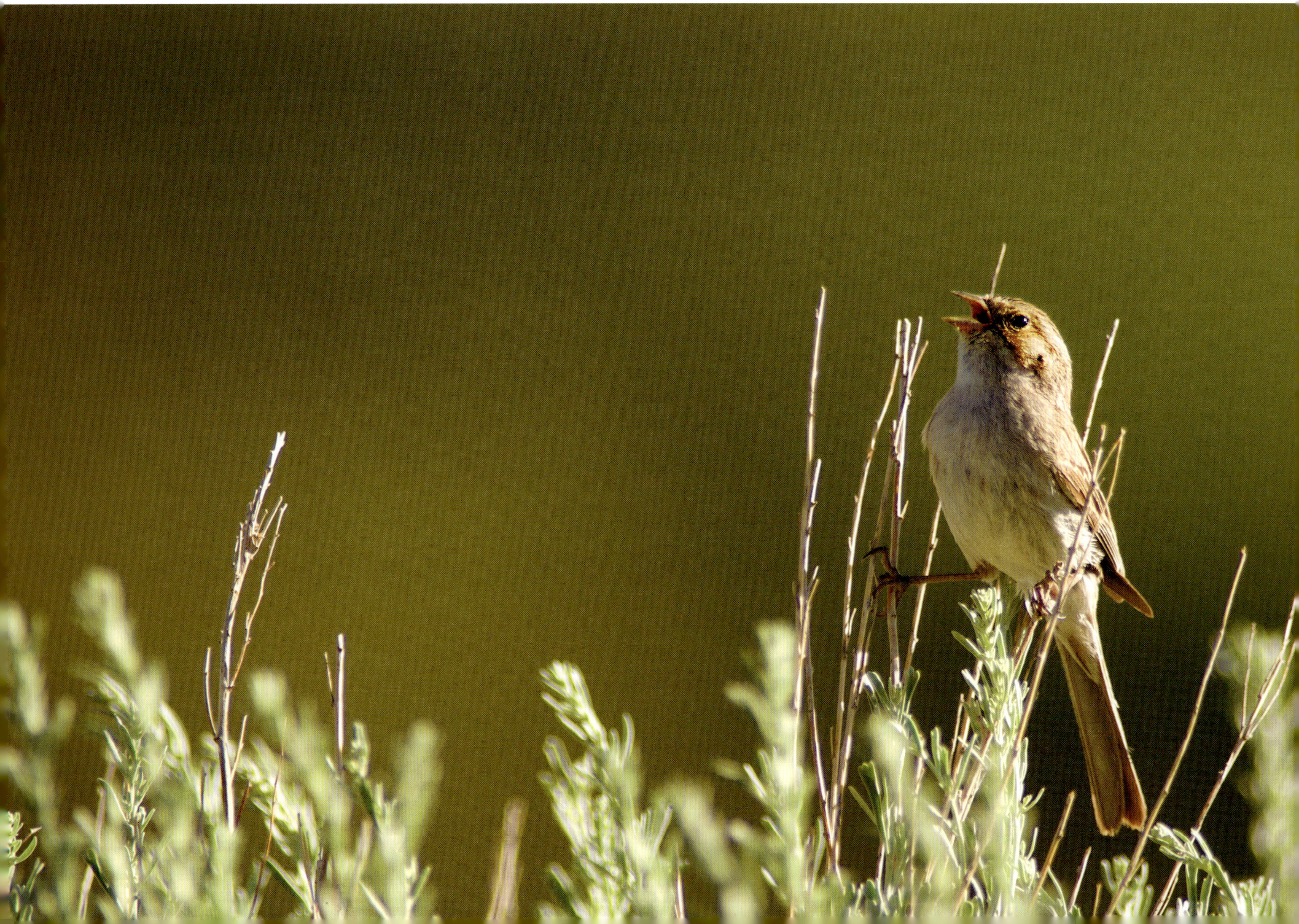

When I started the Sage Spirit project, a friend told me that it just happens that the largest natural-gas deposits are directly under high-density sage-grouse habitat, which is a pretty accurate oversimplification. Sage-grouse were already in decline and occupying 56 percent of their historic range at the start of the fracking boom, while Gunnison sage-grouse were occupying roughly 10 percent of their historic range. Sage-grouse and other species were paying a price for 150 years of bad grazing practices, excessive hunting, development, and sagebrush eradication. Still declining, they are now one of the most-studied birds in the world.

Mule deer winter range overlaps sage-grouse habitat, and mule deer populations are also dropping precipitously. The Pinedale, Wyoming, and Piceance Basin, Colorado, mule deer herds have declined considerably in the last decade. Piceance Basin used to be called the mule deer factory for both the abundance of deer and for its trophy-size bucks.

I've heard people say that the West is lost, just give up. I don't believe that, but we're at a crossroads in the West with much more to learn, and it's important to challenge our assumptions.

For most of the Sage Spirit project, I held the opinion that the vast natural-gas fields are wastelands void of wildlife, undeserving of our best conservation efforts. After nearly six years of work, I'm challenging my own assumption.

I recently had the opportunity to work with Anna Chalfoun and Tracey Johnson, both biologists studying sagebrush-obligate songbirds in natural-gas fields. The term "obligate" means the birds are *obligated* to breed and raise their young in sagebrush; they are so biologically dependent on sagebrush habitat that there is nowhere

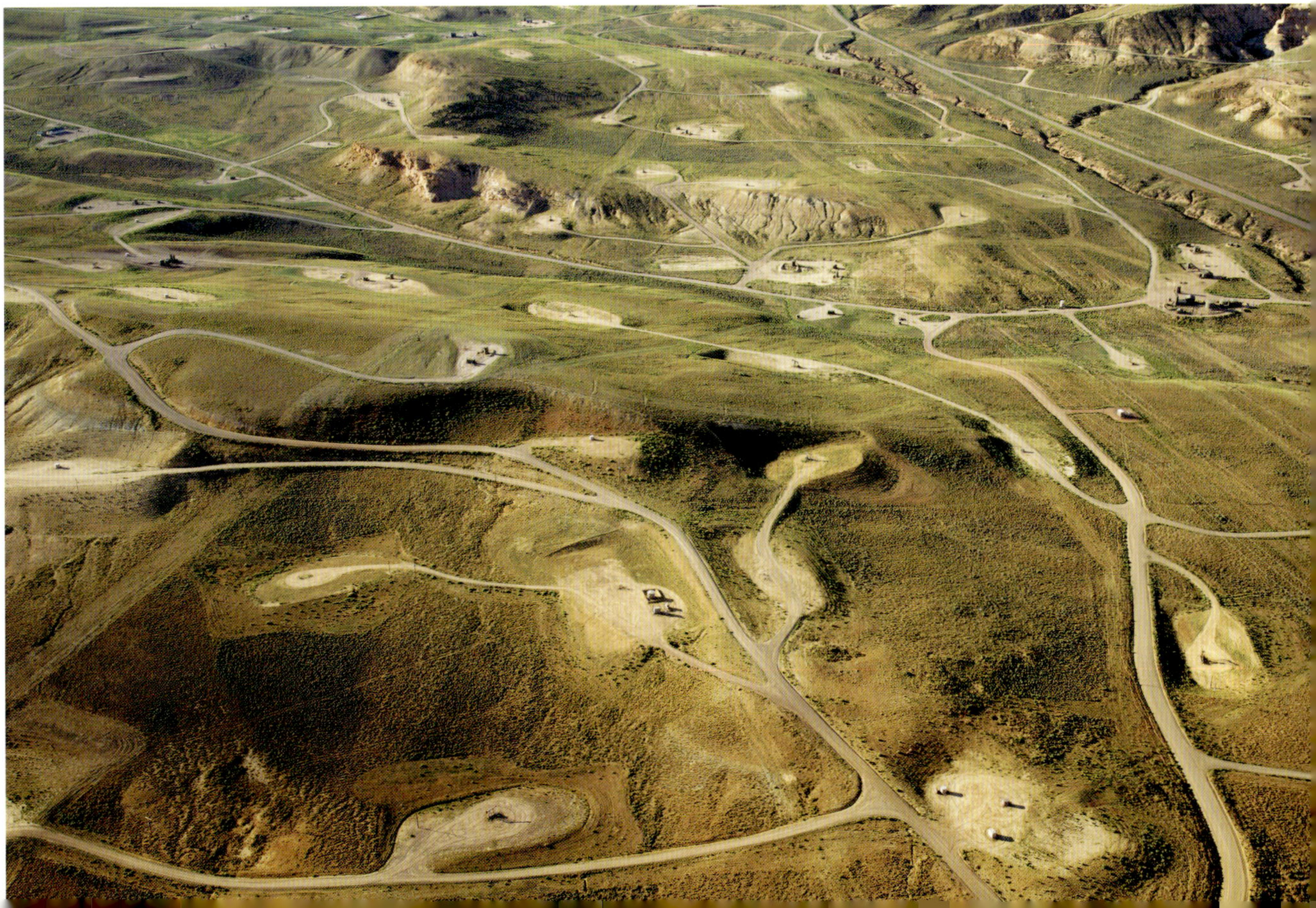

Drill pads and a tangle of service roads in the LaBarge oil and gas field (Sublette County, Wyoming)

else they can breed. The three primary obligate songbird species are Brewer's sparrows, sage thrashers, and sagebrush sparrows. All are in steady decline—Brewer's sparrows are losing 2 percent of their population annually.

I followed Johnson, Chalfoun, and their small team of student biologists for a few days in the Pinedale Anticline and Jonah natural-gas fields and was astonished by the numbers of songbirds. Even in the Jonah, the highest-density gas field in the world, obligate songbirds were singing and attending nests in swaths of sagebrush between drilling rigs. These aren't functioning ecosystems but flickering hope lives in these small patches of a sagebrush mosaic where biologists seek answers in the transformed industrial landscape.

Golden balsamroot flowers above the Noble Basin in the Upper Hoback region of the Wyoming Range. Once planned for an industrial natural-gas field, the area is now protected after the Trust for Public Land purchased and retired natural-gas leases here (Bridger-Teton National Forest, Wyoming).

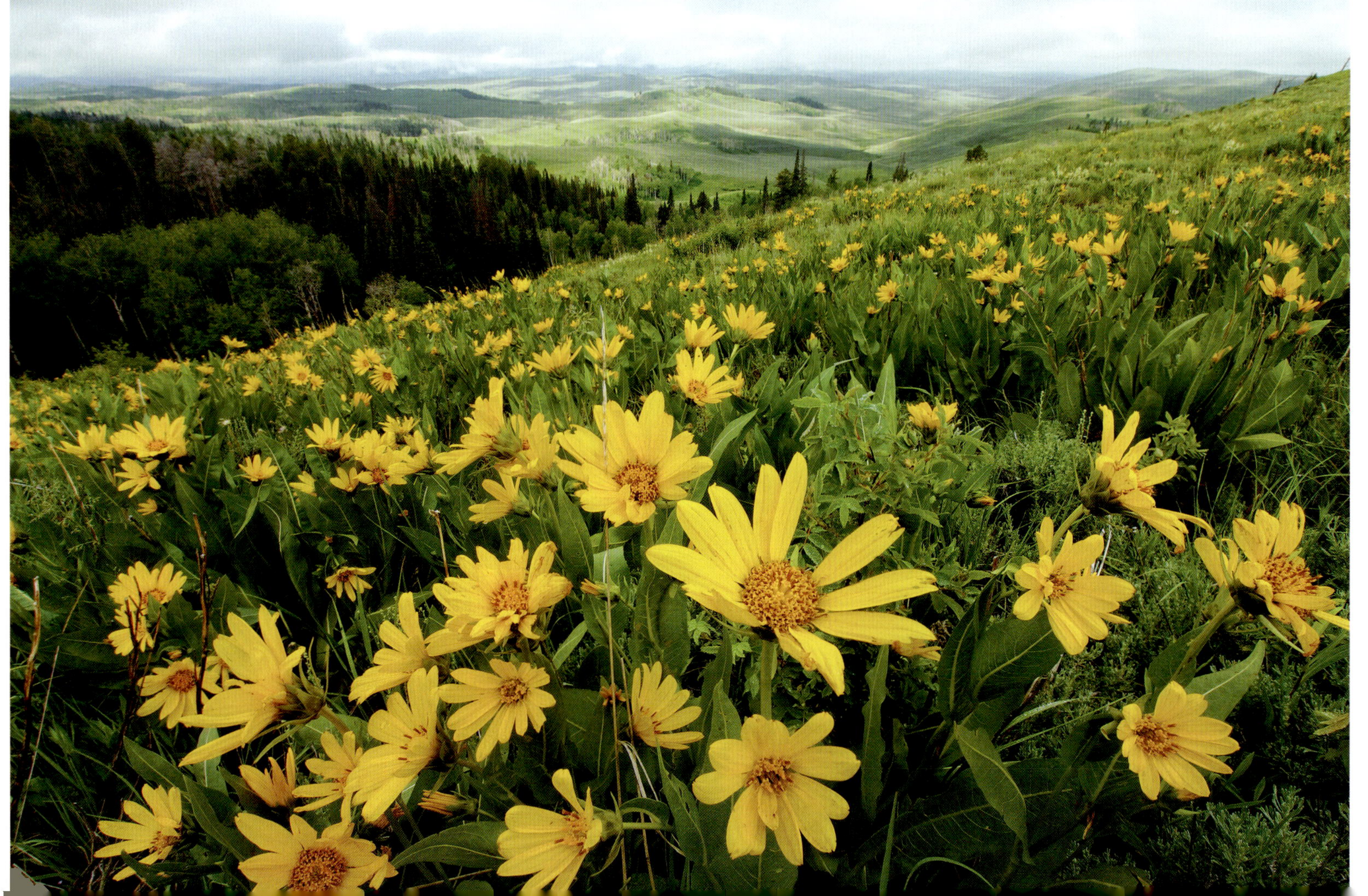

How we view the West often depends on our vantage point. For example, take a satellite image of our eastern sagebrush landscape and the overlay of towns, urban areas, residential development, roads, powerlines, dams, agricultural conversion, and oil and gas fields: our best conservation opportunities, the promise of the American West, live in the remaining wild. In a shrinking West, every new development project must be carefully scrutinized; even green wind farms must be sited properly to avoid killing golden eagles, hawks, and bats.

The greater sage-grouse Priority Areas of Conservation strategy being adopted range-wide by the the US Fish and Wildlife Service and federal and state land management agencies is a way forward that may reverse sage-grouse declines and benefit other wildlife species. It's a historically important initiative to protect the highest-density sage-grouse populations, yet no one really knows if sage-grouse will rebound in a dynamic landscape with limited open sagebrush habitat.

In the course of this Sage Spirit project, I've had the pleasure of working with the most dedicated conservation-minded folks from all walks, some drawn to issues that directly impact their lives, land, and communities, many just trying to make a difference. Grassroots conservation and collaboration saved the Upper Hoback River Valley south of Jackson, Wyoming, and the Roan Plateau and Vermillion Basin in northwest Colorado. The extended Gunnison, Colorado, community came together in an attempt to save their namesake Gunnison sage-grouse; collectively they're making a difference for grouse while improving sagebrush habitat. Many more of us will need to be engaged for any meaningful change toward lasting recovery in the West.

Dave Burke, Dan Smitherman, Heather Stirling, Alison and Matt Holloran, and Roland McCook—all profiled in this book—chose to be a part of making the West better. Famed author, ornithologist, conservationist, and artist David Allen Sibley, who produces *Sibley Guides*, has contributed the foreword of *Sage Spirit*. Patrick (Pat) Magee, who has spent his career educating at Western State Colorado University in Gunnison, and is a staunch defender of Gunnison sage-grouse and their habitat, provides an important historical perspective. Todd Wilkinson, a renowned western conservation writer based in Bozeman, Montana, recently published a biography of Ted Turner, and here in this book, Todd's authentic words show us what's at stake in his essay "Shaping the New West." Award-winning author and activist Rick Bass crafted the epilogue, striking a hopeful tone for the sagebrush ecosystem of the American West. Each of these remarkable individuals and all of these words add up to this: they decided to make a difference and contributed their voices.

These sagebrush lands hold a promise. Simply take that first step, then another, and keep going until you find the Sage Spirit waiting for you. Roll sagebrush leaves in your fingers and breathe in the aroma, be surrounded by birdsong harmonies, hear the huff of a pronghorn attending a nearby fawn, surprise a badger, flush a sage-grouse, listen to the chorus of coyotes as you feel the chill under a diamond-filled sky.

That first step into the sage is also a metaphor for taking action, getting involved in the discussion of what we want *our* West to be. The Sage Spirit vision—connecting wilderness, public, and private lands to conserve the heart of the West—and passing on a proud western legacy are still within our grasp. In the pages that follow, you'll see wildlife that, however imperiled, still dance, sing, migrate, and inspire. We can celebrate our natural inheritance in these sagebrush valleys, spread our own wings, and, if we do things right, it will still be possible to pull over, set up camp, and catch a mess of fish with your dad. §

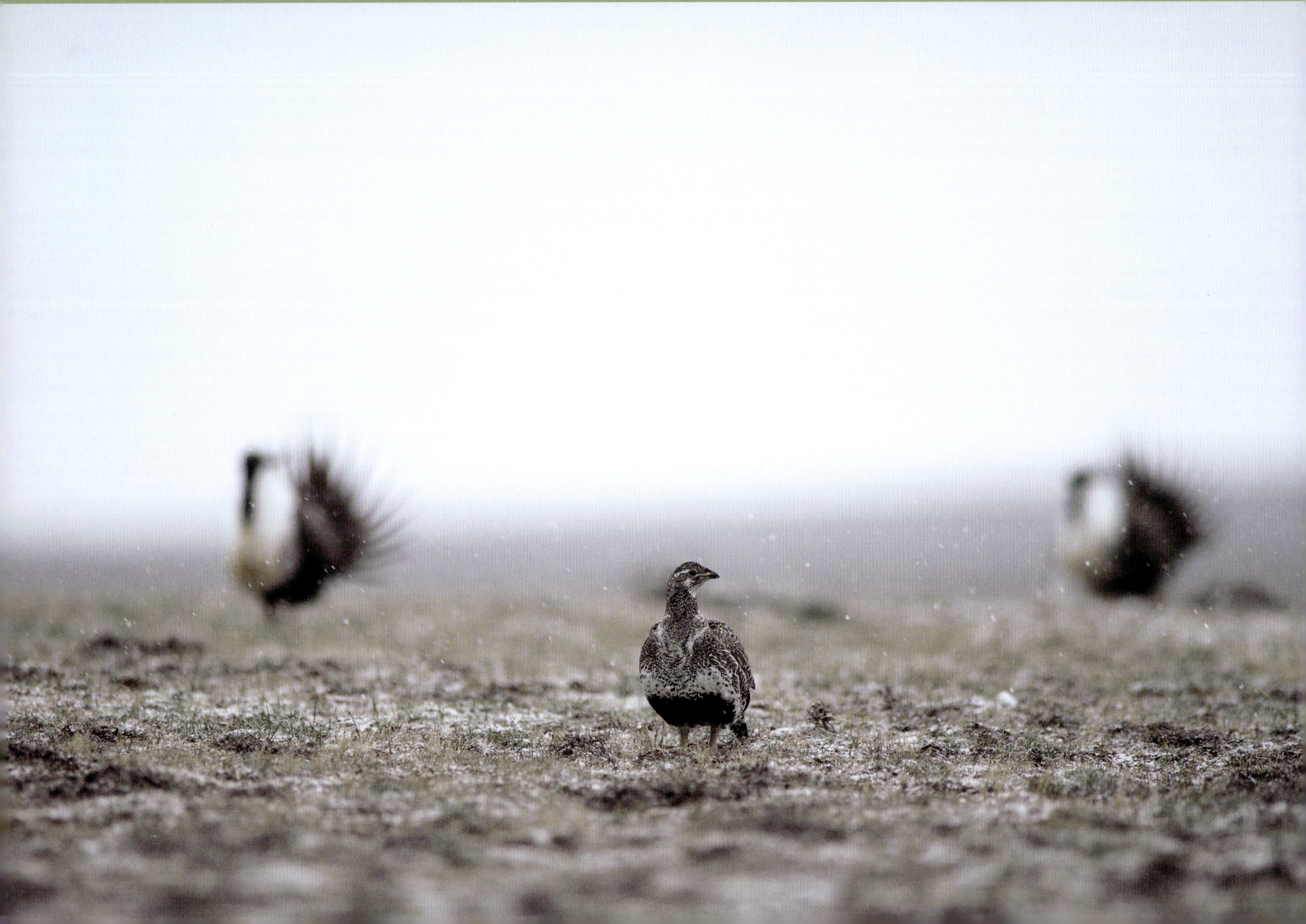

THE SAGEBRUSH ECOSYSTEM

REVERSING A HISTORY OF OVERUSE

Patrick Magee

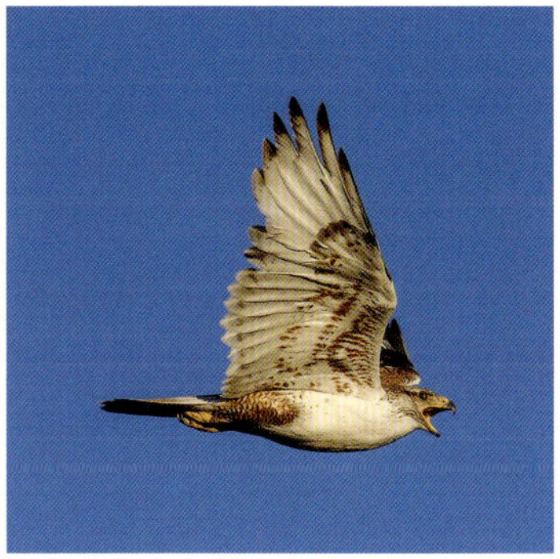

IN THE GUNNISON BASIN, Colorado, in late March, winter's unyielding grip begins to slacken, leaving behind months of long, icy, forbidding nights and frigid days. I'm on an annual predawn rite of passage of sorts, winding through sagebrush on a frozen two-track road. I drive up a steep, loosely graveled slope, and with my lights off I reach the crest of a rolling hill covered with sagebrush. About a quarter mile below me, the dense shrubland opens to an avian dance floor, called a *lek,* where the ancient mating ritual of the Gunnison sage-grouse unfolds each year around this time.

For now, all is quiet as I sit in my eighteen-year-old midsize pickup, engine off, window rolled down, waiting in the dark, chilly stillness, hoping the birds will perform their dance again this year. If they do, I won't see them at first in the dark; instead, I'll hear a sound like bubbles popping. This low-frequency, resonating vocalization carries just above the canopy of the sagebrush more than 3 miles (4.5 kilometers) into the distant sea of sage, attracting females and neophyte males to this vital ritual.

I wait; still no sound. Last year I saw six males, down from the twenty males I observed fifteen years ago at this same lek. Decades ago, higher numbers of males undoubtedly gathered at this same dance floor. Will they show now? Will they persist another year?

ABOVE Ferruginous hawk in flight (Red Desert, Wyoming)

OPPOSITE A female greater sage-grouse stands in front of two displaying males on a snowy April morning (Sublette County, Wyoming).

Decade after decade, since the western American frontier was first explored by Europeans, the integrity and connectivity of the sagebrush ecosystem has eroded. Called the sagebrush sea, this massive sagebrush-dominated shrubland historically covered more than 307 million acres (124 million hectares), a large chunk of the American West. To a traveler driving through the sagebrush sea, this ecosystem may appear as a homogeneous and endlessly abundant landscape. But appearances can be misleading. The cumulative impact of more than a century of wounds has pushed this once vast and diverse ecosystem to the edge of collapse. Sage-grouse, which depend inextricably on resources within sagebrush habitats, are at a threshold as well, and as the habitat continues to unravel, they face a dubious fate.

To the east, Venus perches on the horizon. Somehow it seems appropriate that the goddess of love rises in the morning to overlook this scene, perhaps blessing the lek or cueing the grouse. Both greater and Gunnison sage-grouse have mastered not only passionate courtship toward the opposite sex but also fidelity to their lek—they are *philopatric,* lovers of their home. They return to the same lek, year after year, generation after generation; these sites are sacred altars of rebirth. Since the start of American western expansion, forty or more generations of sage-grouse have returned to their leks, loyal to tradition and geography. But across those generations, they've faced persistent and insidious foes that have undermined their ability to thrive in their sagebrush home.

Sitting here beneath Venus and a dazzling starry sky, I see one light, very low and yellowish, just beyond the lek. This noncelestial porch light indicates the presence of a house; two more glimmer within a half mile. In addition to these homes, dozens of roads, several fences, and a looming powerline immediately surround the lek. Much of the sagebrush country all over the West looks similar. These human changes to the landscape have a price: sage-grouse populations are dwindling and, lek by lek, appear to be sliding toward extinction.

Not so long ago many leks were less lonely; indeed, they were crowded, with males intensely competing through their bizarre and unique mating display.

THRIVING IN THE SAGEBRUSH SEA

What is now known as North America has supported grouse for 25 million years, and during the ice ages (2.5 million to 12,000 years ago) sage-grouse roamed portions of the continent alongside cheetahs, woolly mammoths, and cave lions. About 850,000 years ago, Gunnison and greater sage-grouse diverged evolutionarily, and since that time each species has developed an obligate, specialist relationship with big sagebrush (*Artemisia tridentata*). The bird and the shrub are ecologically and geographically intertwined.

Called *sisk-a-dee* by the Shoshone and *sagagah puch* by the Utes, sage-grouse coexisted with Native Americans across much of the sagebrush sea for thousands of years. Native Americans hunted sage-grouse for food and used their feathers for ornamentation and ceremony. In fact, Native people mimicked the sage-grouse mating display in their own dances.

By the time of the American Revolutionary War, between 2 and 16 million sage-grouse may have occupied 500,000 square miles (1.29 million square kilometers)—approaching half the land area of the western United States—of sagebrush-dominated lands. Upon Lewis and Clark's discovery of sage-grouse in Montana in 1805, the cock of the plains was on the brink of a bewildering transformation: they were about to unwittingly experience European exploration and settlement of the new West.

COLLISION WITH MANIFEST DESTINY

In 1803 the Louisiana Purchase added 530 million acres (215 million hectares) to the American public domain, and for the first

Historic and Current Range of the Greater and Gunnison Sage-Grouse

The greater sage-grouse occupies 56% and the Gunnison sage-grouse occupies roughly 10% of their historic ranges.

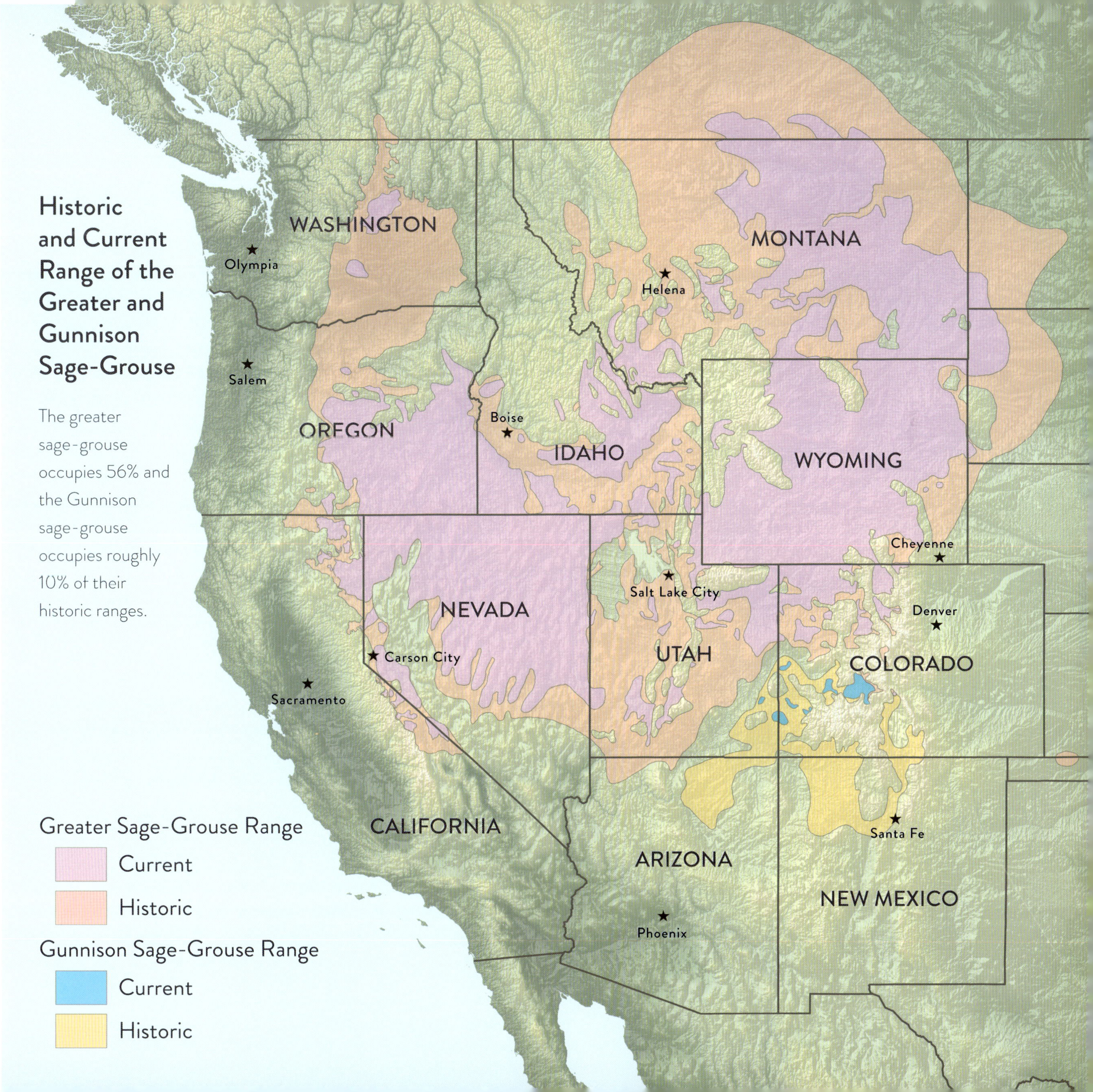

time portions of the sagebrush sea became part of the United States. The US government began to dispose of this land with an orderly vision fueled by Thomas Jefferson to sell 640-acre (259-hectare) parcels to farmers. And on this land, sage-grouse would very soon meet white humans, who had a novel and dangerous idea: the vastness of the western continent, including the sagebrush ecosystem, was their destiny.

Trappers and buffalo hunters, along with military expeditions seeking routes across the continent, were the first Europeans to exploit the rich wildlife resources of the American plains. They paved the way for settlers who advanced west, a migration facilitated by hordes of gold seekers during California's 1848 gold rush and the 1862 Homestead Act, which lured waves of farmers and ranchers, including German and Irish immigrants. These immigrants, along with Civil War veterans, sought redemption and respite, and the western frontier offered a new life.

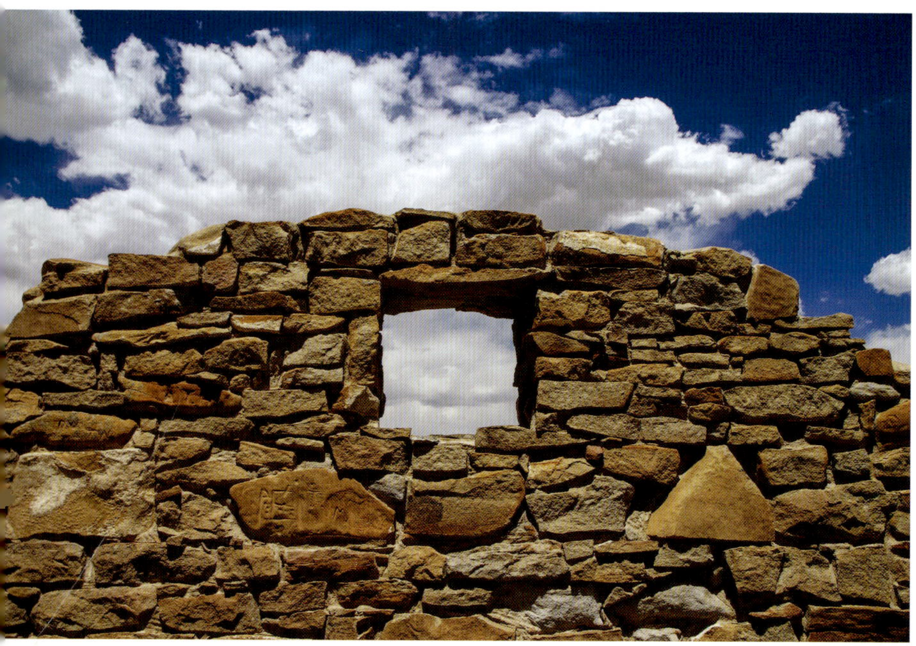

Built in 1862 by the Overland Stage line, Point of Rocks Stage Station in Sweetwater County, Wyoming, served stagecoach travelers making the arduous Overland Trail journey. The station building has been restored, one of few still standing on the Overland route.

They moved farther into the frontier, beyond the midwestern tallgrass prairie states, into the western Great Plains dominated by shortgrass prairie and, farther west, deep into the Rocky Mountains, where they first encountered the sagebrush sea. This immense, arid shrubland extended from the Rocky Mountains across the western states to the major mountain ranges of Washington, Oregon, and California. Here in these dry landscapes traditional farming became marginal, but livestock grazing seemed feasible. In 1870, 4.1 million beef cattle and 4.8 million sheep grazed America's western public domain, and by 1900 the industry grew to 19.6 million cattle and 25.1 million sheep. The unclaimed public land was up for grabs, driven by provisions in the Homestead Act that allowed settlers to claim 160 acres (65 hectares) if they built a residence and farmed the land for five years.

The settlers and stock growers regularly hunted a large, abundant game bird and aptly called it the sage chicken, as it fed many a hungry rancher, miner, railroad worker, and settler in the late nineteenth century.

HUNTED TO THE EDGE OF EXTINCTION

No one knows precisely how many greater and Gunnison sage-grouse existed at the end of the nineteenth century, but they were ubiquitous throughout the sagebrush ecosystem and immensely abundant. On one early morning in 1886 in central Wyoming, the eminent naturalist and conservationist George Bird Grinnell scribbled in his field notes (excerpted here from Robert Patterson's 1952 *The Sage Grouse in Wyoming*): "The number of grouse

Fog and golden autumn cottonwoods along the Gunnison River, Colorado

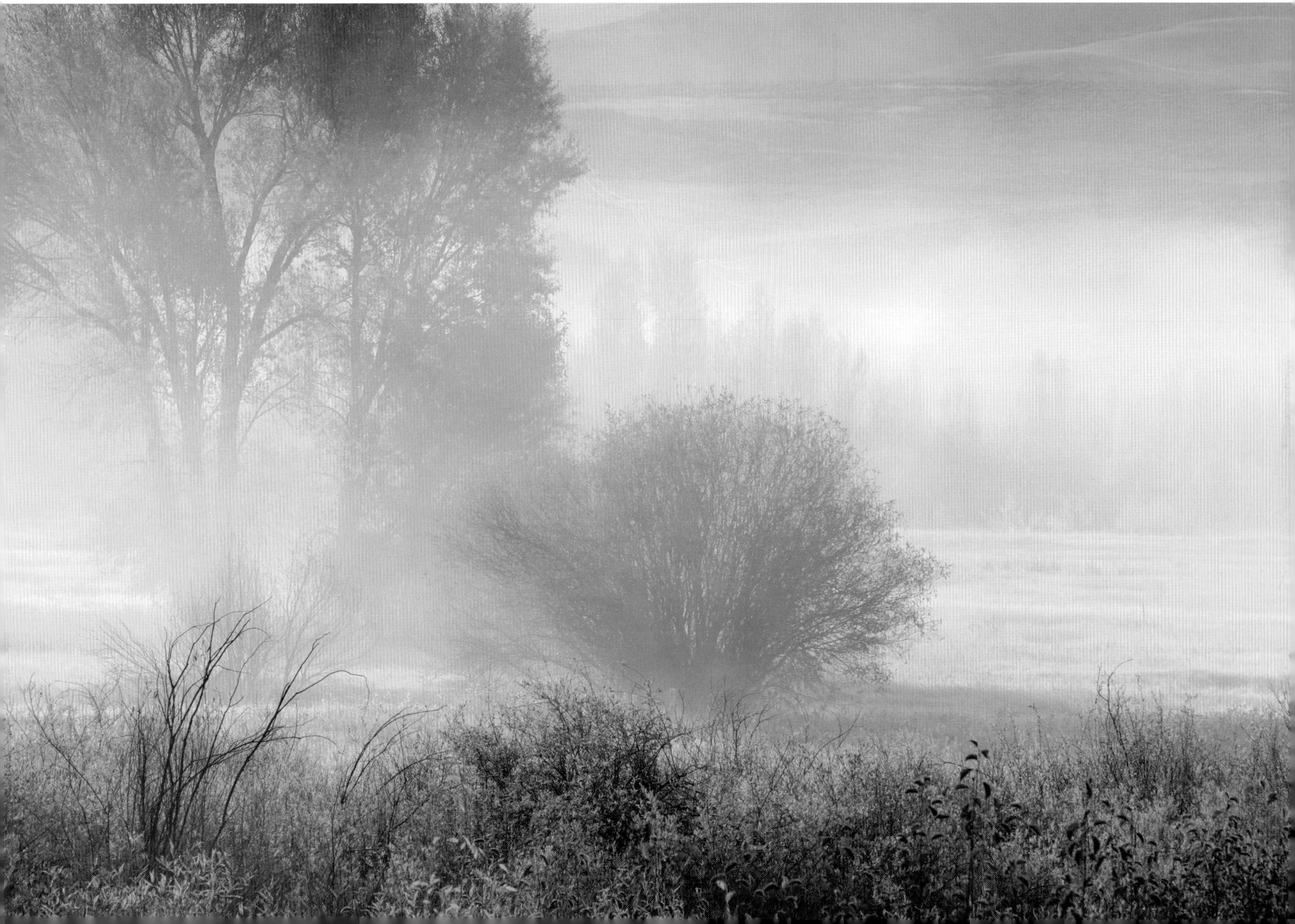

which flew over the camp reminded me of the old time flights of passenger pigeons that I used to see when a boy. Before long the narrow valley where the water was, was a moving mass of gray. I have no means what ever of estimating the number of birds which I saw, but there must have been thousands of them."

But even before the turn of the last century, sage-grouse populations began their decline as the nation expanded westward. Residents of Lake City, Colorado, opened the *Lake City Times* on September 13, 1894, to read, "Billy Green, a genial fireman on the 64, bagged 863 grouse while out hunting on Sapinero last Sunday." To put this "feat" into perspective, in 2013, the total number of male Gunnison sage-grouse counted across their global range was 848 individuals—15 fewer than Billy Green harvested on one September Sunday.

While most people living in the West did not recognize the mounting imperilment of the sage-grouse, one outspoken crusader for wildlife protection, William Temple Hornaday, listed the sage-grouse as the next candidate for oblivion in his 1913 book *Our Vanishing Wild Life*: "In every state, without a single exception, the bag limits are far too high, and the laws are of deadly liberality," raged Hornaday. His chief goal was to severely restrict legal hunting carried out by more than 2.5 million "men and boys warring upon the remnant of game."

Hornaday formed the Permanent Wild Life Protection Fund with backers such as Henry Ford and Andrew Carnegie, along with many other members of the affluent early twentieth-century business community. In 1916 he focused his attention on the protection of the sage-grouse. In that year, he published a manifesto entitled "Save the Sage Grouse from Extinction: A Demand from

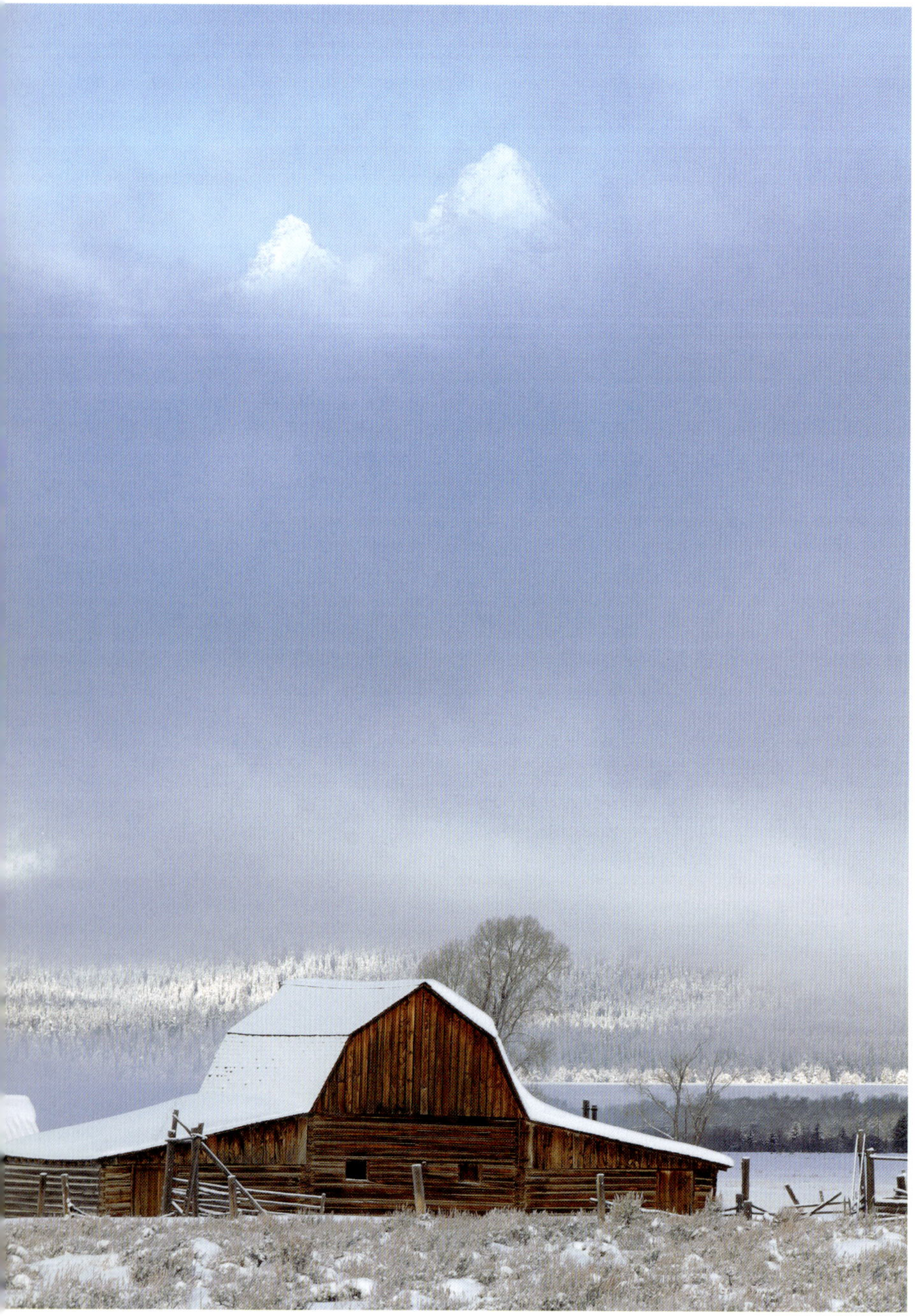

The Grand Teton just breaks through clouds above the historic Moulton Barn in Grand Teton National Park, Wyoming.

Civilization to the Western States," in which he outlined the horrific excesses in legal hunting and illegal "murder" of sage-grouse and predicted its demise as a result of automated firearms, automobiles, and roads allowing easy access by an "army of destruction" to previously inaccessible lands.

Eventually, the western states heeded Hornaday's call for restrictions on legal hunting and increased enforcement on the ground to curb illegal poaching. A conservative estimate of 2 million sage-grouse (and some say up to 16 million) inhabited the vast sagebrush sea prior to western expansion, but the population likely declined by orders of magnitude during the time of unregulated hunting and subsistence and commercial harvest . . . and would never fully recover.

After at least 1.5 million years of occupancy in the sagebrush shrublands of North America, enduring prolonged, severe droughts and bitter-cold winters, had the sage-grouse finally met its match? White folks who claimed it was their destiny to live in this once-lonely sagebrush desert had just begun to move in and reorder the place.

RANCHING RESHAPES THE LAND

When post–Civil War emigrants moved west and became cowboys, they had a fairly simple strategy: release their livestock on the vast open rangeland and round them up occasionally to brand or to sell at market. There may have been a home ranch, but at first there

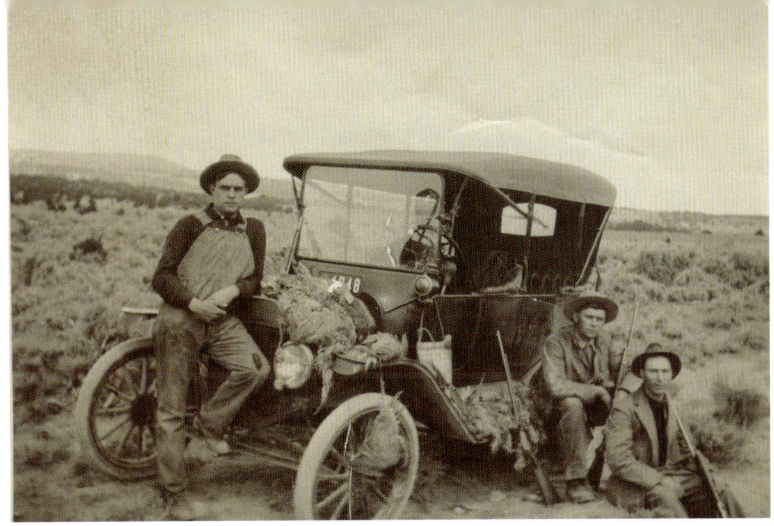

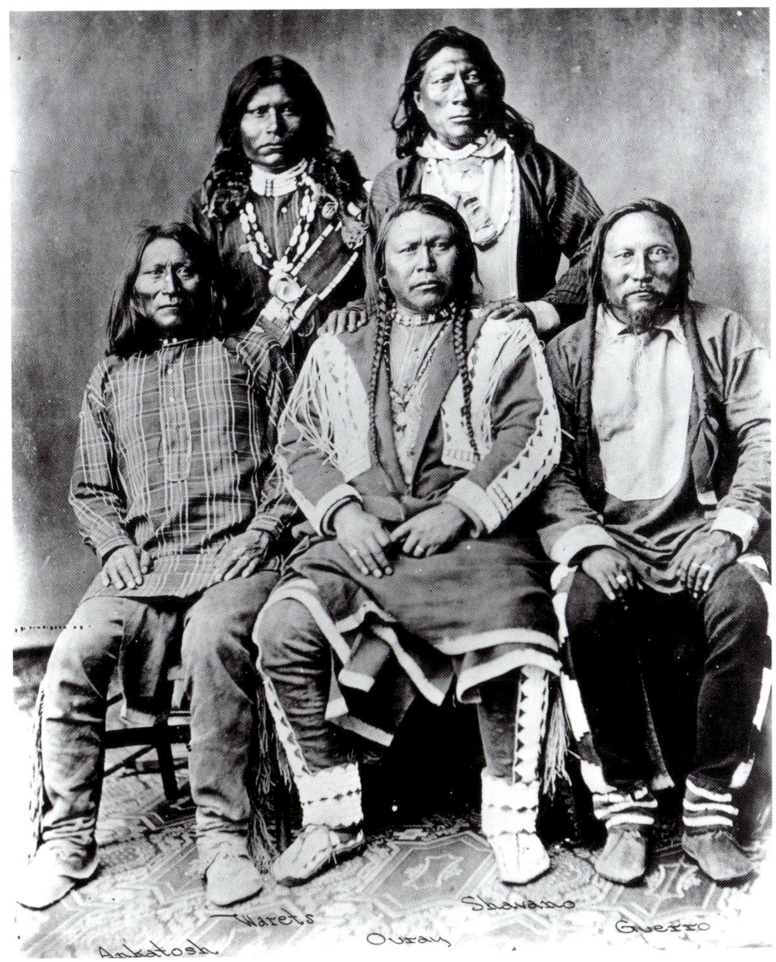

TOP RIGHT Sage-grouse hunters Frank Viox, Charles Viox, and Ed Peverly pose with grouse draped across their vehicle in 1916 in Sweetwater County, Wyoming.

BOTTOM RIGHT Ute Indian Chief Ouray (center) with sub-chiefs in 1868 when the Utes still occupied the western third of Colorado. At home in the sagebrush, the Utes coexisted with sage-grouse for thousands of years.

was no hay production or irrigation. However, the bitter cold and blizzards of the mid-1880s convinced many cowboys that cattle needed to be kept close to home and fed hay during winter, then set out on open range the rest of the year.

Thus began the process of converting sagebrush and riparian habitats to hay, but in this desert landscape, dryland agriculture was not an option; irrigation was required. Small-scale projects sprang up, and in 1902 the Bureau of Reclamation was created to build large dams and divert water to the parched landscape, altering the flow of rivers and flooding miles of riparian and upland sage-grouse habitats.

Millions of cattle and sheep grazed the public domain hard, largely in unregulated fashion, from the 1870s until the 1930s compacting and eroding soil, depleting nutrients, reducing plant cover, and altering plant communities. Their repeated trailing in riparian habitats led to channel incision and eventual lowering of water tables, effectively drying out the moist, productive biological havens within the sagebrush sea. By the early 1930s significant damage was done, and the livestock industry was reeling.

The solution, championed honorably by cattlemen, was the 1934 Taylor Grazing Act, which would "stop injury to the public grazing lands" and "stabilize the livestock industry." Rather than disposing of public land, now the federal government would divide it into districts and allotments that ranchers with a home base would be allowed to lease for a fee. This allocation of the public land brought forth the finite reality of the landscape, thus reducing the number of stock growers and ending nomadic ranching practices.

On these same degraded lands, the Civilian Conservation Corps built fences, wells, and erosion control structures, but while reforms slowed the devastating effects of decades of overgrazing, the ultimate goal to improve grazing land would eventually do even more damage to the sagebrush ecosystem.

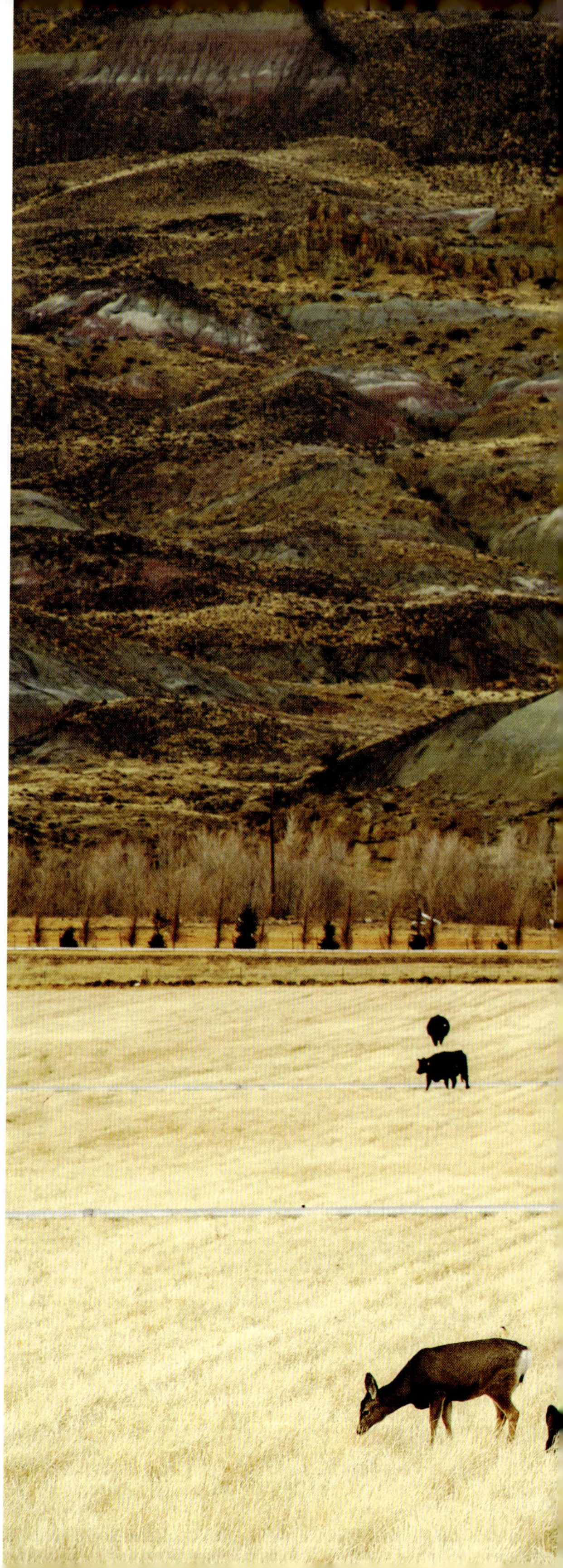

Mule deer graze with cattle in a hay meadow on the North Fork of the Shoshone River in Park County, Wyoming.

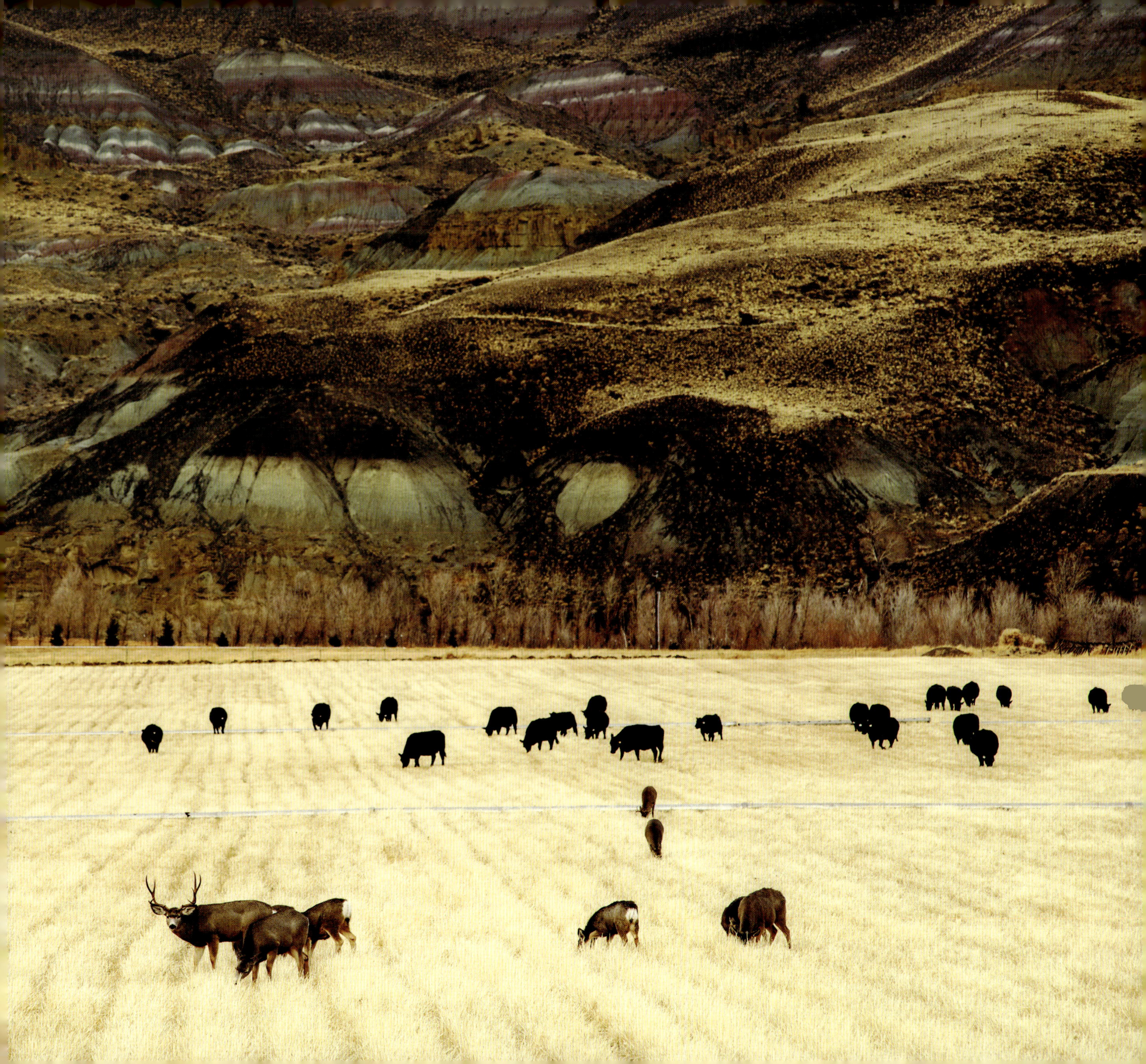

Headed by the Bureau of Land Management, a new federal agency created in 1946, and an ambitious group of land managers and scientists, range improvements were underway. These improvements translated to a narrowly focused mission to increase forage (grass) for livestock. In a 1958 article published in the *Journal of Range Management,* scientists Walter Mueggler and James Blaisdell wrote, "Control of big sagebrush (*Artemisia tridentata*) on heavily infested areas has long been recognized as an effective range improvement practice. By killing this one undesirable species the grazing capacity of rangelands can often be increased many fold."

In a monumental effort to "improve" rangeland, the sagebrush sea was burned, sprayed, railed, chained, rotobeaten, and mowed, and many of these treated areas were then seeded, usually with nonnative species. It is unlikely that any substantial parcel of sagebrush escaped large-scale treatment initiated in the 1930s and aggressively applied through the 1960s. The cumulative impact of ecological transgressions took their toll on sage-grouse, which for the last third of the twentieth century declined at a rate of 2 percent annually—a population decline of 50 percent or more.

LANDSCAPE FRAGMENTATION

From a few wagon trails to new railroad lines to the development of the modern road system, the fragmentation of sage-grouse habitat has been an insidious threat, slowly slicing apart the once-vast oceans of sagebrush. The sky above and the earth below are decorated with human-generated lines, lines that fragment the sagebrush sea, crippling its capacity to support not only sage-grouse but a diverse group of sage-dependent species.

Today 83 percent of the United States is within 0.6 mile (1 kilometer) of a road; within sage-grouse range, less than 5 percent of their habitat is more than 1.5 miles (2.5 kilometers) from a road. And the sky is a matrix of vapor trails recording the daily comings and goings of jet planes, contributing to the now-pervasive anthropogenic noise that disrupts communication and behavior of sage-grouse and many wildlife species. In addition to a landscape fragmented by transmission lines, pipelines, fences, and roads, the outright conversion of sagebrush to hay and row crops, reservoirs, well pads, landfills, tailings, open-pit mines, homes, other buildings, and parking lots has multiplied.

Further, fire suppression has led to an invasion of juniper, converting sage shrublands to woodlands. Stealing into the country from Asia in grain imports, invasive cheatgrass has overthrown vast

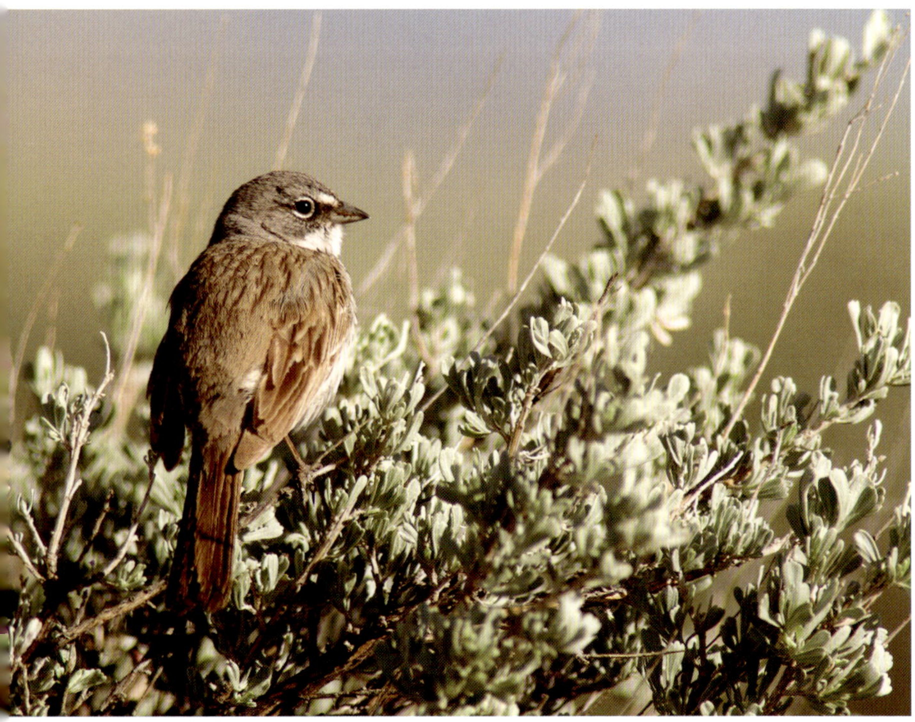

A male sagebrush sparrow perches on a Wyoming big sagebrush during spring breeding season (Gunnison Basin, Colorado).

landscapes of sage, converting them to annual grasslands. On a larger scale, human contributions to rapid climate change may result in the transformation of up to 80 percent of sagebrush into drier desert ecosystems.

Even human recreation has spread into remote portions of the sagebrush sea. Routes exploited by off-road vehicle users, snowmobilers, motorcyclists, mountains bikers, runners, and hikers penetrate deep into the heart of the sage. Today, greater sage-grouse reside on 56 percent of their original range, and Gunnison sage-grouse persist on just 10 percent. Of the millions of sage-grouse that once populated the Wild West, less than 10 percent exist today. In general, across the Rocky Mountain region, the quantity and quality of sagebrush has shrunk dramatically as human presence is increasingly felt in the remotest corners of this sea.

FAITH AND HOPE

In the dim light of morning above the lek, abruptly I hear the flutter of wings, and the first *pop* advances through the cold air. At least one male pumps up his air sacs, braces his wings, throws his ponytail forward, and lets loose a liquid *plop,* his breath rising—as if the display needs any more pageantry. Through the spotting scope I count the birds: one, two, three, four, five males. They have returned for another season.

In some Native American cultures, sage-grouse represent the sacred spiral, the turning of time and life, the eternal return. Together sagebrush and sage-grouse evolved, survived, and adapted.

Today the vulnerability of sage-grouse to extinction represents the widespread havoc that humans have perpetrated across the West. The sagebrush sea has been attacked from all sides; it has been degraded and fragmented, losing its connectivity that is vital to the diverse and productive natural communities within it—such as big-sagebrush shrublands, black sagebrush, shrub-steppe, riparian wet meadows, ephemeral wetlands, springs, seeps, and many, many more. And, for as long as humans have lived alongside these natural communities, we have benefitted from an abundance of food and water that nourishes wildlife populations and livestock, healthy soils that generate forage and retain precious nutrients, the production of oxygen and sequestration of carbon performed by billions of tiny sagebrush leaves, and an array of other roles that the sagebrush ecosystem fulfills.

Now we are called to forge a new relationship with the sagebrush ecosystem in which we use and value the natural resources of this vast western region in a way that sustains the underlying structure and resources that flow toward us. We have the opportunity to reimagine our place in the once-vast sage—to rethink how we extract energy, how we expand our urban dwellings, how we adventure and play. Many concerned advocates of the sagebrush sea have begun to reverse the historic trend of depleting this ecosystem, and now it is time to focus our energy, finances, knowledge, and passion on restoration. The issues around protecting this ecosystem are infinitely complex, requiring creative, collaborative, long-term solutions that will be rewarded not only by the annual return of sage-grouse each spring but also by rescuing something central to the identity of the West: the sagebrush sea. §

A WALK THROUGH THE SAGEBRUSH SEA

Kris Fulsaas

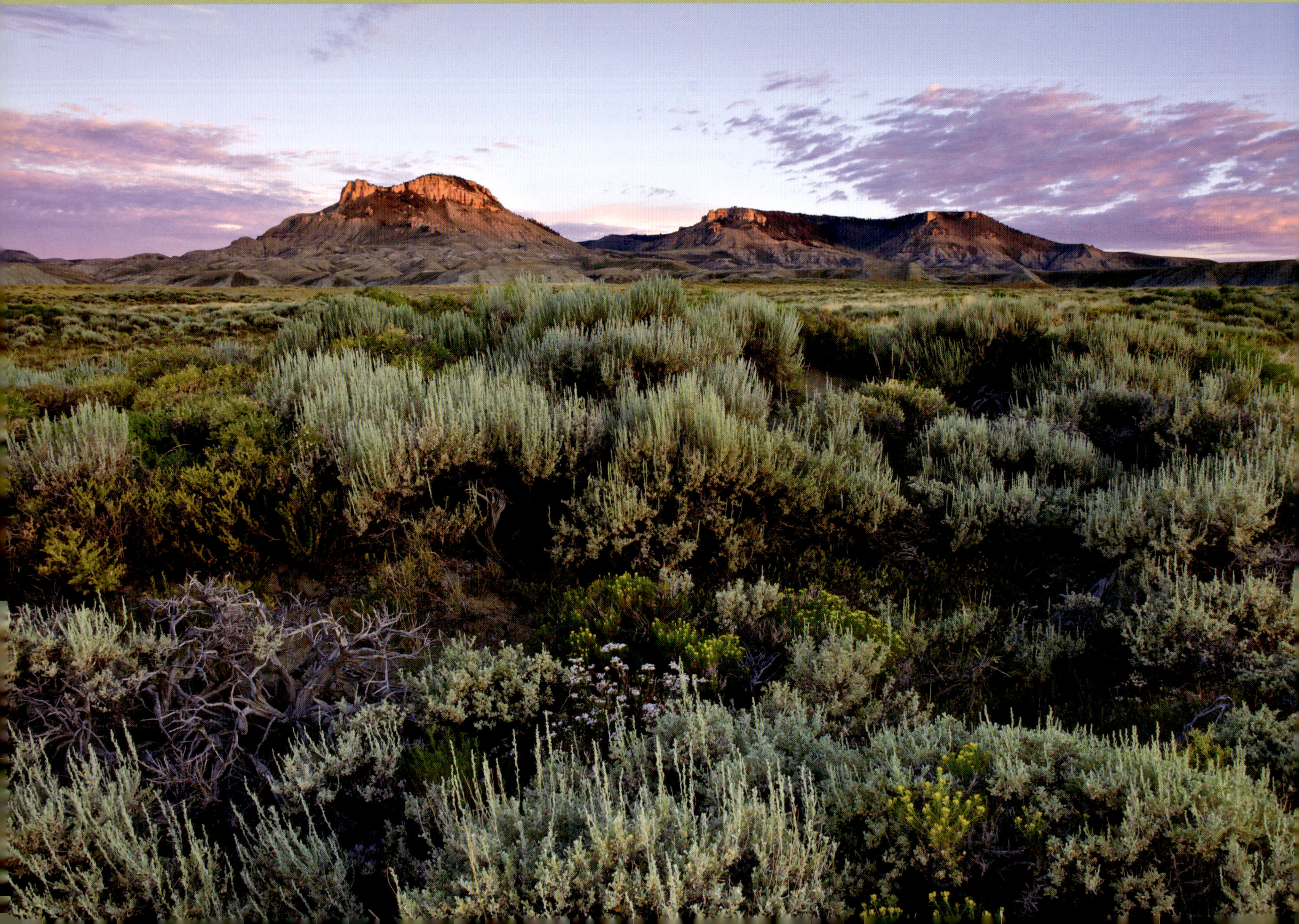

THE VAST ECOSYSTEM known as the sagebrush sea wears many faces across eleven states: from the sagebrush-steppe, where grasses and shrubs are co-dominant, to grasslands and southern desert shrublands, from riparian areas to mountain woodlands. The sagebrush-steppe, perhaps the sagebrush sea's most recognizable form, is a mix of big sagebrush, antelope bitterbrush, horsebrush, and rabbitbrush, with perennial bunchgrasses providing horizontal cover for ground nests and dotted with balsamroot, lupine, Indian paintbrush, pasqueflower, saxifrage, and shooting stars.

This is the realm of the greater sage-grouse, Gunnison sage-grouse, sagebrush sparrow, Brewer's sparrow, and sage thrasher—obligate species that require sagebrush for cover, nest protection, and food.

A visit to the sagebrush sea's desert shrubland reveals big sagebrush mixed with black greasewood, spiny hop-sage, and saltbush. Keep an eye out for tiny ball cactus, hedgehog cactus, prickly pear, and stonecrop sparsely scattered among desert parsley, rock penstemon, and bitterroot. Raptors perch in wait for mice and voles to venture out; nighthawks swoop for bugs through the evening air with a buzzy, nasal *peent*.

Yet the sagebrush ecosystem also encompasses riparian areas lush with creek dogwood, willows, rushes, sedges, and monkeyflower. The Intermountain West's "gray sea" includes the upper watersheds of the Colorado and Missouri rivers and watersheds in the Great Basin. Red-winged blackbirds make throaty, reedy calls from cattails; the songs of yellow and MacGillivray's warblers, song sparrows, and Bullock's orioles resound. The scent of water draws creatures at dawn and dusk.

On the fringes of the sagebrush sea, woodlands of aspen, cottonwood, pinyon-juniper, mountain mahogany, and serviceberry are the haunts of red-tailed and ferruginous hawks, western scrub jays, black-billed magpies, and Virginia's warblers, as well as mammals on the hunt.

In any of these habitats, the sagebrush ecosystem is dominated by nearly thirty species, subspecies, and hybrids of the iconic plant that gives this habitat its name. According to Leila Shultz in the *Pocket Guide to Sagebrush*, the best known is big sagebrush, which has more biomass than any other shrub species in the world. It and its four subspecies have multiple races—hybridization that results from habitat-specific specialization.

Basin big sagebrush is found in Nevada's Basin and Range country and beyond. Higher-elevation mountain big sagebrush is the most palatable to wildlife, and thus very important for animals in the Rockies, Cascades, and Sierras. Parish big sagebrush is found in California's Mojave Desert. Wyoming big sagebrush, despite its state-specific name, grows throughout the lower-elevation eastern sagebrush sea. Bonneville sagebrush, a hybrid of mountain and

Sunrise lights the Oregon Buttes in the Jack Morrow Hills region of Wyoming's Red Desert. The Oregon Buttes were a major milestone for nineteenth-century travelers on the Oregon Trail. The undeveloped Oregon Buttes Wilderness Study Area is home to many endemic species of the sagebrush ecosystem.

Wyoming big sagebrush, is also palatable to wildlife. Any of these plants has an unmistakable shape, tall and pale grayish-blue, with a strong scent evoking the wide-open spaces of the Old West.

Lesser-known sagebrush species include fringed sagebrush, the most widespread (ranging from the Arctic to the High Plains to central grasslands); California sagebrush (found only on that state's Pacific coast); and scabland or stiff sagebrush (rooted in the volcanic soils of the Columbia River Basin in Oregon, Washington, and western Idaho). Other species endemic to particular areas include birdfoot sage (Wyoming and Montana), fuzzy or Owyhee sagebrush (Oregon, Idaho, and Nevada), Porter sage (only in the Wyoming badlands), and the rare pygmy sage (Nevada, Utah, Colorado, and northern Arizona and New Mexico).

Each sagebrush species has a unique mix of bitter terpenoids as well as tasty compounds attractive to browsing animals. Most sagebrush species are evergreen, with two kinds of leaves: large, irregular early spring leaves drop off by midsummer—gather these aromatic sprigs for sage bundles—but tight little evergreen bundles of perennial leaves photosynthesize during winter, providing important forage when most other woody plants are dormant.

In spring, sagebrush leaves are at their most bitter and aromatic, to discourage insects and grazing animals while flowers are forming. Its many small flowers start opening in late summer through early fall, producing huge amounts of pollen; sagebrush is primarily wind pollinated. In fall the bitter compounds that deter insects and predators break down, so the microscopic seeds become more palatable; animals that eat them assist the wind in dispersing them. At 20 percent protein, sagebrush seeds are a significant winter food source, according to Leila Shultz.

Among the many animals that forage on sagebrush are Rocky Mountain elk, mule deer, pronghorn, and pygmy rabbits. Sage-grouse eat scabland sage, low (or little or dwarf) sagebrush and its two subspecies, silver sagebrush and its three subspecies, and black sagebrush, as well as the three primary subspecies of big sagebrush: Wyoming, basin, and mountain, the most palatable of all. Greater sage-grouse and Gunnison sage-grouse eat sagebrush foliage year-round, supplementing it with insects in summer—but in fall and winter, sagebrush is their sole food source.

The sagebrush sea supports around a hundred bird species, ninety mammal species, more than a hundred plant species in addition to sagebrush, plus more than three hundred species of insects, fish, reptiles, and amphibians. In the *Pocket Guide to Sagebrush Birds*, Melissa Pitkin and Laura Quattrini define sagebrush's iconic species as greater sage-grouse, Gunnison sage-grouse, sage thrasher, sagebrush sparrow, Brewer's sparrow, pygmy rabbit, mule deer, Rocky Mountain elk, moose, black bear, pronghorn, mountain lion, coyote, and gray wolf. No matter which sagebrush habitat you walk through, you'll see clear blue sky and big views, hear a unique bird chorus, and breathe in the unmistakable scent of sagebrush: the vital, life-giving shrub of the Intermountain West. §

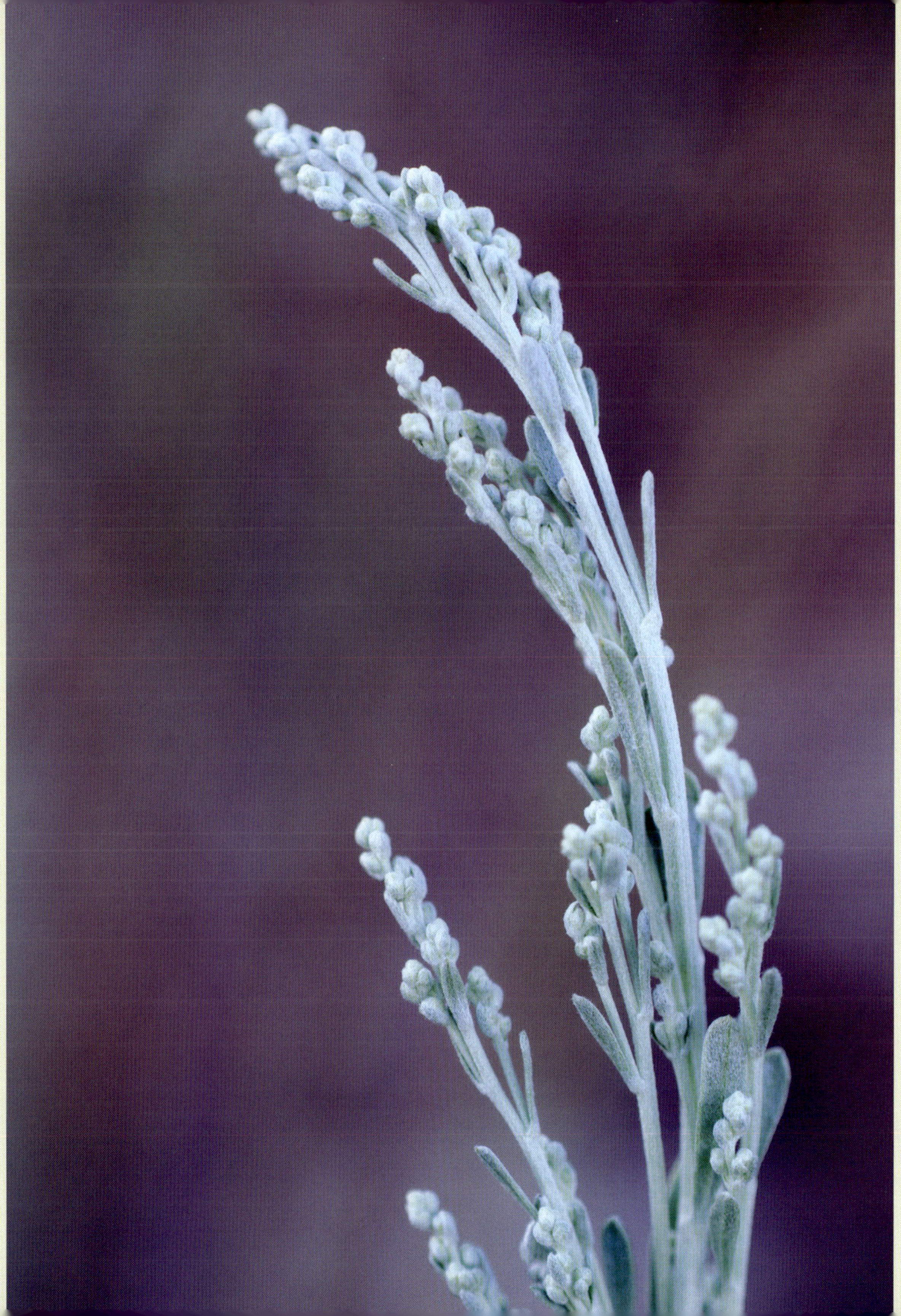

Sagebrush in bloom
(Roan Plateau, Garfield County,
Colorado)

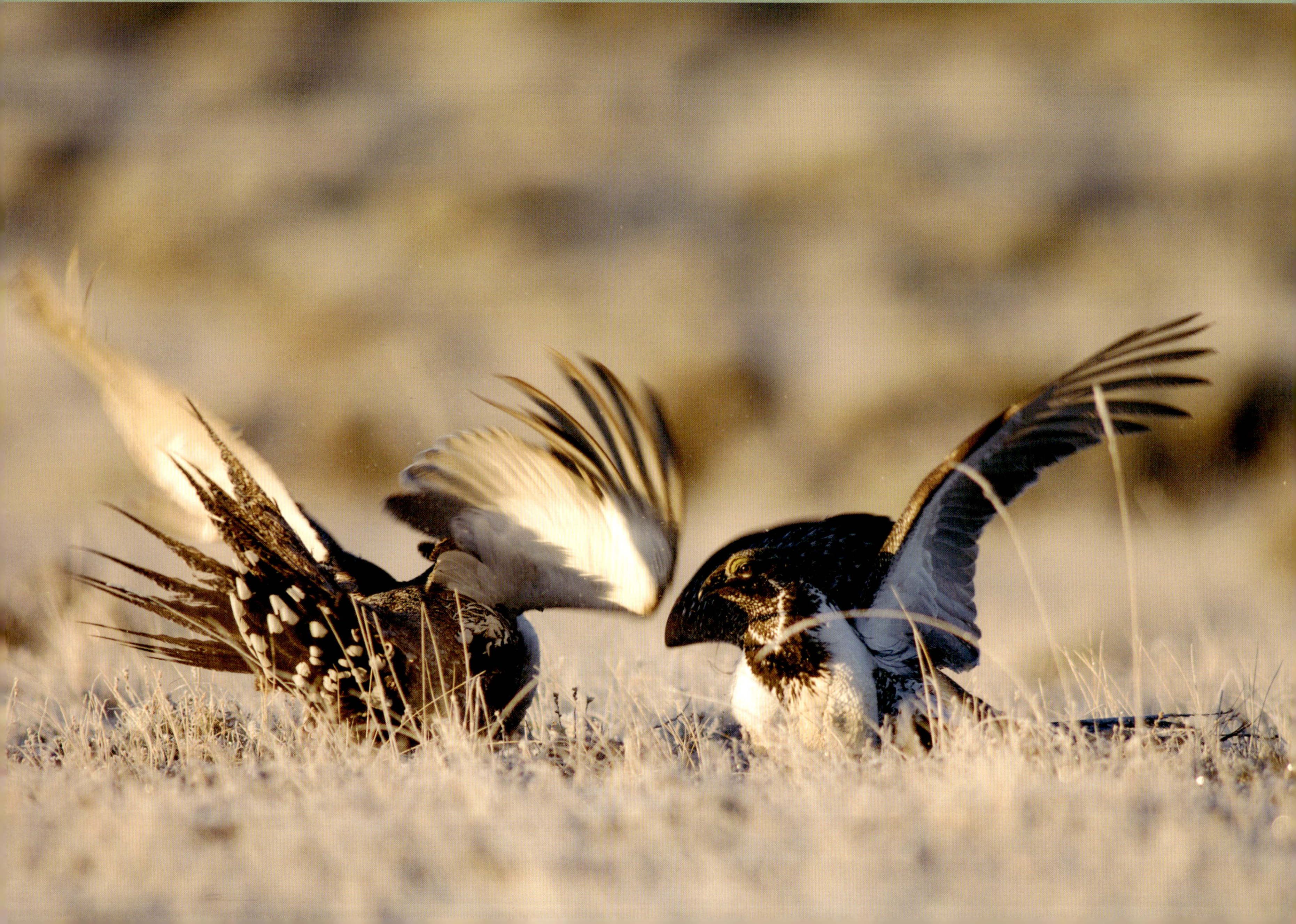

RHYTHM OF THE SAGE

OUR CONNECTION TO THE LAND

Dave Showalter

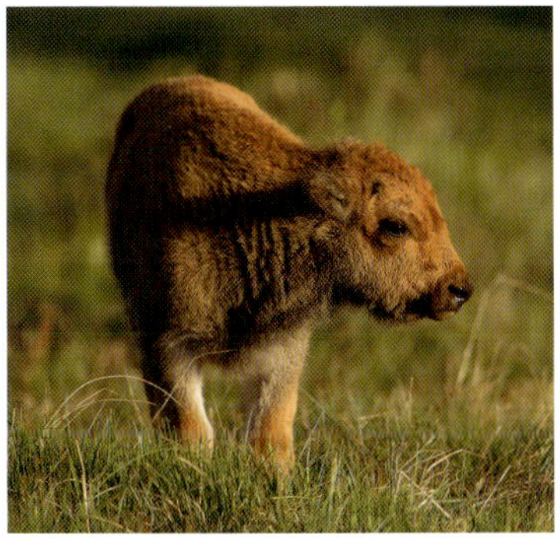

AS I SIT IN A COLD, DRAFTY, portable blind on a moonless April morning, it's so dark that I can't see the nylon walls that will keep me hidden from greater sage-grouse. The stillness and quiet are pure, the stars like diamonds. I woke at 2:30 a.m. for the privilege of a few hours with these iconic western birds—birds that are getting a lot of attention these days.

During the wait for grouse and light, I quiet my body, but my mind doesn't follow. My fatigue-addled state yields a stream of consciousness, like that on a sleepless night, but with senses alert and thoughts shifting rapidly: "My toes are cold . . . there's a breeze . . . I hope it doesn't get windy . . . coffee would be nice."

Then the birds arrive in total darkness with a few wing flaps. One, then another, then more start making the *bah-looping,* popping, swishing sounds until my piece of sage country, hard against western Wyoming's Wind River Range, sounds like an off-kilter popcorn popper. This place, at this moment, is as wild as it was in the 1800s, and I have a front-row seat. I will sit here impatiently another two hours before I can see the grouse.

Steely-gray predawn reveals silhouettes of strutting, fighting males chasing one another. I can see the birds displaying as their wings brush against inflated air sacs that impress

ABOVE Bison calf in Lamar Valley (Yellowstone National Park, Wyoming)
OPPOSITE Greater sage-grouse males fight during *lekking*, or mating, season (Upper Green River Basin, Sublette County, Wyoming).

females. Males square up chest to chest and sometimes fight, the rapid wing slaps of a territorial conflict lasting less than a second.

A pale magenta sunrise is followed by a warm glow in the short grasses on the mating ground, or *lek*—cinnamon light for birds and photographer. No golden eagle on the hunt flushes these birds today, and I take photos and watch and smile as two hours pass like a thought—that is one of the rhythms of the sage. The grouse face the breeze and fly from the lek a few at a time until just a handful of males are displaying for one another and the sage. They'll rest and eat in sagebrush, then return tomorrow morning and the next day, as they have every spring, on this lek, for thousands of years. That, too, is the rhythm of the sage.

My small camouflage blind is a speck in the eastern sagebrush sea, an apt moniker for the Intermountain West's dominant shrub species. Here, halfway between Grand Teton National Park and Rock Springs, in Wyoming's Green River Basin, is a stronghold for greater sage-grouse and a migration corridor for mule deer and pronghorn moving between winter and summer range. On the east side of the basin, the jagged Wind River Range, pure white with winter snowpack, stretches nearly 100 miles (160 kilometers) northwest, spawning the Green, Big Sandy, and New Fork rivers on this side of the mountains. The Wyoming and Gros Ventre ranges form the basin's western and northern boundaries. In between is rolling terrain covered in sagebrush in a classic, spectacular intermountain valley rich in wild diversity and rugged beauty.

In mid-April, the peak of grouse lekking season, pronghorn and mule deer are migrating north from this basin hundreds of miles to fawn and spend the summer in the mountains, and sandhill cranes are arriving and dispersing throughout the Greater Yellowstone Ecosystem after a 1000-mile (1600-kilometer) migration from New Mexico. In a month, all of the sagebrush will be alive with the sweet melodies of songbirds singing from the tallest sagebrush tops to protect mating territory.

By mid-June, memories of a harsh winter fade when showy yellow balsamroot, lupine, and scarlet Indian paintbrush splash color across the land. Broods of mottled sage-grouse chicks awkwardly trail behind mothers as they grow by eating succulent forbs and high-protein insects. A few months of spring seem to span winter to summer until the sage is pulsating with life's rhythms.

If I were dropped into this outsize landscape in mid-June, I would think the world is absolutely perfect, but the truth is this ecosystem is collapsing under the weight of human development. The estimated quarter to half million greater sage-grouse in eleven states (Montana, Wyoming, Colorado, Utah, Idaho, Nevada, California, Oregon, Washington, North Dakota, and South Dakota) spanning 166 million acres (75 million hectares) represent a small fraction of the species' historic population. In four decades from 1965 to 2003, greater sage-grouse declined at a rate of 2 percent per year and a lost half their population; a bird that has thrived in the West for millions of years vanishing in plain sight. Imperiled and declining steadily describes the status of sage-grouse, Brewer's sparrows, and pygmy rabbits. Sage thrashers and sagebrush sparrows are declining slowly. All these are sagebrush *obligate* species—wildlife that must breed and raise their young in sagebrush, dependent on healthy expanses of sage for survival. Mule deer are struggling too—down 44 percent since 2001, when drilling began on Wyoming's Pinedale Anticline, and down 66 percent in northwest Colorado, which includes the Piceance Basin.

My idyllic lek location in the Upper Green River Basin is only a couple of ridges and a dozen miles from the Jonah Field, a natural-gas megadevelopment that transformed more than 30,000 acres (12,000 hectares) into a fragmented, single-use industrial zone. Humankind's penchant for development without concern for long-term impacts is displayed in the Jonah, Pinedale

Portable hunting blind (used for photography) on a greater sage-grouse lek in Sublette County, Wyoming

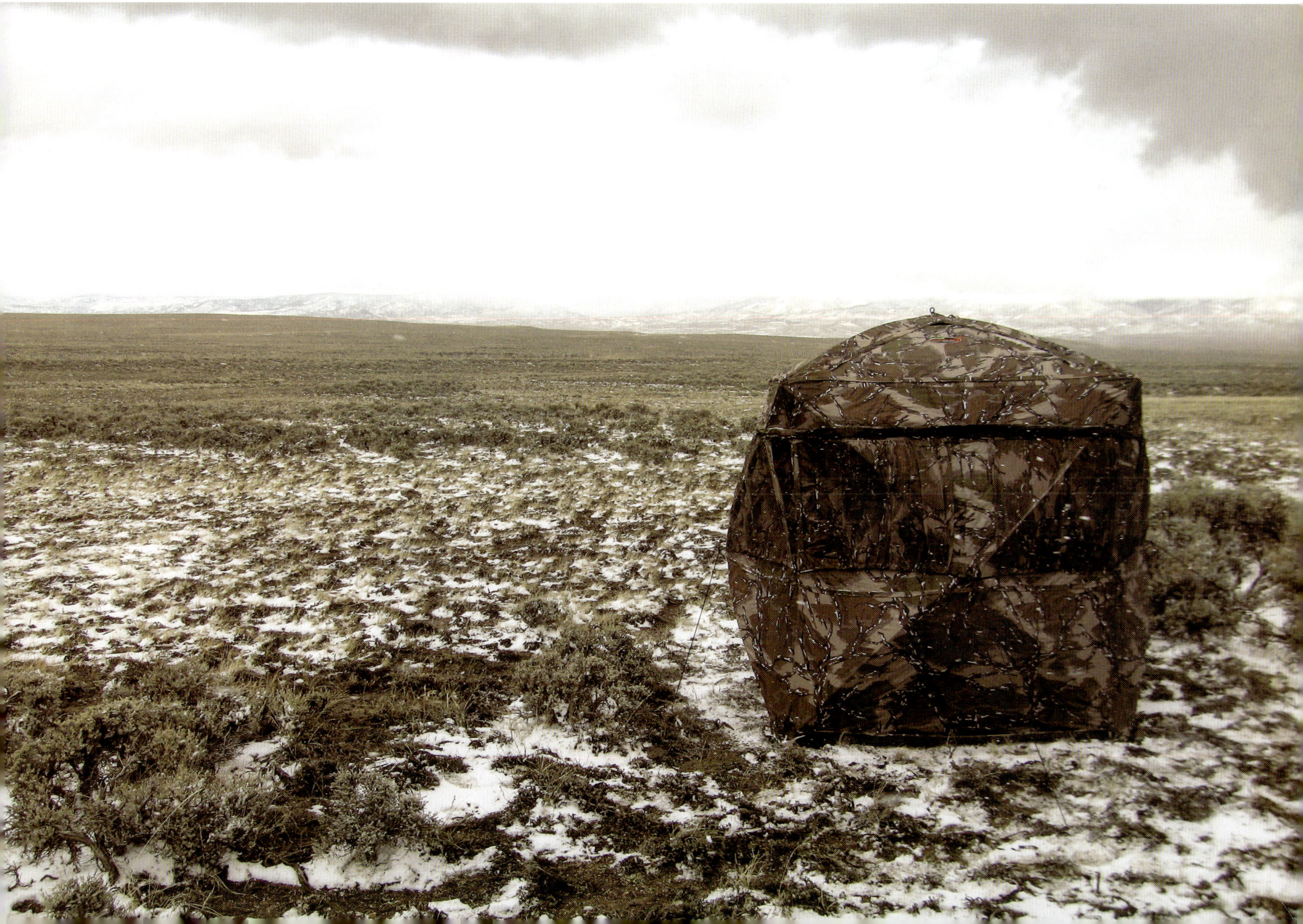

Anticline, La Barge, Continental Divide–Creston, and Piceance megafields: each is interlaced with roads, five-acre (2-hectare)well pads, powerlines, and compressor stations attended by thousands of heavy trucks. A gas field is never dark or quiet; semitrucks laden with toxic fluid for hydraulic fracturing and by-product water from fracked wells crisscross the gas fields day and night, dust clouds billowing across the sage. In 150 years of remaking the West, this energy-development rush epitomizes a determination to bend the land to human will.

Greater sage-grouse are often described as a brownish chicken-sized bird that lives in sagebrush, and they are compared to the spotted owl because of the nexus of an imperiled species and burgeoning development. In the 1980s the endangered northern spotted owl, clinging to the last 10 percent of its habitat—Pacific Northwest old-growth forest—was pitted against a dying timber industry. The parallels to greater sage-grouse are undeniable—but while the energy-development rush is booming in the Intermountain West, with courage and political will, westerners can choose to leave critical sage-grouse core-area habitat and other sensitive areas undeveloped.

Evolved over eons, the sage-grouse date back 25 million years. They are perfectly suited for their environment, for sage-grouse spend their entire lives in sagebrush. They eat, mate, nest, and raise their young in sage, migrating to find good grass and wet meadows for forbs and insects, typically returning to the same lek every spring to perform one of the most spectacular mating rituals in the natural world. Gunnison sage-grouse, which occupy a much smaller area than the greater sage-grouse, share the same dependence on healthy sagebrush habitat—and they are far more vulnerable than greater sage-grouse, currently inhabiting about

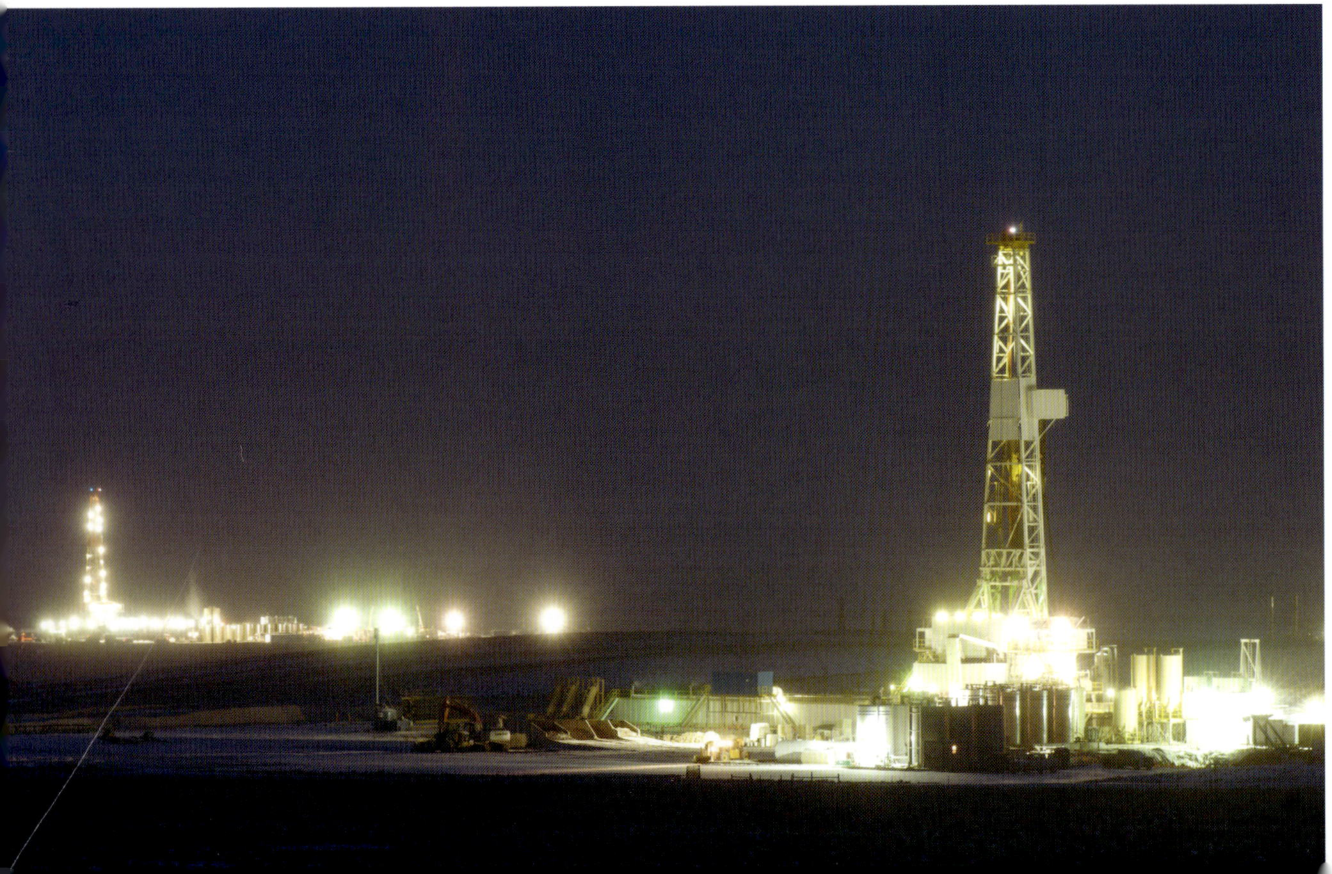

Drilling on the Pinedale Anticline never stops—it is never dark or quiet. Natural-gas rigs run through winter on this critical mule deer winter range in Sublette County, Wyoming. The deer population, which migrates 40–50 miles (64–80 kilometers) to winter here, has declined by 44 percent since 2001.

This greater sage-grouse female blends in with her sagebrush environment. Sage-grouse spend their entire life in sage (Pinedale Anticline, Sublette County, Wyoming).

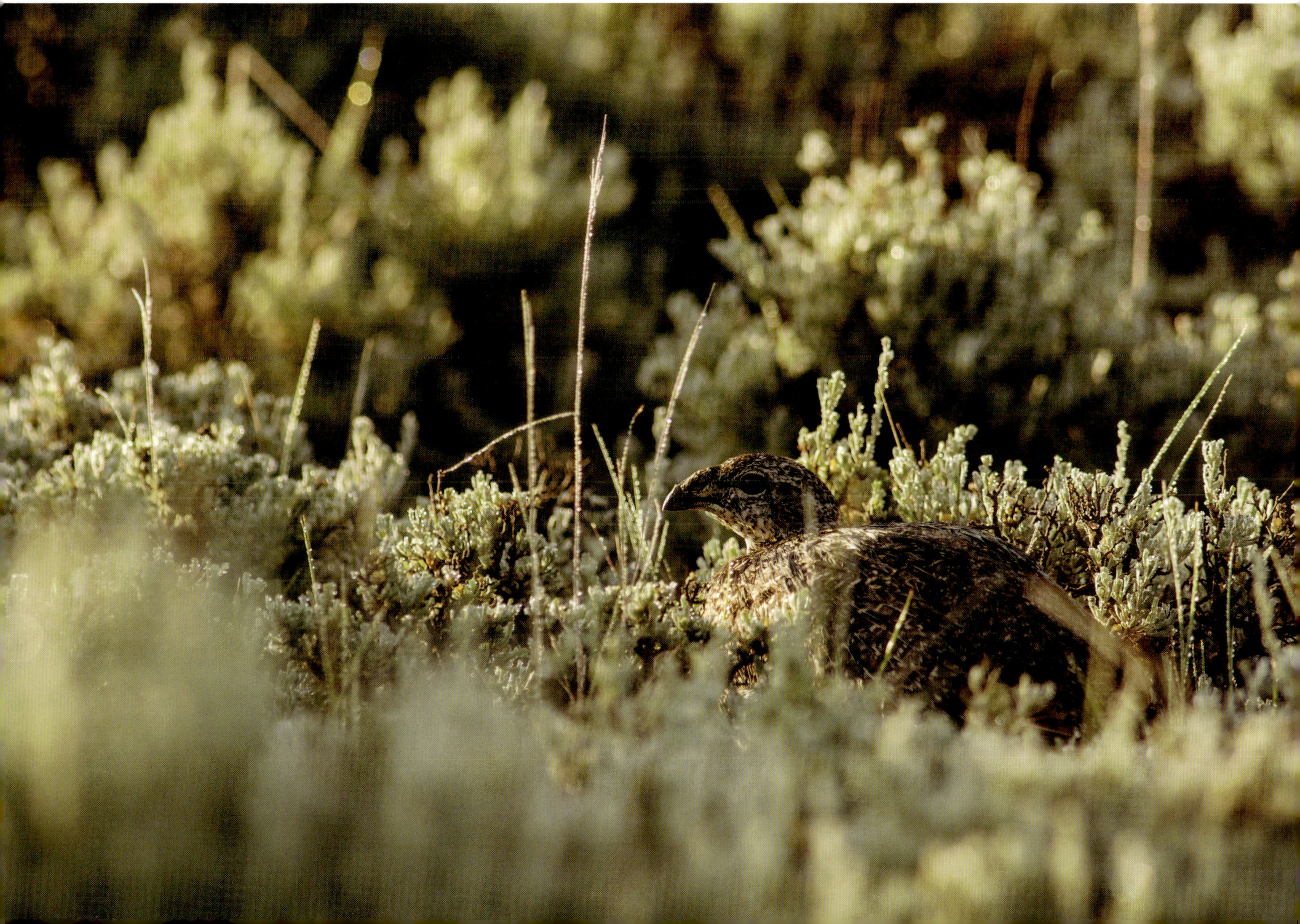

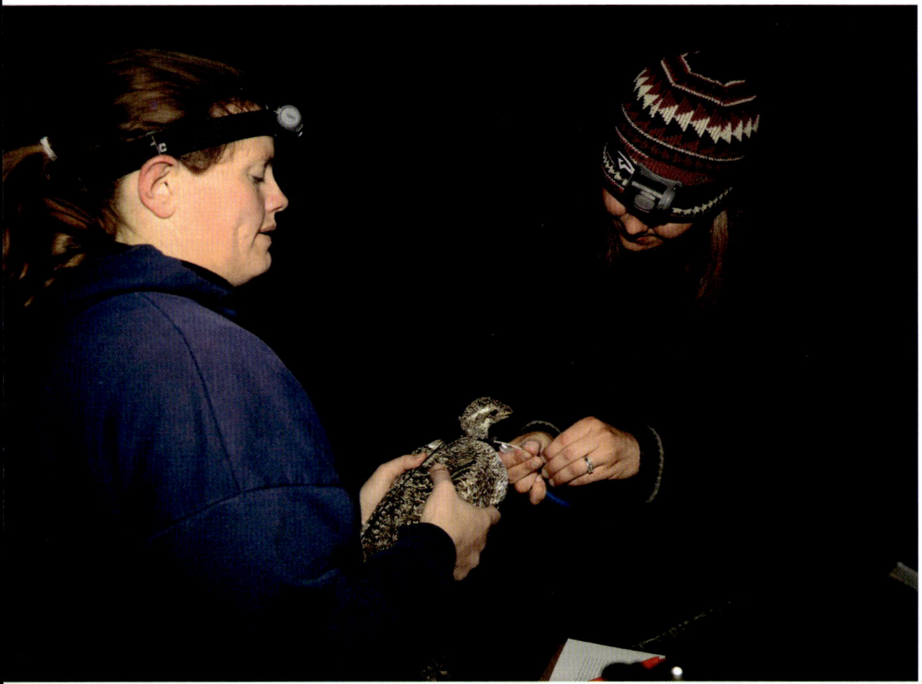

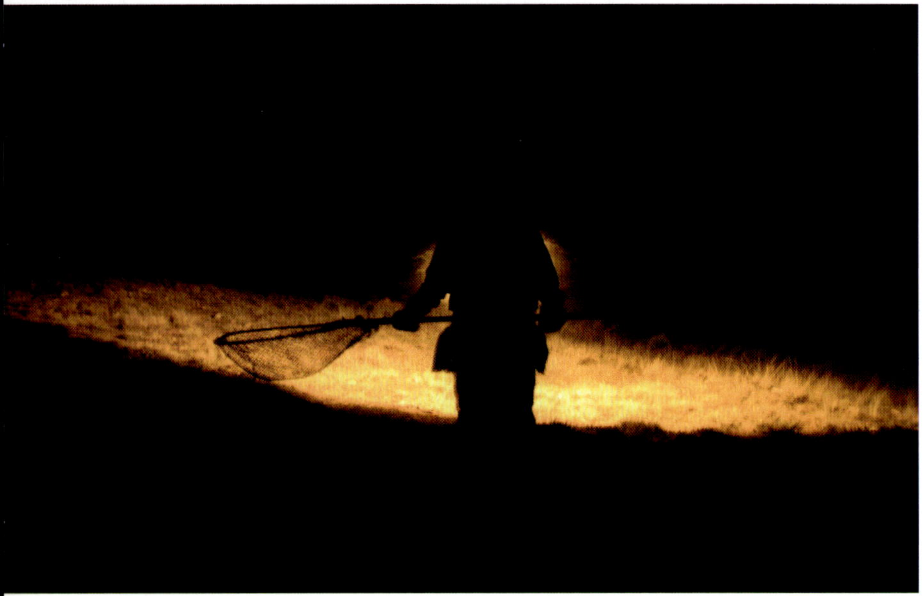

10 percent of their historic range in west-central and southwest Colorado and eastern Utah. Habitat loss and fragmentation is the primary cause for alarming population declines in both species of sage-grouse.

Managing imperiled species often requires catching them to study the creature, attaching a telemetry collar, even relocating animals to other populations to improve genetic diversity or to ensure there are enough birds to continue mating. On a full-moon night in mid-April, I found myself in the back of a one-ton pickup, escorted by Brooke Palmer and Jessica Divens, two of the best Gunnison sage-grouse seasonal trappers with Colorado Parks and Wildlife. We bounced our way uphill on rutted two-track to a ridge above Ohio Creek just outside of Gunnison, Colorado. It seemed too warm for April; the moon, just beginning to wane, peeked in and out of fast-moving clouds; lightning bolts silently danced up the valley near Crested Butte, the scent of rain on a breeze.

I was there to photograph trapping for relocation to the Crawford, Colorado, lek, which was in need of males for the remaining

TOP LEFT Gunnison sage-grouse are trapped, collared, and loaded for transfer to the Crawford lek that evening by Brooke Palmer and Jessica Divens of Colorado Parks and Wildlife. The relocation program, which bolsters declining grouse satellite populations outside of the Gunnison Basin, is intended to improve breeding success where small numbers of grouse and limited genetic diversity threaten local populations (Gunnison Basin, Colorado).
BOTTOM LEFT Jessica Divens, a trapper with Colorado Parks and Wildlife, heads into the sage of the Gunnison Basin to capture a Gunnison sage-grouse for relocation to a satellite lek.

lekking season. Gunnison sage-grouse are the fourth most imperiled bird in North America, behind the ivory-billed woodpecker, California condor, and whooping crane. Numbering around 4500 individuals, Gunnison sage-grouse need help to recover and survive.

"Get ready," says Jessica, at the wheel, while Brooke runs the spotlight in transects across the sage. As we creep around a lek site, some of the males are displaying puffed-out white chests under the moon's brightness. The spotlight stops when Brooke sees the eye shine of a grouse, a tiny blue light in a world of black and silhouettes. Sage-grouse freeze when the light lands on them, and Brooke is the best at locking in on a grouse, even at a few hundred yards. Once Brooke is locked on, Jessica stops, Brooke holding the light steady while noting coordinates in her mind—a juniper, the location of a star, or the moon at midnight.

Gear is readied, vests go on, the boombox is cranked; tonight it's "Crazy Train" by Ozzie Osbourne. The loud music confuses the birds and covers the sound of footsteps while the trappers move in with a giant fishing net. I can see the activity in the field when Brooke and Jessica stop. They've netted a grouse and quickly hold him in hand to calm the bird on the way back to the truck. The music is off, the trappers silent, communicating without sound.

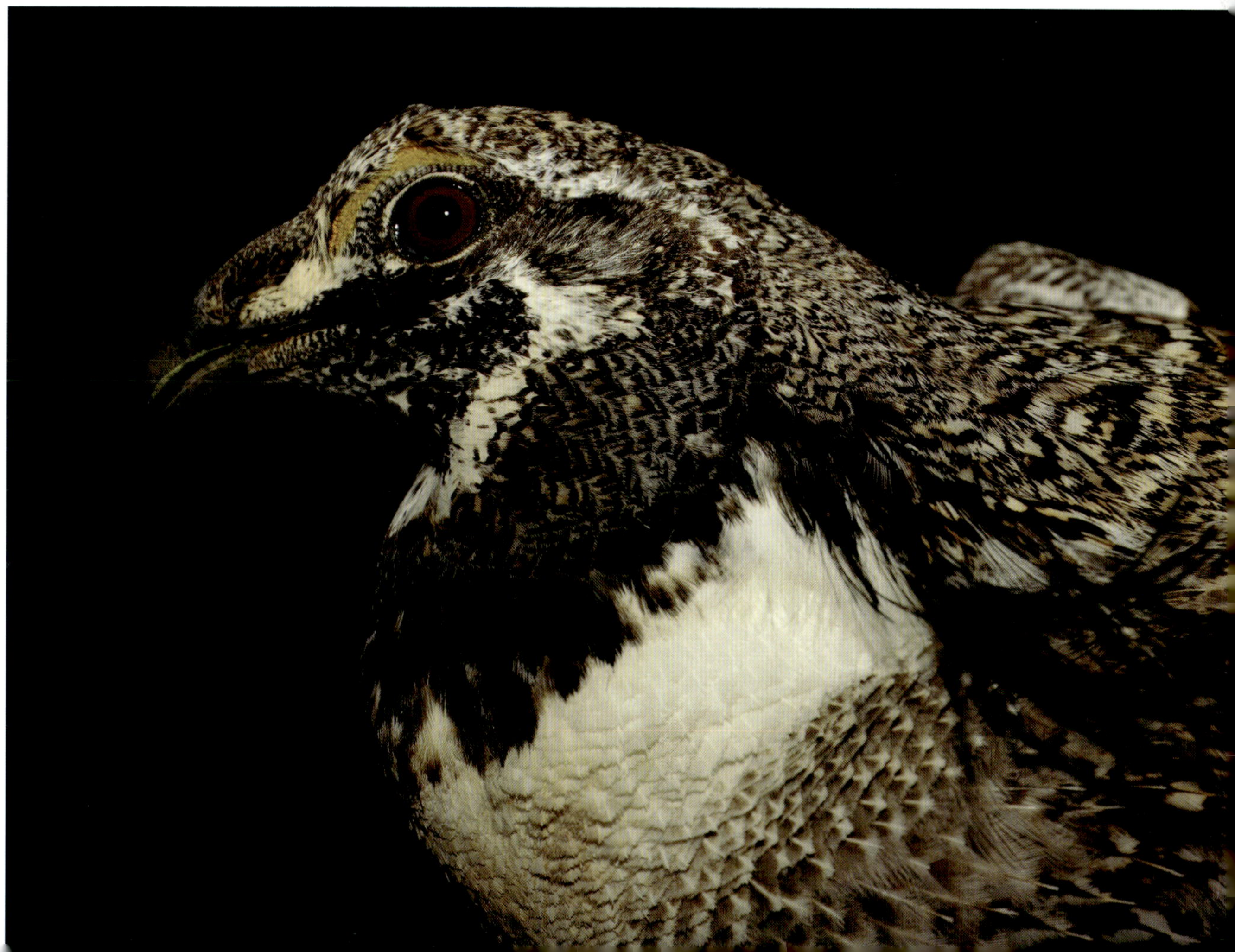

This Gunnison sage-grouse male is one of several that were collared and loaded for transfer to the Crawford lek as part of a Colorado Parks and Wildlife trapping program (Gunnison Basin, Colorado).

When they reach the truck, they set the bird gently in a cotton sack, weigh him, then remove him from the sack, measure him, record the sex and maturity, and put on a telemetry collar with short antennae attached. All of the information is recorded, with few words exchanged. When Brooke holds up a mature male for a close-up photograph, I'm awestruck by the remarkable beauty of this vulnerable creature. Brooke then carefully places the grouse in a custom wood box marked "Crawford." He'll be displaying on a new lek before dawn. They caught four grouse, three males and a female—a big night near the end of trapping season.

There almost wasn't a Gunnison sage-grouse. Until 2000, when Gunnison sage-grouse were recognized as a separate species, there were only *sage-grouse,* all lumped together as one species. Jessica Young knew there was something different about the Gunnison sage-grouse and pursued her PhD by working on "sexual selection and physical and behavioral differences in the Gunnison sage-grouse" at Western State Colorado University in Gunnison. Countless predawn mornings gathering data in cold blinds by listening to and observing grouse ultimately led to designation of the Gunnison sage-grouse in 2000.

Gunnison sage-grouse are smaller than greater sage-grouse, with wide white bands on tail feathers and filo-plumes that are often called a "ponytail" on the top of their heads. Males amusingly shimmy their rear end at the end of a display. Their vocalizations, while similar to greater sage-grouse's, are distinctly different. Had Jessica not followed her instincts and spent more than a decade studying Gunnison's namesake grouse with a team of talented scientists, these extraordinary creatures might never have been known.

Species that are officially designated endangered require changes in land management practices, with government oversight, and so not everyone was excited about the recognition of a new

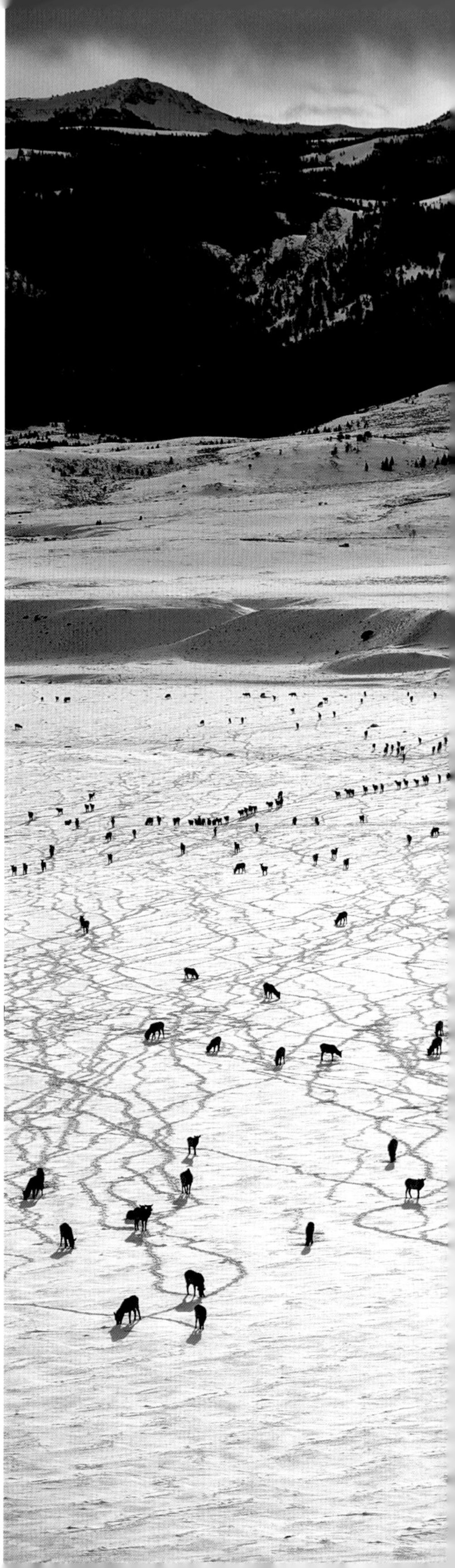

Elk seem to pour from the Gravelly Range to windblown open range in Montana's Madison River Valley. Elk are among the western ungulates that use sagebrush lands for critical winter range, later following snowmelt as they return to higher elevations in spring.

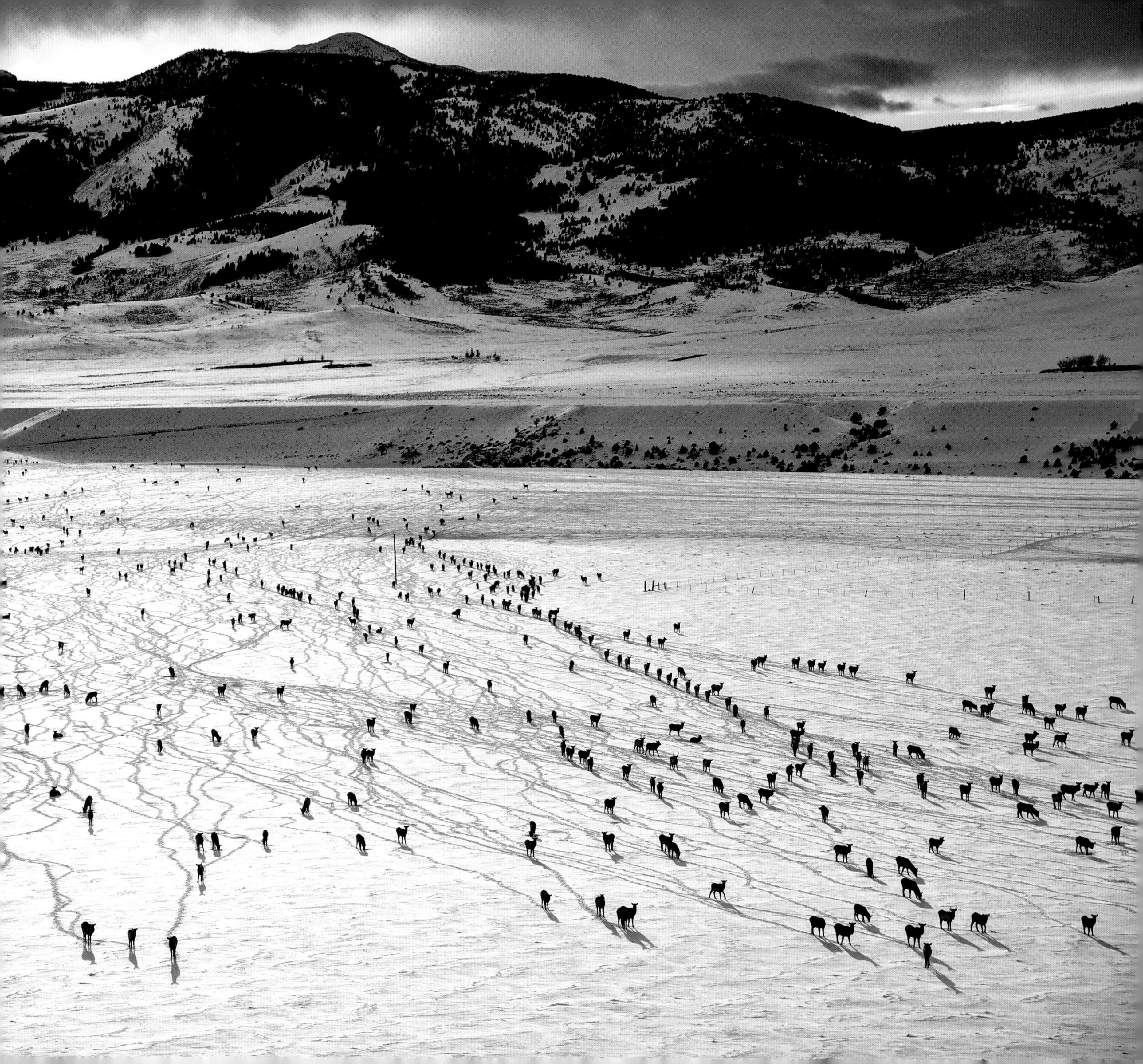

species that was already imperiled and surviving on 10 percent of its historic range, much of it ranch land. So much habitat had already been lost to powerline construction, a tailings pile and tailings-removal road from a Gunnison uranium mine that split an active Gunnison sage-grouse population, and lands flooded by Blue Mesa Reservoir near Gunnison and Miramonte Reservoir near Telluride, Colorado. Locals tell heartbreaking stories of grouse displaying on Blue Mesa Reservoir ice after it was flooded to create Colorado's largest body of water—a last dance as close as they could get to where their lek used to be. Sage-grouse are so hard-wired to their lek that they simply die out after being displaced from it. Even relocating them doesn't guarantee success.

But there was no going back, and the extended Gunnison community of Colorado Parks and Wildlife, Western State Colorado University, the National Park Service (managing Black Canyon of the Gunnison National Park), ranchers, conservation groups, community members, and recreationists came together. Working groups were formed; planning for the recovery of the newest (unofficially endangered) species in the West was underway. Gunnison sage-grouse were designated as "threatened" under the Endangered Species Act in November 2014. Today, Gunnison sage-grouse habitat is guarded for seasonal nesting and brood rearing, many popular public access roads are closed seasonally, every lek is counted, and sound is monitored near leks, since even seemingly innocuous sounds mask communication between grouse and disrupt male displays on the lek, potentially resulting in fewer opportunities to breed. Experimental habitat-improvement projects in sage and riparian areas on private lands have multiplied. Birders from around the world flock to Gunnison every spring, viewing from a trailer blind operated by Sisk-a-dee, a local nonprofit, as grouse display on the Waunita lek.

From Greater Yellowstone to Telluride, Colorado, blobs on maps that look like spilled ink mark "sage-grouse core-area habitat," areas of highest population density and greatest conservation priority. The boundaries aren't square, for animals move, and grouse need roughly 4 miles (6 kilometers) of good sagebrush habitat around each lek to thrive. In theory, protecting core-area habitats gives sage-grouse the best chance to rebound—of course, the habitat has to be healthy, the boundaries respected and fiercely protected. In addition to having those core-area habitat maps, we also know where hoofed mammals (ungulates) travel: mule deer, pronghorn, and elk move between summer and winter range, often over great distances, migrating and birthing in sage.

It matters little whether people like or dislike sage-grouse; what's important is its role as an umbrella species that must be embraced to see a much bigger picture. Protecting grouse habitat benefits the sagebrush ecological community: the songbirds chirping from sagebrush tops and laying eggs beneath; bears, wolves, mountain lions, coyotes, ravens, golden eagles all following the cycle of life and death. Prairie dogs, ground squirrels, and badgers bulldozing and tunneling the understory, which shelters toads, lizards, snakes, burrowing owls, and insects that feed grouse chicks and songbirds. More sagebrush rhythms.

Humans have also been a part of this web of life for thousands of years. In Greater Yellowstone, Shoshone Indians act out greater sage-grouse courtship in dance. Ute Indians in southwest Colorado dance as Gunnison sage-grouse, the ancient rhythmic movements of each tribe passed down over generations out of reverence for grouse, sagebrush, and all living things.

If I wanted to see the West on a grand landscape scale without borders, a bright red Cessna tail-dragger with the door removed seemed like a good way to do it. Chris Boyer is a pilot and photographer out of Bozeman, Montana, who runs Kestrel Aerial Services and contributes part of his time to LightHawk, an organization that donates flights to elevate conservation. Some things simply must be observed and photographed from the air, so we prepared to fly a cold mission over elk herds on winter range in southwest Montana's Madison River Valley. Removing the plane's door in February spices things up and gives me the perfect platform for photography.

"There are the Spanish Peaks," Chris says. We're headed directly toward Ted Turner's Flying D Ranch over a world of white in flat, overcast light. Small details such as aspen forest in a crease catch my eye, then a few bison, like pepper flakes. Turner is the second-largest private landowner in America. The Flying D's 2 million acres (809,000 hectares) sprawling across southwest Montana is one of sixteen Turner ranches, all of them managed for sustainability. All the creatures of the Greater Yellowstone Ecosystem are welcome on the Flying D Ranch, wolves and grizzlies too—*especially* wolves and grizzlies.

Ten minutes later, we're still flying over Turner's ranch when we see the Madison River cut through Bear Trap Canyon, our cue to hang a left and head north to follow the Madison toward West Yellowstone. North of Ennis Lake, trumpeter swans grace curves in the river in wide spots where the current slows. In the distance, we spot elk; the thousands in the valley have split into groups of hundreds, their hooves pounding an elk highway on the snow-covered open range. A sun break lights one herd and their tracks, elk pouring from the Gravelly Range. This is the image I'd hoped for: big herds moving en masse. Wildlife need more than national parks and alpine wonderlands—they go to windblown lower elevations in the sage to survive a harsh winter, a race against time until spring green-up. We chased evening light all the way back to Bozeman on cold, still air, never quite catching up to it.

If we could keep going in Chris's red plane and maintained a southern heading, we could fly the entire east side of the sagebrush ecosystem and see how the land flows, follow great rivers and migration paths, and marvel at red-rock desert in the rain shadow of the mountains. Ranch lands, gas fields, mountains poking the clouds, forest lapping sagebrush at the edges: we could see what we want to see—or, better yet, see it all, and realize that this sagebrush sea is finite. There is just this one.

We Americans have claimed all of this sageland we see stretched out before us and we've used it, but it's not used up. Our legacy will be not how we profited from the land, but whether wildlife will have freedom to roam and whether future explorers can find their own rhythm in this sagebrush sea. §

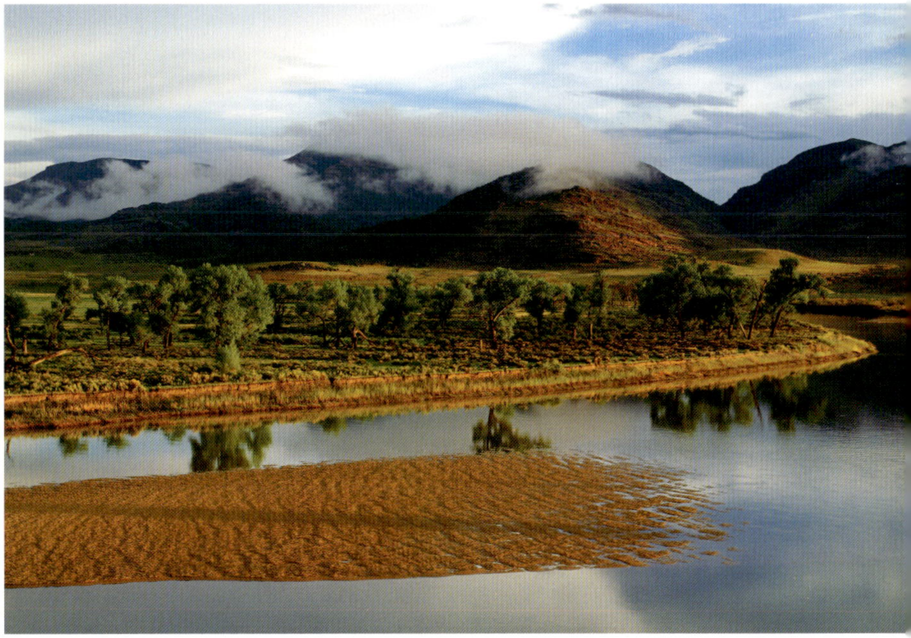

Warm sunset light on a bend in the Green River after an intense summer thunderstorm (Browns Park National Wildlife Refuge, Colorado)

SAGE MIGRATIONS
AMERICAN SERENGETI
Dave Showalter

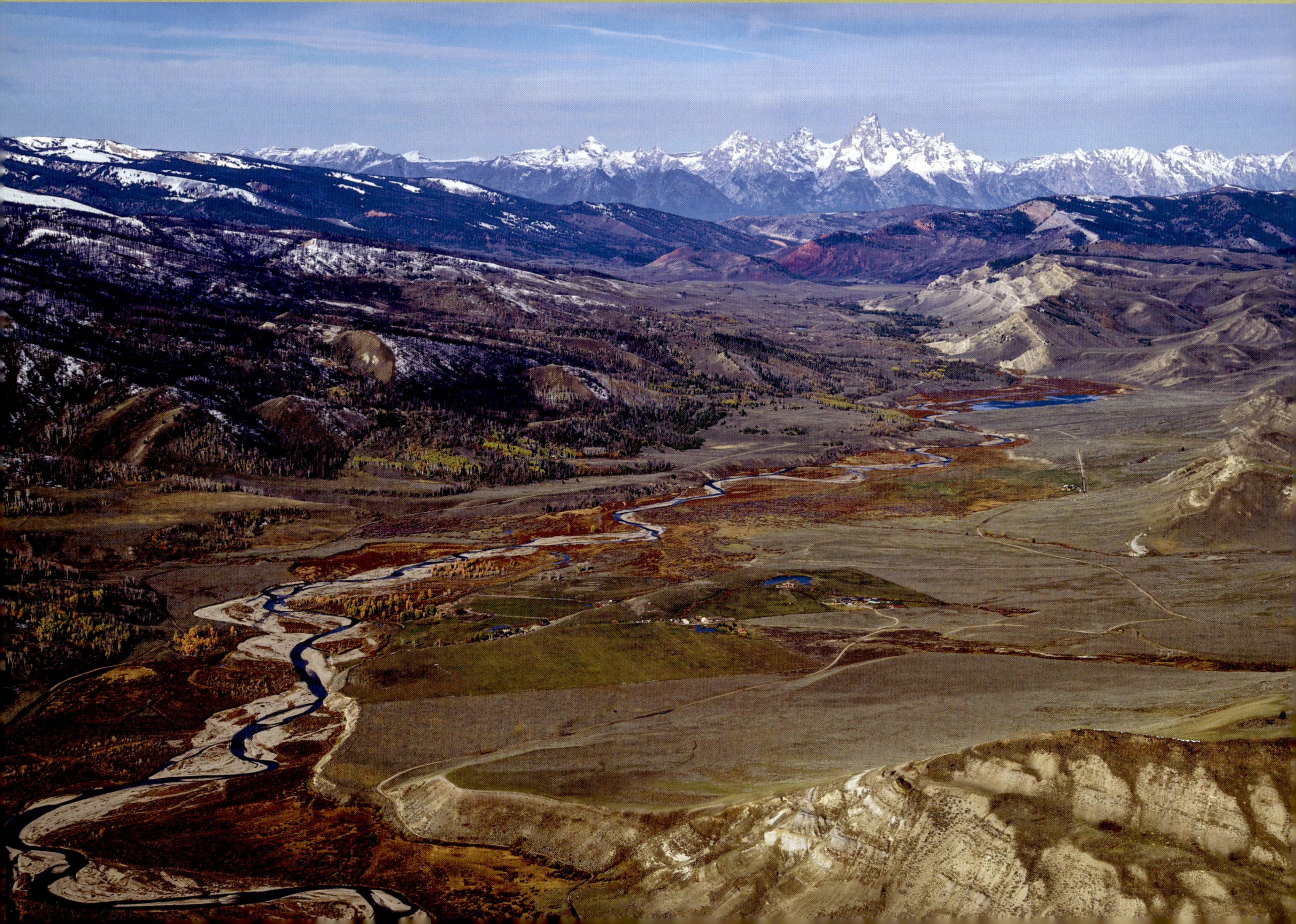

LARGE HERDS OF MAMMALS moving across an immense landscape are one of the most awe-inspiring sights in the West. With movements that reveal the terrain on a grand scale, migration tethers the entire West, from high sagebrush desert to mountainous areas. Migration is about an internal clock telling wild creatures to leave one seasonal home for another. Migrations are the arteries in the heart of the West, lifeblood for sustainability: migration means survival. Connecting the remaining fragments of the sagebrush sea is critical to the migrations for a wide variety of western mammals, particularly pronghorn and mule deer. If conservation efforts focus solely on sage-grouse, we miss this big picture.

A fresh snowfall blankets Grand Teton National Park, in Wyoming's northwest region, in early October: a disruption of golden autumn days, forewarning that winter is near. Pronghorn on Antelope Flats bunch up, grazing on snow-covered sagebrush. The rut is still on, males chasing females, guarding harems, thrashing the sage with pointed horns. I watched a buck mount a doe, his antlers full of tangled sagebrush, just their heads poking above the sage. Another doe fed on sagebrush with her pair of juveniles, fueling up for the journey ahead.

The next day they were all gone from Antelope Flats.

Why would the Teton pronghorn leave in early autumn while foraging on grass and sagebrush leaves is still good? Locals know that the second major snowstorm of the season pushes the animals out of their sagebrush summer range on their migration south to winter range in the Upper Green River Basin 120 miles (190 kilometers) distant, as they've done for thousands of years.

This aerial view of the Gros Ventre River Valley shows a large swath of the "path of the pronghorn" migration corridor, with the Teton Range as a backdrop (Bridger-Teton National Forest, Wyoming).

North America's fastest land mammal, pronghorn prefer open country where they can spot predators from a distance and use speed to escape. Yet the need to survive winter, by migrating through wooded mountains to reach wind-scoured range where grasses aren't buried in snow, supersedes habitat preference. The Grand Teton "path of the pronghorn" herd migrates up the Gros Ventre River Valley to rugged, forested subalpine terrain at the river's headwaters before dropping down into the expansive Upper Green River Basin and its sage-covered, corrugated ridges and mesas. The Teton pronghorn disperse south of Pinedale, as far as the Jack Morrow Hills in the Red Desert, joining other migrating pronghorn herds in search of windblown ridges in the sagebrush sea to forage through the long Wyoming winter.

For western ecosystems to function and to prevent national parks and wilderness from becoming biological islands, migratory wildlife corridors need to be open. Migrating wildlife—from pronghorn to elk, moose to bighorn sheep—can and do pass through developed areas, but the next badly sited road, housing development, fence, or energy field may be the tipping point that obstructs a migration. Although three major bottlenecks and the vagaries of weather make their migration precarious, the Teton pronghorn are using a newly constructed wildlife bridge at Trapper's Point that allows safe passage over Highway 191 near Pinedale. Forty-five miles (70 kilometers) of US Forest Service land on the "path of the pronghorn" is now federally protected, the only designated migration corridor in the nation.

In southwest Wyoming just north of Interstate 80 near Rock Springs, a herd of about five hundred mule deer also spend winter in the Red Desert, situated in the Great Divide Basin, which is wrapped by the Continental Divide. In 2011 research biologist Hall Sawyer set out to learn about the movements of these mule deer. Sawyer, a specialist in western Wyoming's ungulate (hooved mammal) migrations, has studied many of the state's great migrations, yet he wasn't expecting to discover that the longest mule deer migration ever documented was "happening right underneath our noses."

Sawyer put telemetry collars on forty mule deer to learn about their migratory behavior and where they go to summer range. He discovered that in late winter or early spring, the Red Desert mule deer herd begins slowly moving up in elevation as they forage on emerging grass. Traveling due north through the Red Desert and along the west side of the Wind River Range, these mule deer join thousands of deer from other herds, resting and refueling at known stopover points, the way people use rest areas along a road trip.

The mule deer migration may take up to three months to cross multiple highways, fences, lakes and rivers, and housing and energy developments to reach the Upper Hoback River Basin at the base of the Wyoming Range 150 miles (240 kilometers) from where they started. Here the Red Desert mule deer herd, on summer range along a wildlife migration superhighway, spends the season fawning

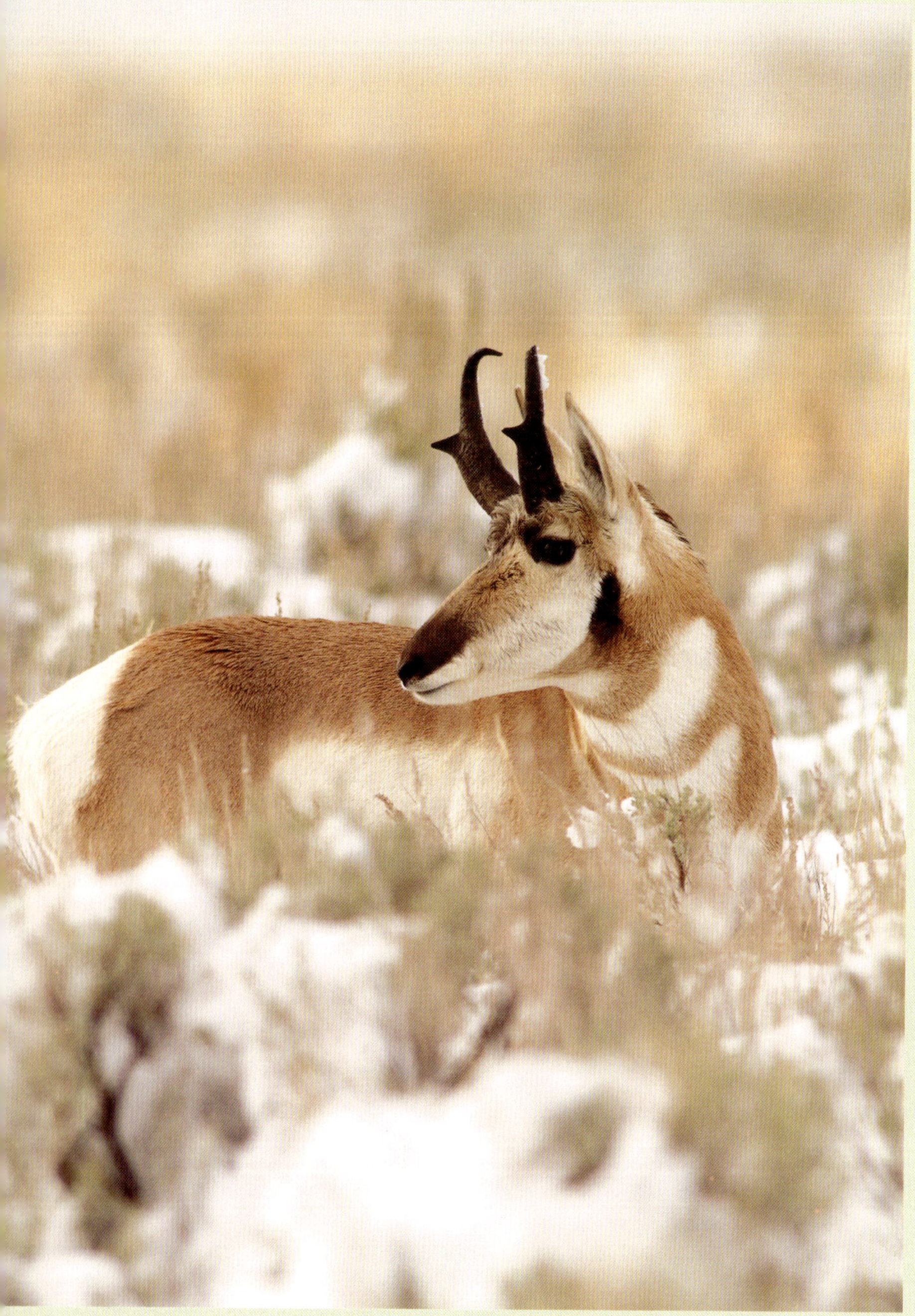

A pronghorn male observes another male during rut in October snow on Antelope Flats in Grand Teton National Park, Wyoming. The famed Teton herd was gone the following day, on their "path of the pronghorn" migration to the Upper Green River Basin.

TOP RIGHT Pronghorn of the Teton herd migrate from the Upper Green River Basin in mid-April north toward Grand Teton National Park, Wyoming. The roughly 120-mile-long (193-kilometer-long), migration route that has crossed the rugged Gros Ventre Range for thousands of years has been designated a National Migration Corridor, providing passage to the three hundred to four hundred animals that make the journey. Yet the pronghorn still face a gauntlet of unnatural challenges in addition to unpredictable weather.

BOTTOM RIGHT A pronghorn buck feeds in a strip of sagebrush in the Jonah Field, Sublette County, Wyoming.

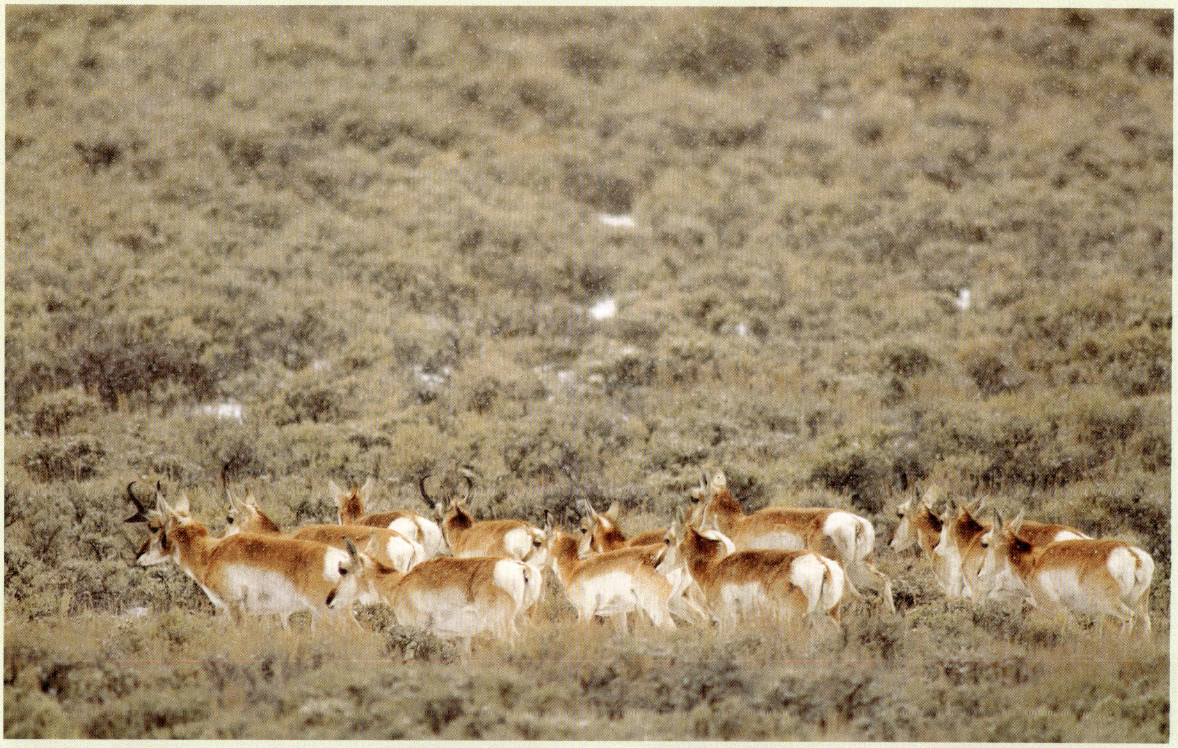

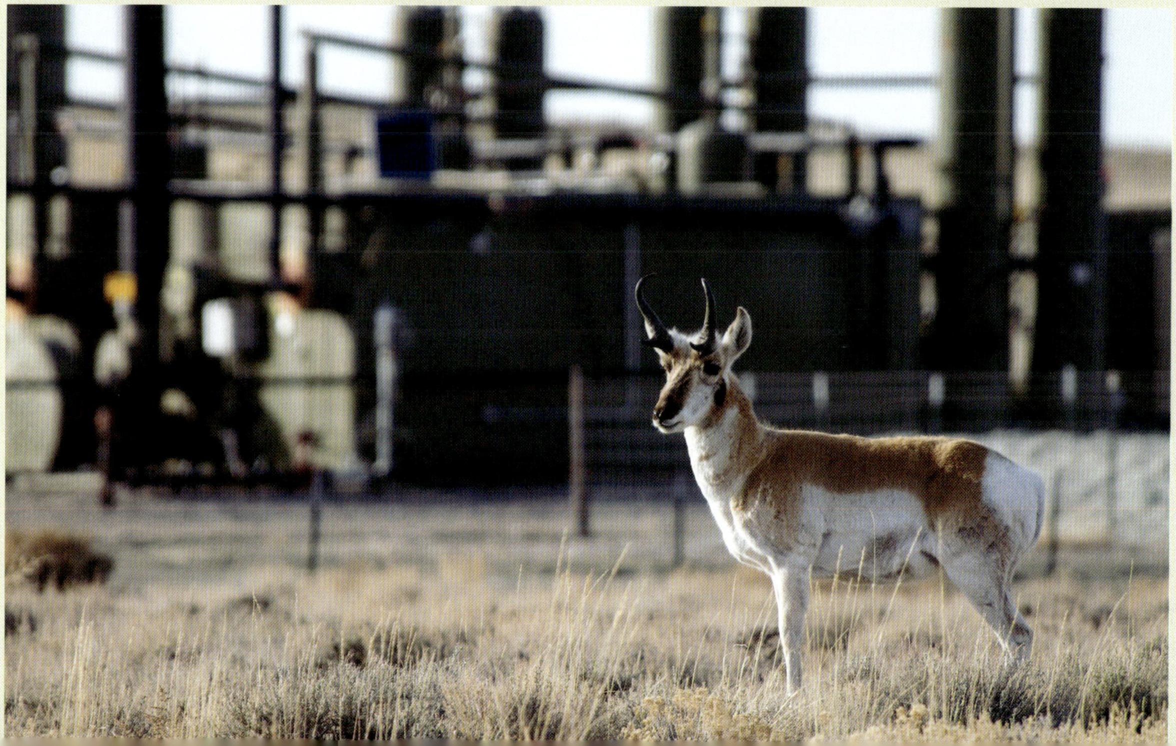

and grazing in alpine meadows and on mountain slopes as high as 10,000 feet (3048 meters).

It is astonishing that this herd goes only to the Hoback and that they make the reverse journey on the exact same route again in autumn, back to their sagebrush winter range in the Red Desert—a path passed on generationally from mothers to offspring.

Mule deer pay a heavy price if winter range is developed. The Pinedale mule deer herd relies on the sagebrush habitat of the Pinedale Mesa for critical winter range. Yet in the fourteen years since drilling began on the Mesa, the herd's population has decreased by 44 percent.

In an increasingly developed and fragmented West, wildlife face daunting challenges as they migrate between seasonal habitats. The migrations of mule deer in Colorado's Piceance Basin, called the "mule deer factory," cover between 20 and 35 miles (30–55 kilometers) dotted with natural-gas drill pads, rigs, and roads. Sawyer's research shows that mule deer populations will migrate through gas fields, but they move through quickly, rather than resting and refueling. As the Piceance Basin's development grows, mule deer have declined significantly there in recent years.

Migrating ungulates spend 95 percent of their migration resting and refueling at stopovers used year after year. Sawyer, who considers protecting stopovers a way forward for corridor conservation, explained: "If we want to continue to enjoy and sustain the deer herds we have today, we need to take care of these migration corridors."

Many pronghorn and mule deer migrations travel over sage-grouse core habitat. Although conserving core habitat is critical for sage-grouse survival, it is inadequate for protecting corridors

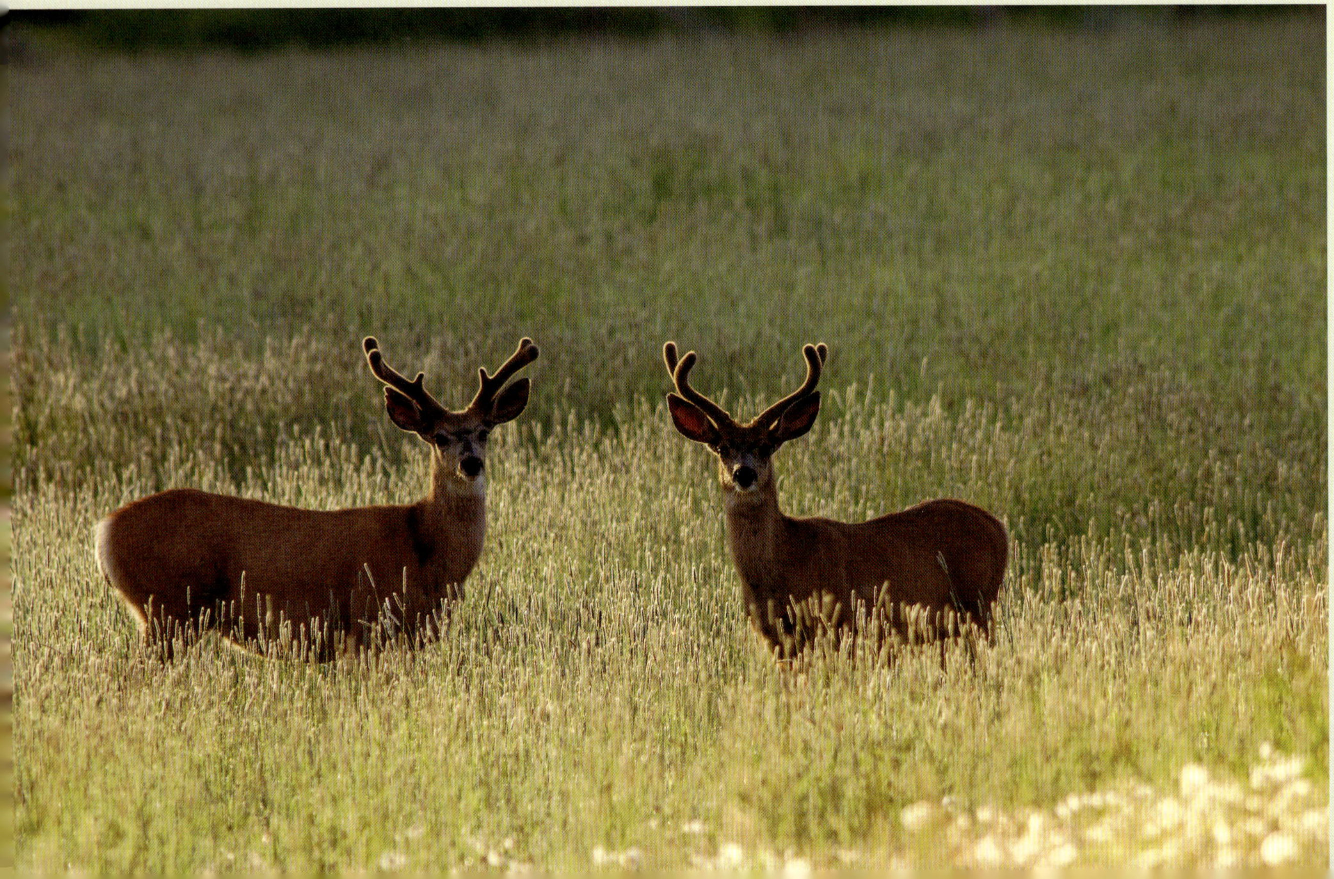

A pair of mule deer backlit by morning sun in a Hoback River Basin meadow (Sublette County, Wyoming)

for migrating mammals. For the big herds to survive and continue moving freely across the land, the West—and conservation opportunities—must be viewed on a landscape scale in the way wildlife have moved across it for thousands of years.

It's midsummer now, an easy time of year for hooved mammals raising young and grazing on succulent green grass. In two months they'll be moving again, passing on their migration path to juvenile successors. Keeping these corridors open so wildlife may continue roaming freely is up to us. §

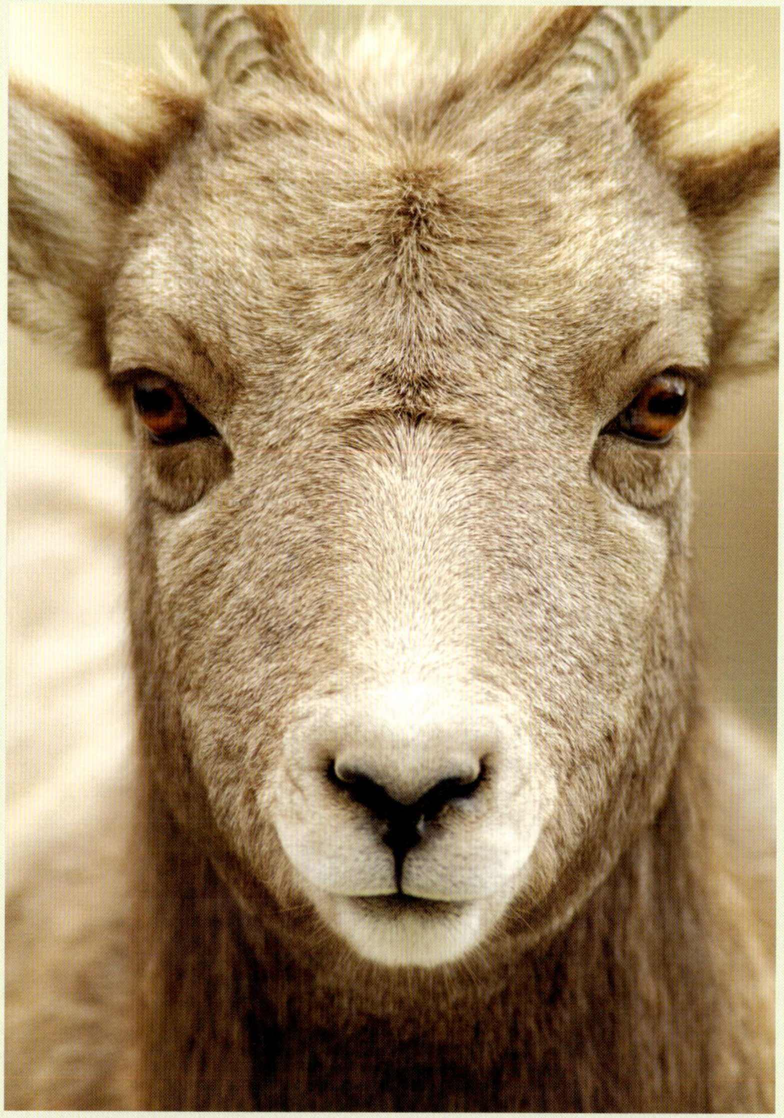

LEFT AND BELOW Bighorns along the North Fork of the Shoshone River migrate from Yellowstone National Park for autumn rut, with rams battling for supremacy and the right to mate, then spend winter grazing at lower elevations before returning to Yellowstone in spring (Shoshone National Forest, Wyoming).

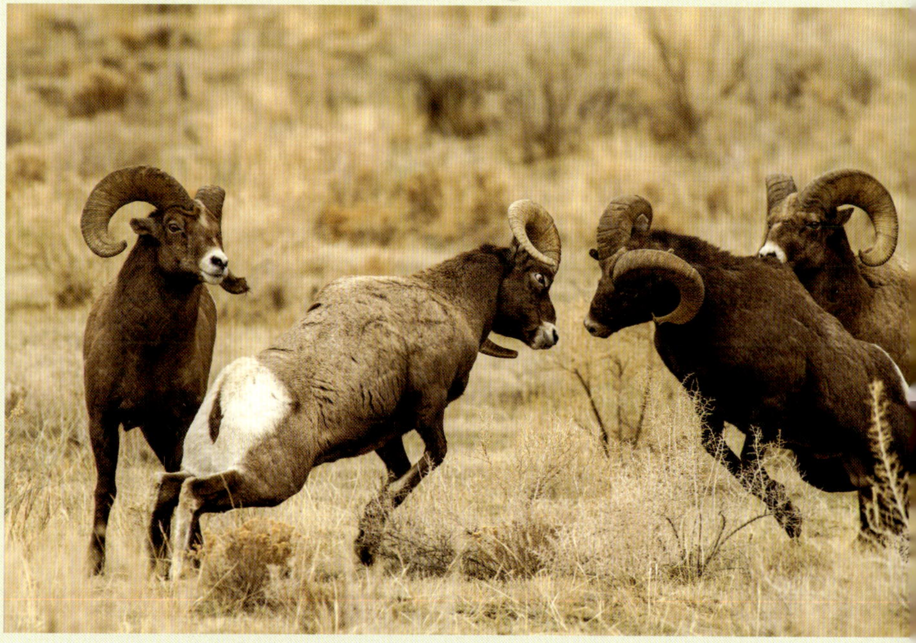

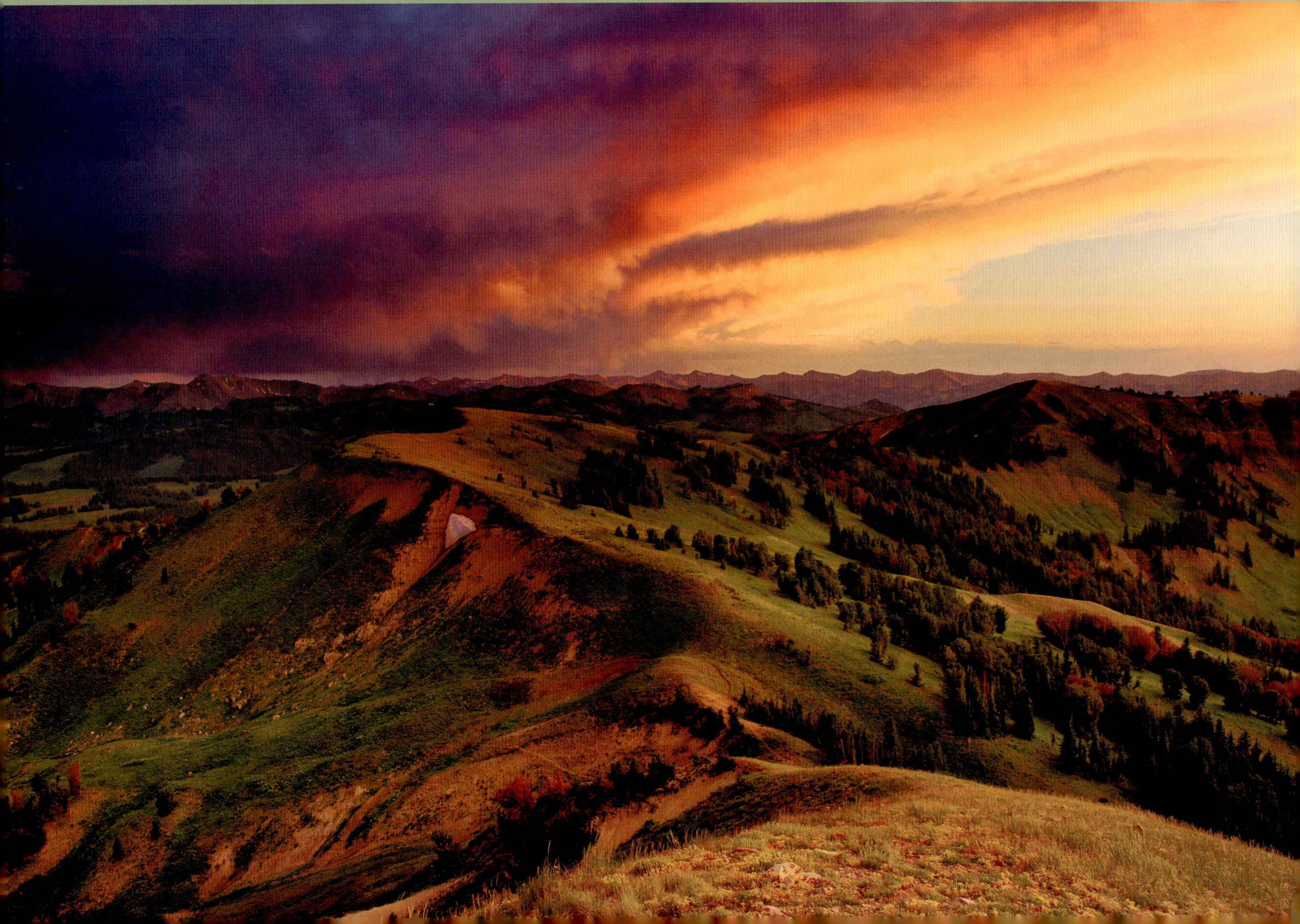

SHAPING THE NEW WEST

MOVING BEYOND THE FRONTIER MIND-SET

Todd Wilkinson

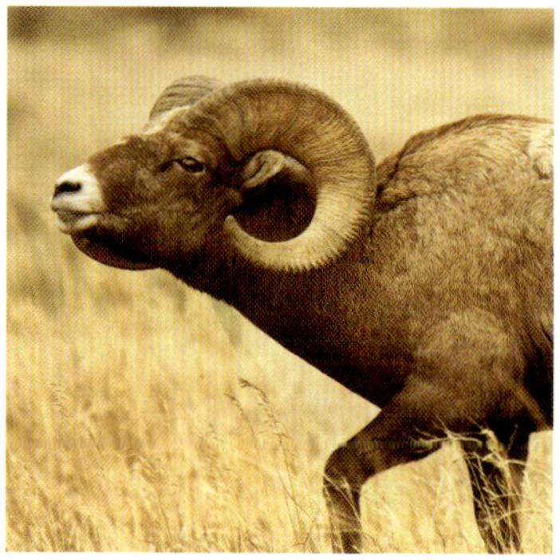

PICTURE: A HOT AND HUMID EVENING in the West, sun slipping fast into ethereal half-light on the horizon, suddenly a booming timbre. It was on a night like this that I had my first encounter with a sage-grouse, though the bird's spirit presented itself in a mystical human form.

Indigenous singers circled around me in the arc of a Lakota powwow ground crooning a grand entrance song, their voices lilting over the thunder of pounding drums, the rhythm like a fast-beating heart. This was not a reverie.

From out of the shadows emerged a dancer. Plumed in feathered regalia and fringed buckskin jingling with bells, he glided forth strutting, mimicking the courtship steps of a male sage-grouse. His appearance met with joyous ululation from the crowd of onlookers.

Mesmerizing, the ritual became a catalyst for me to more fully consider the long, unbroken bond between people and nature in the sagebrush-blanketed American West. From that night until now, over a span of decades, I've seen the homage to greater sage-grouse, *Centrocercus urophasianus,* repeated dozens of times by many different tribes, each one owning a distinct aboriginal name for this largest member of the grouse family.

ABOVE Bighorn sheep ram (Shoshone National Forest, Wyoming)
OPPOSITE From Lookout Peak in Bridger-Teton National Forest, Wyoming, the setting sun drops below a heavy cloud bank to light the Wyoming Range in brilliant color.

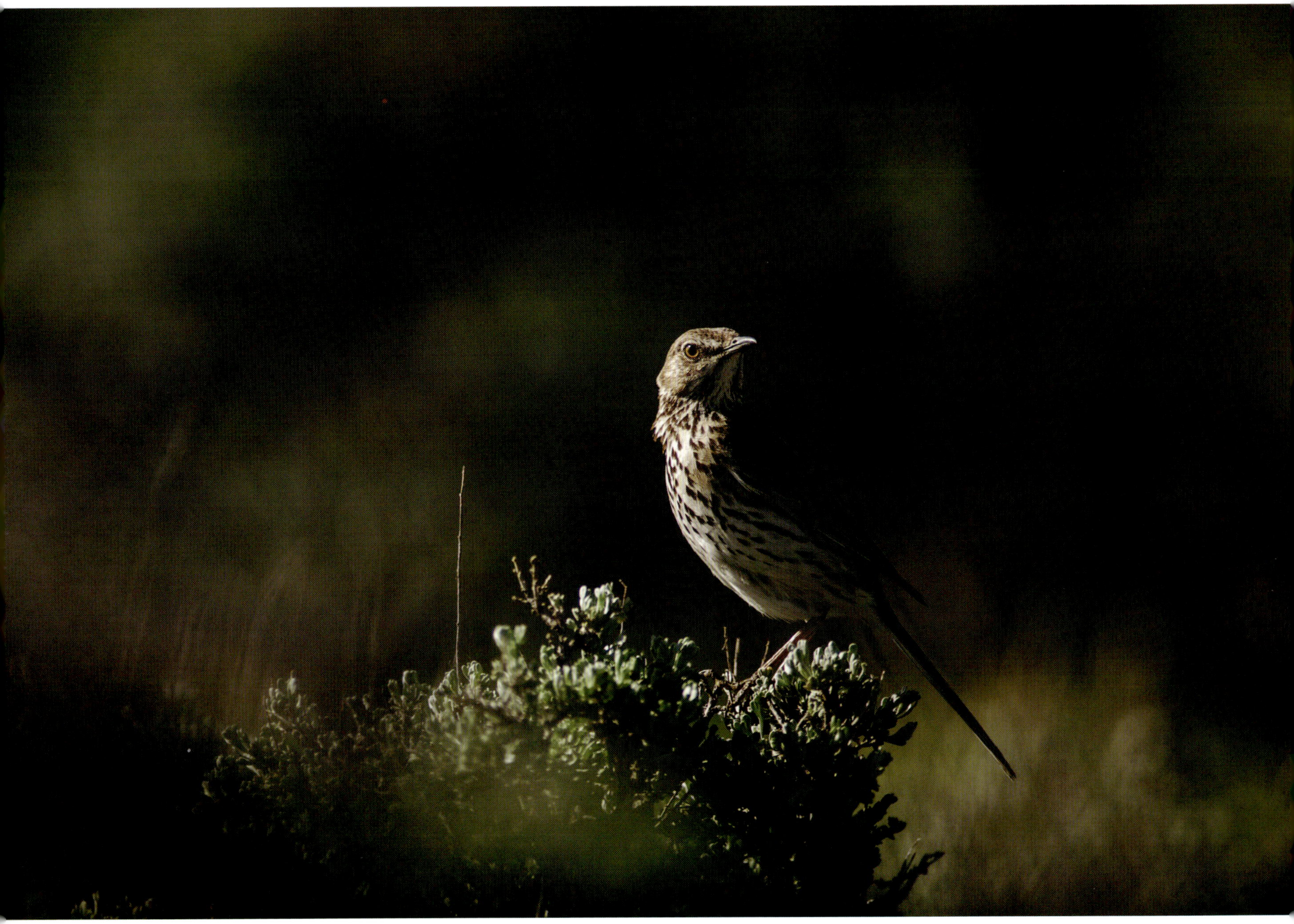

Out here in the geographic center of the West, roughly halfway between the 100th meridian (on the western side of the Mississippi) and the Pacific Ocean, wildlife imagery and symbolism matter because they serve as a window into the sentient soul of the land. This vast expanse dominated by big sagebrush (*Artemisia tridentata*) has come to be known as the sagebrush sea: a mighty inland shrub-steppe environment where one thing is certain: the fates of the greater sage-grouse, the Gunnison sage-grouse, and people are inextricable.

Look around the world and you'll see many earthly landscapes that still exude the charms of changelessness: in the context of human time, they are essentially as they've always been. You can view this timelessness in parts of the sagebrush sea, which stands in sharp contrast to the populous, crowded, artificially engineered settings where most Americans live.

Mind you, this is not a landscape that you can come to know by simply glancing into it or tiptoeing around its edges or flying across its huge, textured surface in the temperature-controlled confines of a jet. To grasp the essence of why the sagebrush sea as a biological homeland is special—and why we are being called to rally in defense of its eponymous bird—requires total sensual immersion. You must wade in ear, eye, and nose deep. You must listen to what this raw environment has to say, visually absorb its subtleties, and breathe in its natural fragrant bouquet.

What is it about this sweep of America that, for most citizens, does not register in their consciousness, yet, once discovered, still possesses the power to inhabit an enduring place in the heart? The sagebrush sea appeals to something deep—archetypal—inside of us.

An adult sage thrasher pauses before tending to the nest woven in thick sage below (Sublette County, Wyoming).

For an adventurous seeker on the trail of a totemic bird like the sage-grouse, the rewards can be sublime. Where the great bird lives—an unending open space—may initially seem lonesome, exposed, and daunting but with a little curiosity turns alive and intimate. What once was invisible can become suddenly revealed.

Grander, bigger-picture thinkers than I have pondered this phenomenon; for example, on July 27, 1869, John Muir made this observation in his journal: "When we try to pick out anything by itself we find that it is bound fast by a thousand invisible cords that cannot be broken, to everything in the universe." Muir might just as well have been referencing the interwoven biological threads and tendrils of the sagebrush sea, where the plight of a bird is a harbinger of the habitat's health. He might have identified us as a cord, too. In Muir's time, the ocean of sagebrush covered approximately 300 million acres (121 million hectares); half of it has now disappeared.

The topography encompassed by sagebrush in the West is astoundingly diverse. The massive terrain for sage-grouse stretches across eleven states. It climbs the sides of rugged snow-clad mountains in the Rockies and plies sandstone arroyos and smooth-worn coulees beneath mesas in Colorado and the same flat-topped buttes in Montana that Charlie Russell painted. It skirts unforgiving deserts as varied as the Red, Mojave, and Chihuahuan. It speckles the high savanna pushing inland from rain forests in the Pacific Northwest and defines the windblown steppe that enwraps the Great Basin, itself a huge area stretching between California's Sierra Nevada and Utah's Wasatch Range, and from southeast Oregon to southern Utah. No other ecotype in the country lays claim to as much terra firma.

And wherever there was sagebrush, there was the real probability that a covey of skittish sage-grouse could be near, piqued to flush, noble whether in flight or on the trot. In steady retreat following the plight of the once-prolific shrub, however, sage-grouse have disappeared from many of their former haunts, their survival inextricably tied to that of their sagebrush habitat.

A WELLSPRING OF LIFE

Sage-grouse are an obligate species of sagebrush, meaning they depend upon it wholly for survival. The plants, a collection of eighteen unique species of sagebrush, provide nutritional sustenance and shelter and shape every element of critical habitat to these avians.

Once upon a time, there were possibly as many as 16 million sage-grouse in the Lower 48 and southern Canadian provinces. Today, greater sage-grouse numbers are somewhere between 200,000 and 500,000. Numbers of birds provide a useful metric for thinking about the arithmetic of demography—and are cited in the debate over whether to give the bird protection under the Endangered Species Act—but what those numbers fail to account for is everything else in the bird's extended web of life. The same ecological factors affecting *C. urophasianus* as an "umbrella species" ripple across a much larger community of 350 species (actually, more than 700 if we include a wider gamut of organisms ranging from vertebrates to plants, lichens, insects, and spiders).

These strands of life are intricately connected to each other in the web of taxa that we call "biological diversity." Here, in an elaborate mixture of readily observable predators and prey, you'll find

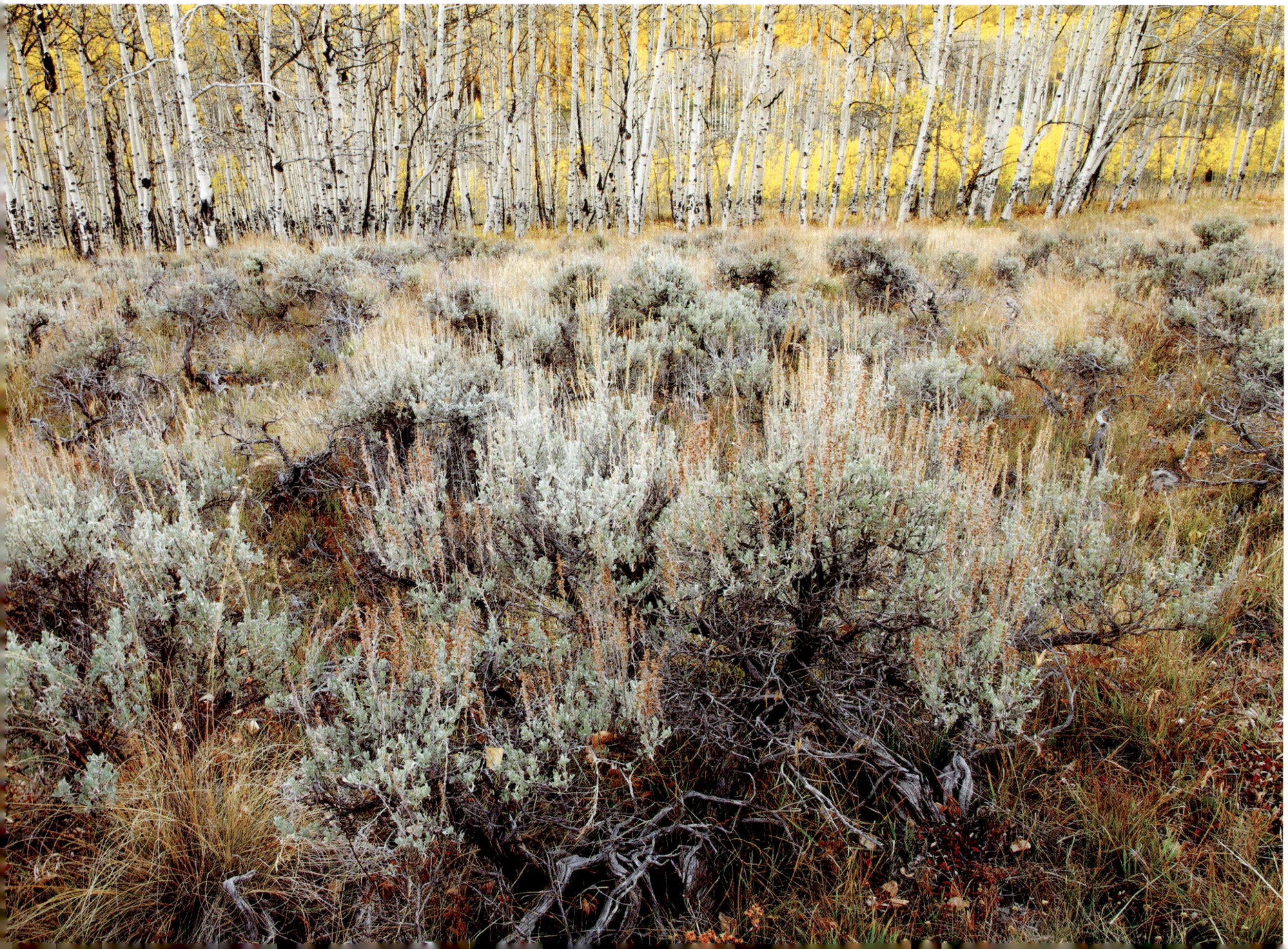

Fall color in the biologically rich interface of sagebrush and aspen forest (Curecanti National Recreation Area, Colorado)

A view across the sagebrush-covered top of Pinedale Mesa in Sublette County, Wyoming, to the magnificent Wind River Range

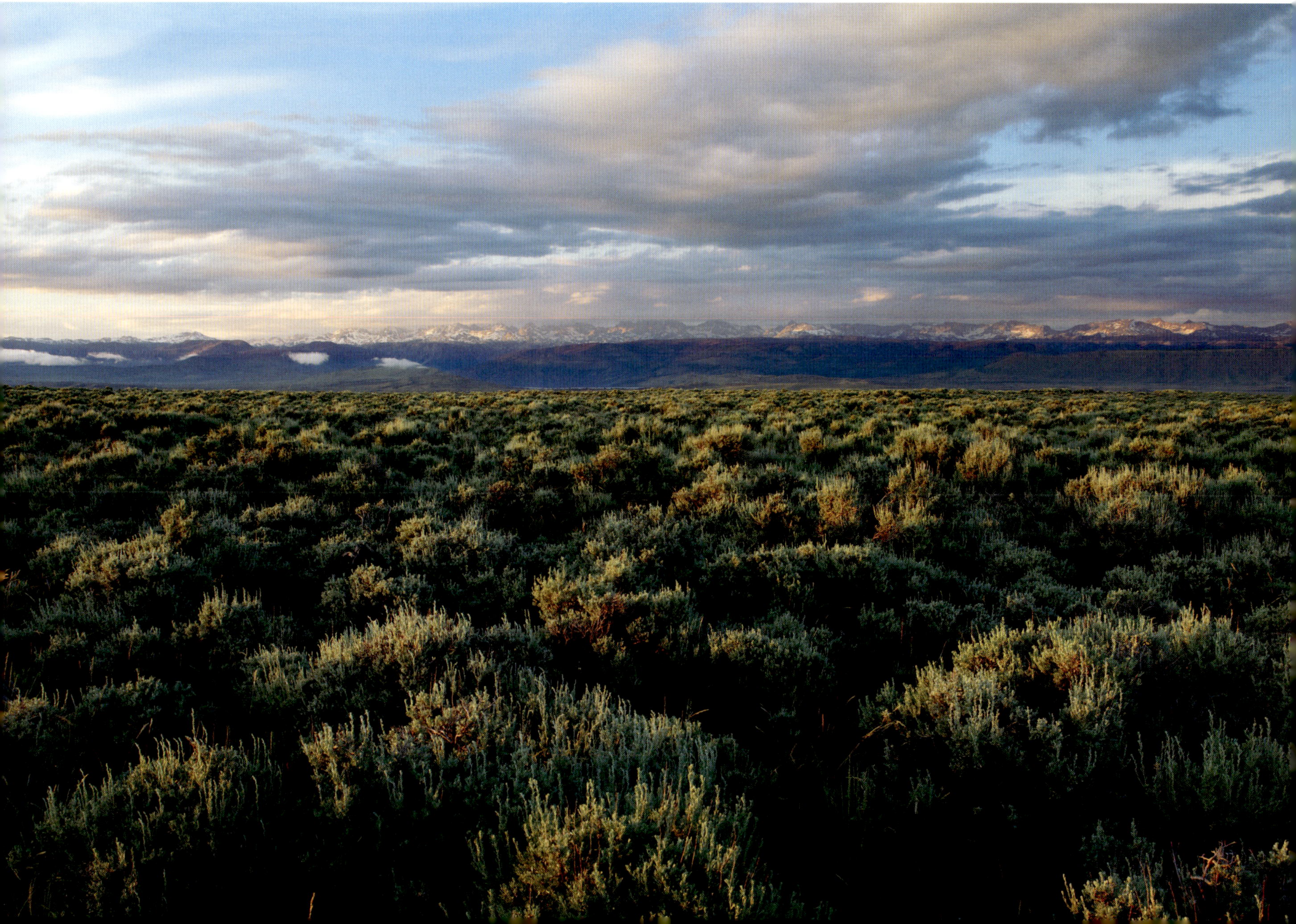

Sagebrush holds many surprises. This newborn pronghorn fawn, maybe a day or two old, was tucked in mixed grass and sagebrush—but the mother was not far away. Within a week of birth, a pronghorn can outrun a coyote (Soda Lake Wildlife Habitat Management Area, near Pinedale, Wyoming).

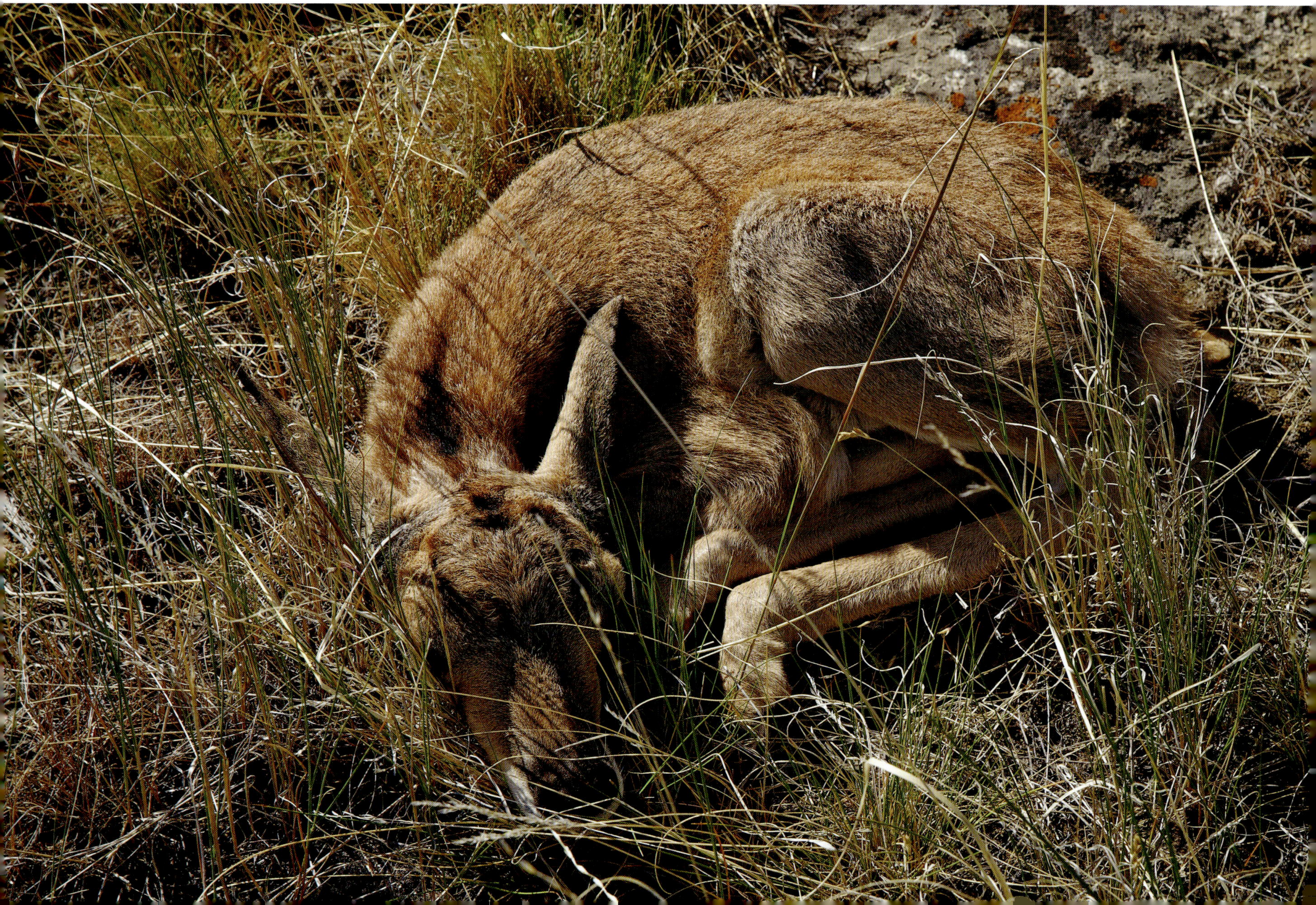

the pygmy rabbit, western burrowing owl, black-throated sparrow, Merriam's shrew, Great Basin collared lizard, loggerhead shrike, kit fox, Wyoming ground squirrel, prairie falcon, striped whipsnake, and chisel-toothed kangaroo rat, among dozens upon dozens of others I could name. The sage-grouse, as an umbrella species, serves as a barometer for considering the fate of the others. Indeed, where does the survival of one begin and another end?

Standing before a vast stretch of old-growth sagebrush can evoke the wonder that wells up in us when entering a disheveled labyrinth. The twisted, misshapen clumps of mature big sagebrush—the most widespread species—can rise more than head high. Around every turn you never know what you might spy, and that's what makes the journey exciting. Usually it's intersecting ungulate trails or scat spoor or animal tracks that offer the first hint of biological richness. Yet pause for a quiet listen, and you'll be amazed that the light breeze gives way to a subtle feast for the ears: the whistling of meadowlarks, the screech of a red-tailed hawk hunting for rodents, a yipping coyote at sunset, or, in the higher country during late summer, a bugling bull elk. There can be wolves in some of these parts, grizzly bears too.

The point? There's always something *happening* here. But it isn't happening as an abstraction apart from our human experience. Remove the sagebrush, and you pull the rug out from under everything else tethered to it. During my hiking and camping forays into roadless sagebrush wilderness, I've ambled close enough to western diamondback rattlesnakes to set them abuzz. I've staked out the ground nests of burrowing owls and startled black bears in a daybed, though in the rouse of their flight it was I who was left

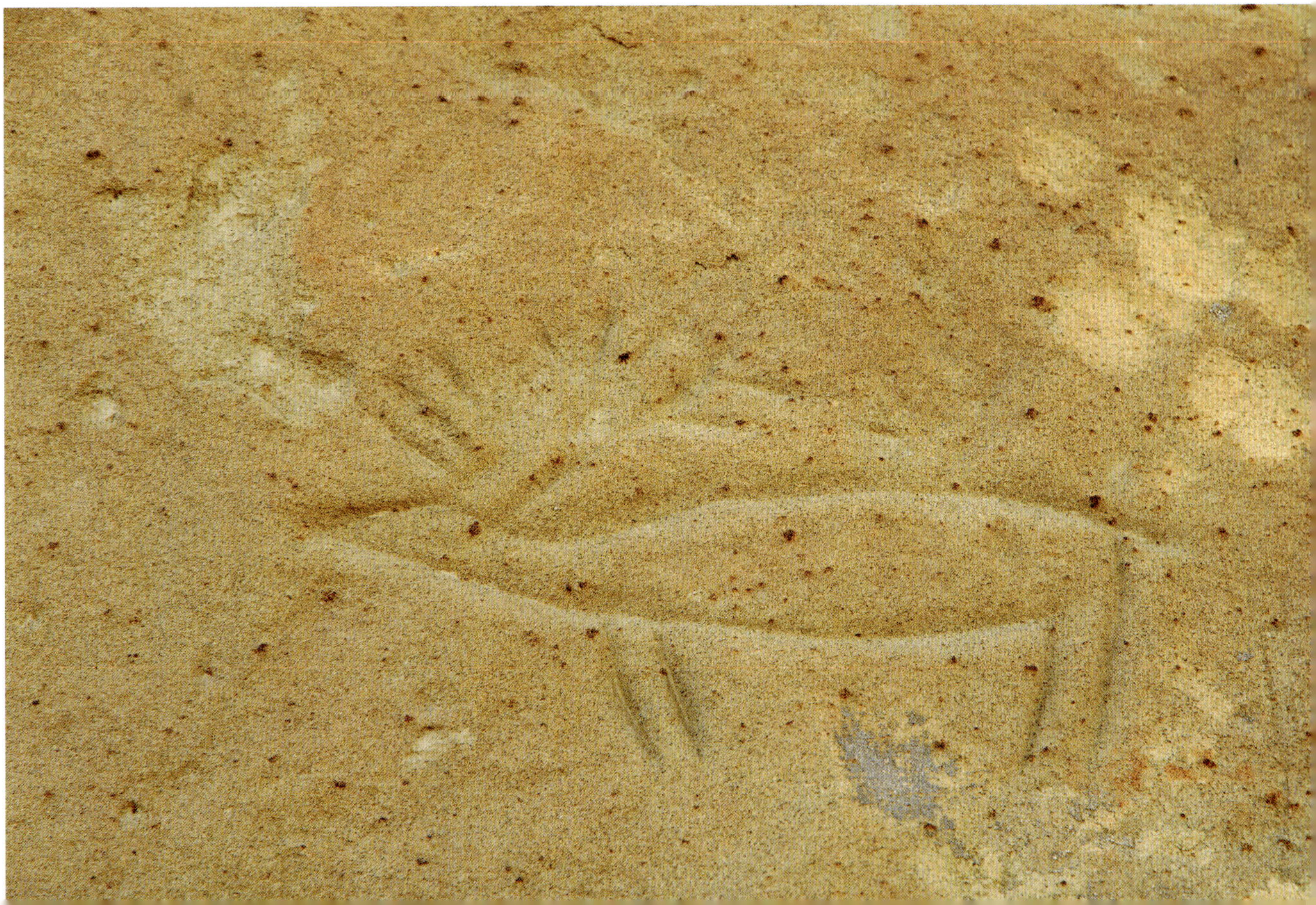

A deer, or possibly an elk, carved in stone by Native hands at the White Mountain Petroglyph Site in Wyoming's Red Desert

RIGHT Pronghorn are the fastest land mammal in North America, reaching speeds up to 55 miles (89 kilometers) per hour on open range. Pronghorn are one of only two mammal families endemic to North America and are highly tied to sagebrush (Sublette County, Wyoming).
BELOW The Pinedale mule deer herd uses the Pinedale Anticline for crucial winter range. The anticline, a long mesa in Sublette County, Wyoming, doesn't hold deep snow on the top, which allows deer to forage through the harsh winter. Natural-gas drilling on the Pinedale Mesa has contributed to a 44 percent decline in the Pinedale mule deer herd population.

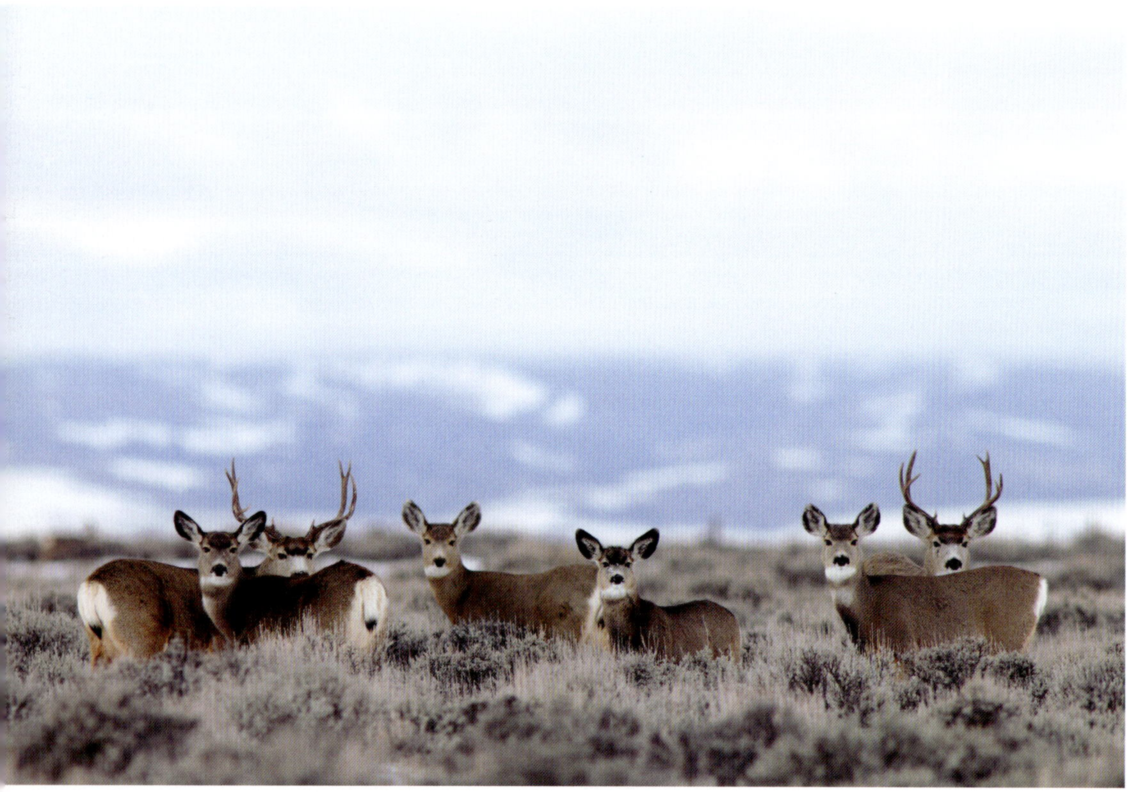

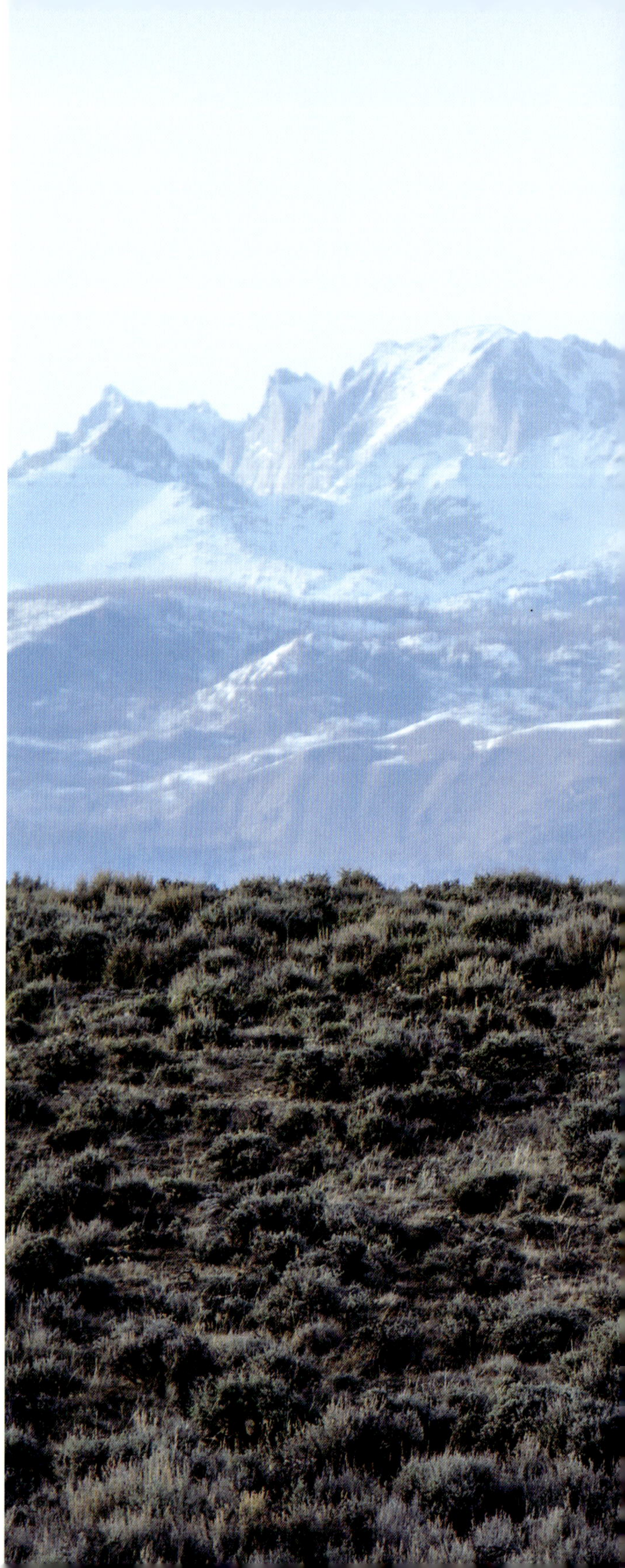

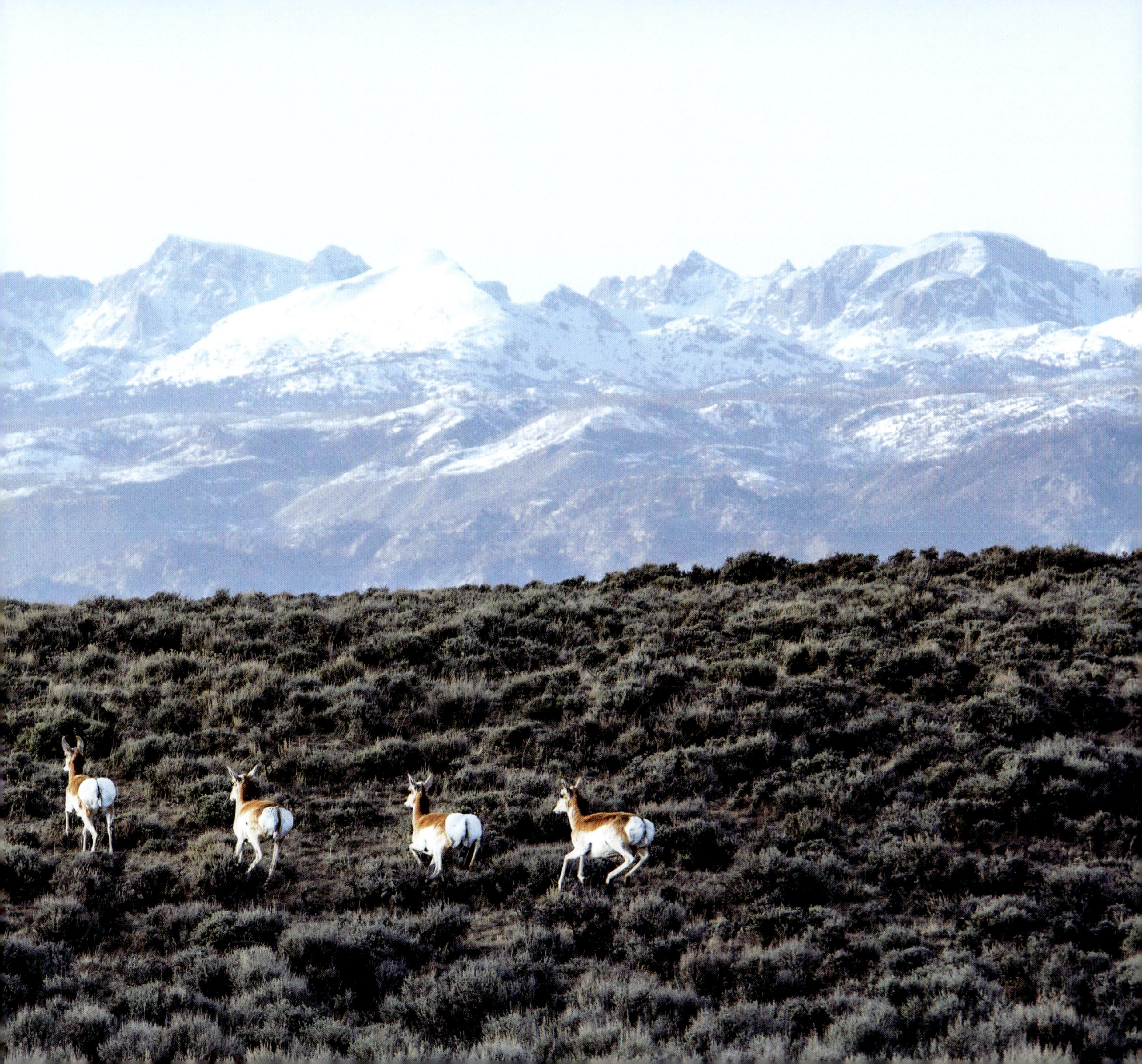

with an adrenaline rush. I've laid out a sleeping bag in the sage-brush outback to watch the stars amid profusions of wildflowers reminiscent of an Impressionist painting. Everywhere you walk, one primal avenue of diversity leads to another, even dead-ending in the natural alcoves of cliff faces where native artists who navigated these parts for millennia have left behind petroglyphs and pictographs confirming the sagebrush-steppe as an ancient homeland and canvas for inspiration.

Where the oceanic allusion of the sagebrush sea's name holds most poignantly is in the fluidity of animal migration. In Wyoming, home to the biggest remaining expanses of undisturbed sagebrush, wildness abounds. The sagebrush sea in the western third of the state serves as a critical undergirding for two of the longest wildlife migrations on the continent. Pronghorn still follow a post-Pleistocene "path" between summer birthing grounds in Grand Teton National Park in the heart of Jackson Hole and winter range in the Red Desert and Upper Green River Valley to the south. Mule deer embark upon similar seasonal treks on ragged carpets of sagebrush, which provides food and concealment. Yet human development pressure on public and private land puts the persistence of those routes in jeopardy. We, as individuals, as a culture, society, nation, and species, have to take responsibility for our own actions.

As the near-annihilation of bison, whooping cranes, and trumpeter swans has demonstrated, there are limits to how much nature can be directly exploited or eroded without consequence. In *Hornaday's American Natural History,* William T. Hornaday (1854–1937)—famed director of the New York Zoological Park, friend of Theodore Roosevelt, and trailblazing zoologist who helped rescue bison from extinction—reserved special admiration for the greater sage-grouse. He wrote fondly of his encounters with sage-grouse in my home state, Montana. "When a whole flock suddenly rises out of the sage-brush and takes wing, it is an event to remember," Hornaday noted, based upon field observations made at the end of the nineteenth and early in the twentieth centuries.

During a round-trip trek by horse-drawn wagon in October 1901 between Miles City, Montana, and the Missouri River to the north—250 miles (400 kilometers) in all—Hornaday saw six flocks of sage-grouse, one a grouping of forty-six birds strong. "I will only add the earnest wish that every one who reads these notes may some day have the pleasure of seeing at close range this glorious bird in its ideal home—on a sage-brush flat in the land of buttes, where the world is big and free, and full of sunshine."

Later, when the birds started to vanish, Hornaday tried to marshal a national public outcry to prevent them from going the way of the extinct heath hen in the East. As the twentieth century demonstrated with the continuous downward slide in sage-grouse numbers, he ultimately was unsuccessful. Today, sage-grouse are still hanging on, by a thread.

DEPLETING THE ORIGINAL WILD WEST

Sagebrush ecosystems fared poorly under the paradigm of the "Old West" when Manifest Destiny fueled the myth of a never-ending frontier replete with boundless, inexhaustible natural resources.

No less a figure than Rachel Carson alluded to destruction of the sagebrush-steppe in her 1962 classic, *Silent Spring.* "In the name of progress the land management agencies have set about to satisfy the insatiable demands of the cattlemen for more grazing land," Carson wrote. "Few seem to have asked whether grasslands are a suitable and desirable goal in this region. Certainly nature's own answer was otherwise. The annual precipitation in this land where the rains seldom fall is not enough to support good sod-forming grass; it favors rather the perennial bunch-grass that grows in the shelter of the sage."

Presciently, Carson questioned the conversion of diverse sage-brush to agricultural monoculture and livestock pasture. "The

A grizzly bear sow, designated Bear 399, sniffs the air with her two subadult cubs just out of view (Grand Teton National Park, Wyoming).

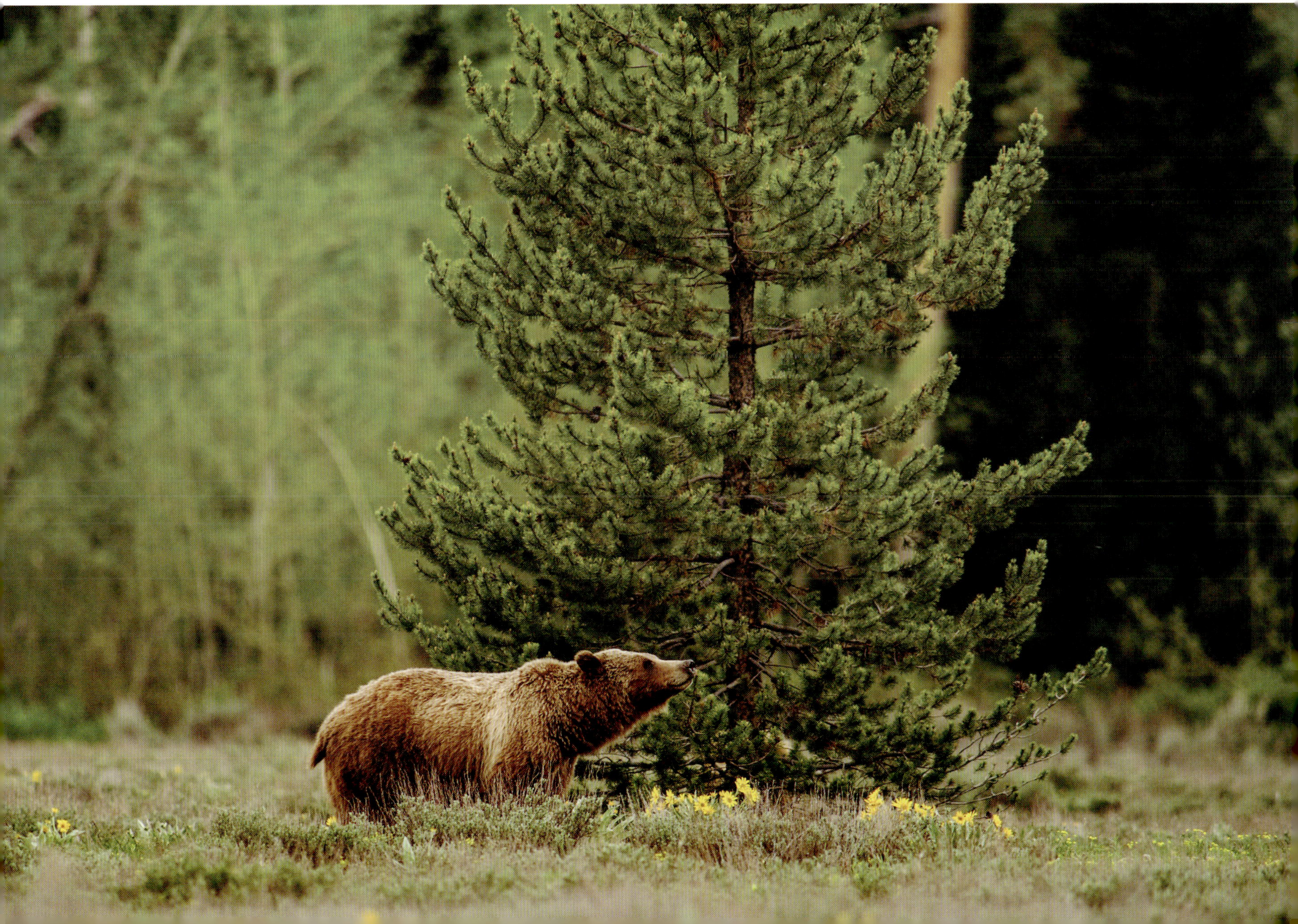

antelope and the grouse will disappear along with the sage," she warned. "The deer will suffer, too, and the land will be poorer for the destruction of the wild things that belong to it."

Indeed, recent studies in Montana have correlated the decline of mule deer to loss of sagebrush, and in Wyoming field researchers have said that energy development in the heart of sagebrush winter range for mule deer has also led to smaller herds. Reams of additional studies illuminate the impacts of sagebrush loss on numerous species. A dozen bird, plant, amphibian, and insect species have either been listed for federal protection or are candidates.

Though rugged and hardy it may seem, the sagebrush sea today is, unbeknownst to many, considered the most endangered ecosystem in America. Steven T. Knick of the US Geological Survey's Forest and Rangeland Ecosystem Service Center joined several colleagues in offering an assessment that appeared in the journal *Ecology and Evolution* in 2013: "The long-term future for this ecosystem is uncertain. Extensive regions of sagebrush have been burned by wildfire or lost to agriculture, energy and infrastructure development, and other resource demands by increasing human populations within the sage-grouse range. Remaining sagebrush landscapes are threatened further by exotic plant invasions leading to altered fire regimes and conversions to unsuitable expanses of exotic annual grasslands."

If that isn't chastening enough, the scientists added this: "Long-term effects of changing climate could result in further loss of sagebrush by the end of this century: as much as *80 percent* (my emphasis added) of the current sagebrush distribution could disappear under extreme projections. Thus, current trajectories and future loss of sagebrush are likely to further imperil sage-grouse and other dependent species."

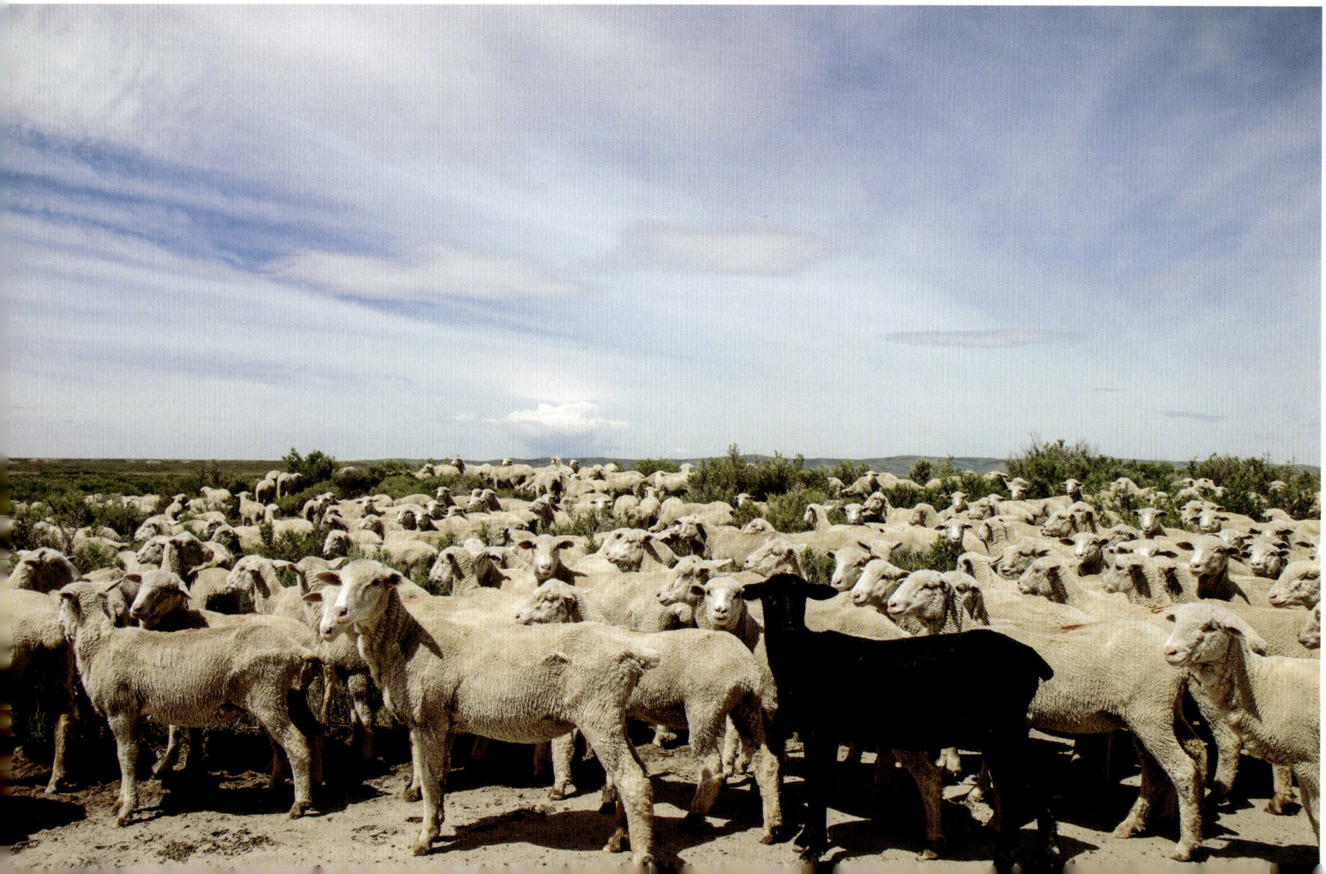

A herd of domestic sheep grazing in the southern Red Desert, Wyoming

Months after that article appeared, international climatologists said extreme future scenarios involving lower precipitation, hotter temperatures, extended droughts, and fires are likely in the region encompassed by the sagebrush sea.

RESTORING WHAT'S LOST, SAVING WHAT ISN'T

We've known that a day of reckoning was coming, when warning signs could no longer be ignored. After decades of deferred action to reverse a downward spiral that began a century ago, conservationists early in the new millennium petitioned the US Fish and Wildlife Service to grant the greater sage-grouse protection under the Endangered Species Act. A separate parallel push to get similar protection for the Gunnison sage-grouse—recognized as a separate species in 2000—was also initiated. And in November 2014 the Fish and Wildlife Service rendered an assessment: The Gunnison sage-grouse was formally listed as a threatened species under the federal Endangered Species Act. The Gunnison exists in a far smaller geographic area and at much lower numbers—fewer than 5000 birds occupying just 10 percent of its historic range—making it in many ways a harbinger of what is in store for greater sage-grouse.

Just as there has been special human intervention applied to save bald eagles, peregrine falcons, wolves, grizzly bears, black-footed ferrets, and other animals, sage-grouse are getting attention that many insist is long overdue.

Auspiciously, most of the species mentioned above have recovered. They've been brought out of triage thanks mostly to citizen support for habitat protection, a political will to make recovery happen, and a mixture of carrot and stick—rewards and regulatory enforcement—to motivate human behavior. Mostly, though, rescue

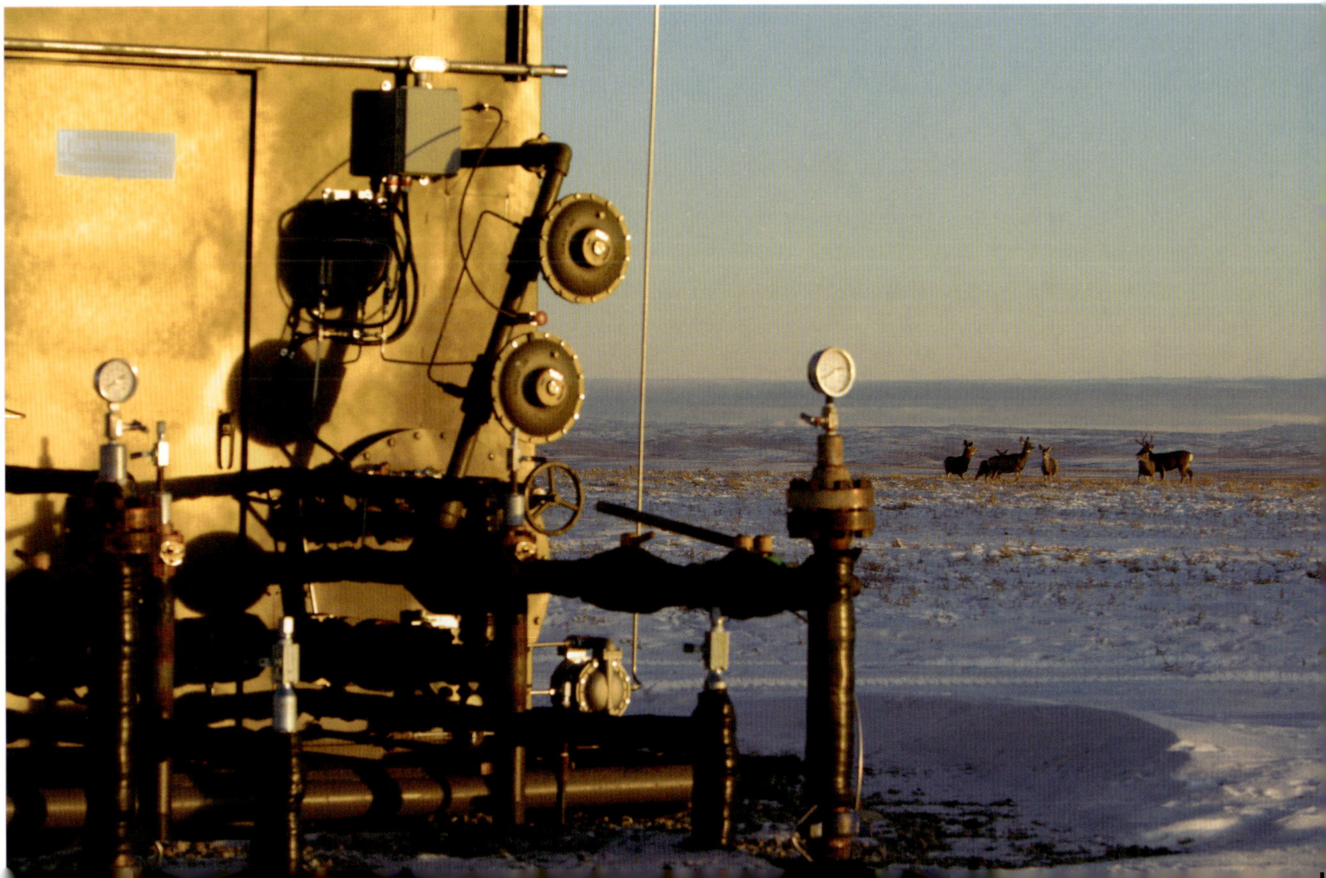

Some of the Pinedale Mesa mule deer herd near natural-gas drilling equipment on their critical winter range (Sublette County, Wyoming)

of those creatures has involved a willingness of human stakeholders to become inconvenienced and to not place their own self-interests ahead, at all costs, of the survival of something else.

In 2004, officials in President George W. Bush's Department of the Interior declined to list greater sage-grouse, and later it was revealed that the decision was owed in part to political meddling overruling the case made by scientists. A federal judge later ordered a second biological review of the greater sage-grouse's status, and it was determined that federal protection is warranted but that other species deserve more priority. The question today is whether the greater sage-grouse will finally and formally be declared endangered.

It is, like many issues in our society, framed as a polemic of humans versus nature—but need it be?

To protect the greater sage-grouse without having to list it as threatened or endangered, unprecedented attention has been galvanized. In 2012, then US Interior Secretary Ken Salazar, appointed by President Barack Obama, joined the governors of the eleven western states of the sagebrush sea and convened the State-Federal Sage Grouse Task Force. Its primary mission has been to identify best practices for protecting sage-grouse habitat that can be incorporated into viable conservation work plans each state is drafting and will present to the US Fish and Wildlife Service for approval. The objective of the states is to show Fish and Wildlife they can caretake the bird without having to contend with the federal Endangered Species Act. Why it didn't happen sooner speaks to an array of similar concerns in our society—climate change being one.

If the goal is protecting a species *population*, experts in "island" biogeography say it is better to have large blocks of interconnected core-area habitat (the best and biggest sweeps of habitat that encompass the most important pieces) than a plethora of smaller pockets of habitat existing in isolation from one another. As has been demonstrated with other species, the faster that core-area habitat shrinks and is fragmented into pieces, the faster the rate of species decline. It's as true for mule deer, pronghorn, and elk as it is for grizzlies, wolves, and sage-grouse.

GROUND ZERO FOR ENERGY DEVELOPMENT

Conserving the sage-grouse is an epic multistate challenge, to be sure, though the stakes are highest in a handful of states that straddle the Rocky Mountains. Wyoming has the largest remaining concentration of greater sage-grouse and, together with Montana, harbors 55 percent of the total sage-grouse population. Habitat destruction associated with ranching, in particular cattle and sheep grazing, has had an undeniable deleterious impact on many millions of acres of sage-grouse homeland across all of the West.

But the issue in Wyoming and Montana is acute, and on top of it, sage-grouse habitat in these states is being inundated by a new bonanza of energy development. In the twenty-first century, this juggernaut of land disturbance is being fed by massive finds of oil and natural gas reachable by technology breakthroughs in drilling techniques such as hydraulic fracturing ("fracking"). There are now aspirations that America could become energy independent and weaned from Middle East petroleum. Moreover, Wyoming is the largest producer of coal in America, and soon Montana, which is sitting upon the nation's largest undeveloped coal reserves, could, like its neighbor, be opening mega–strip mines that come replete with roads, powerline corridors, and railroad tracks. In the aftermath of the Great Recession, economists tout the role that energy development plays in putting people to work. In a case of déjà vu, we are again hearing the argument that safeguarding nature stands in the way of progress.

Resource extraction does bring good-paying jobs. It generates tax dollars for schools, roads, and community services; it keeps shop owners in business down on Main Street and casts a hopeful

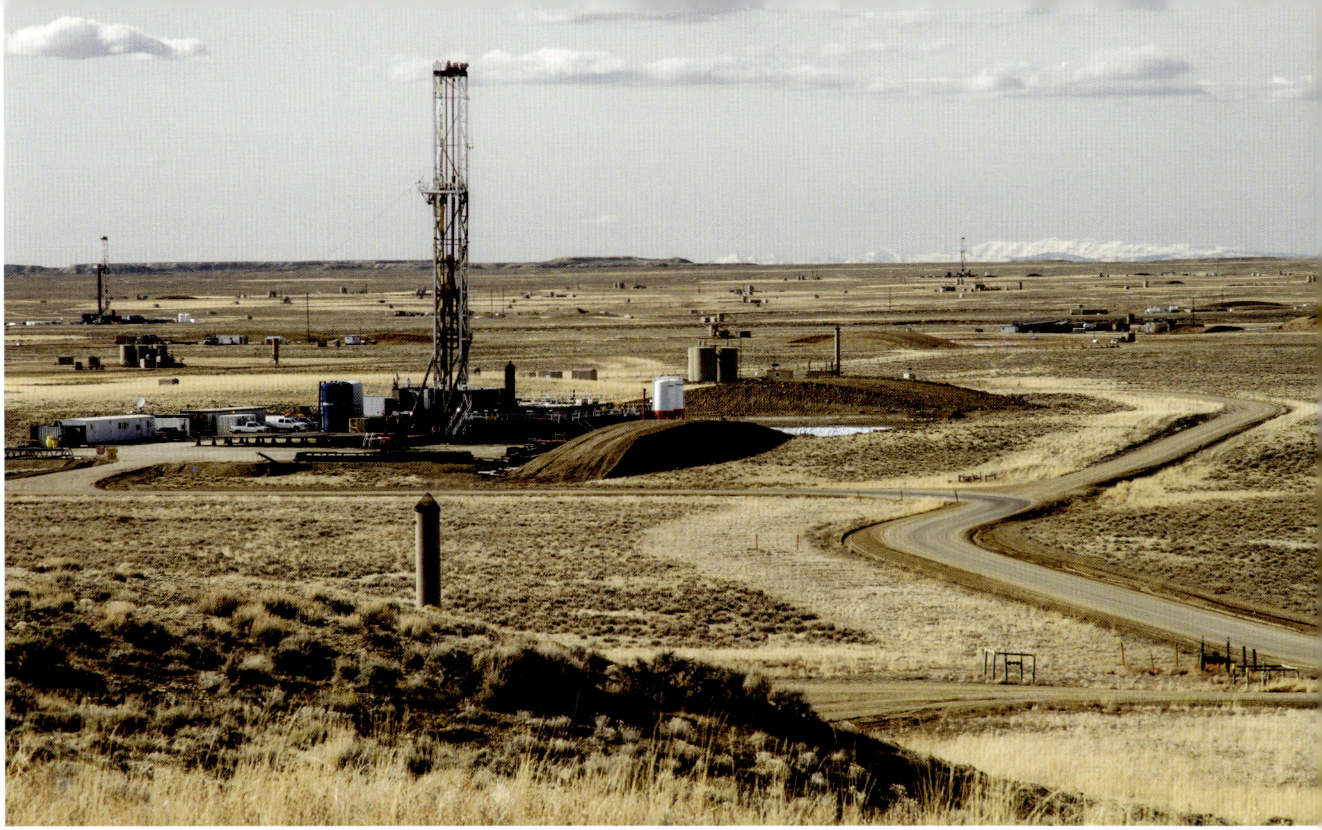

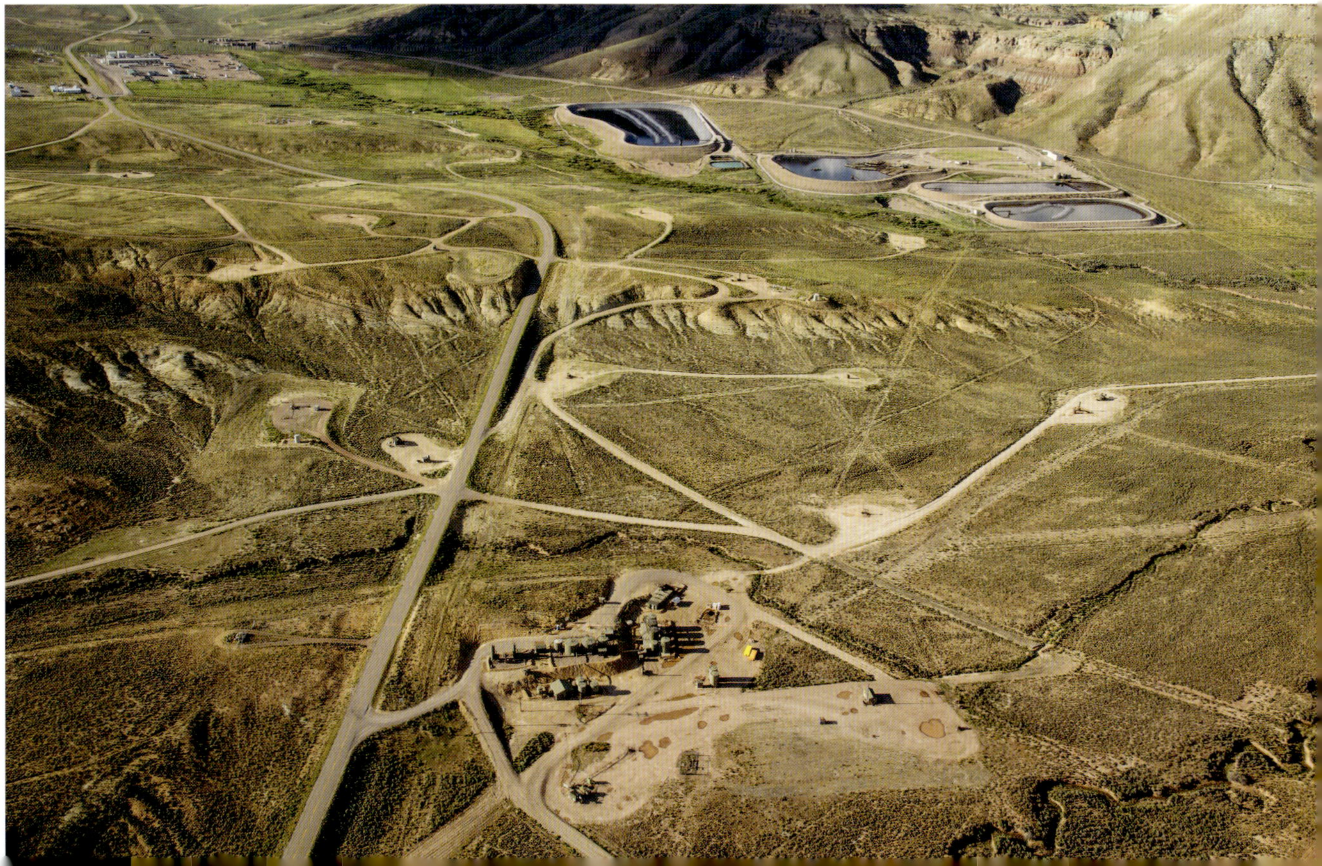

TOP RIGHT Jonah Field, in Sublette County, Wyoming, is one of the most densely developed natural-gas fields in the world.

BOTTOM RIGHT Full-scale industrial development in the La Barge, Wyoming, oil and gas fields has left a fragmented landscape, complete with produced water waste ponds, that is now suitable only for energy development (Sublette County, Wyoming).

Aaron Pruzan carves a turn in the Upper Hoback River near its confluence with the Snake River. Pruzan, owner of Rendezvous Sports, was a staunch advocate for the conservation of the Hoback River.

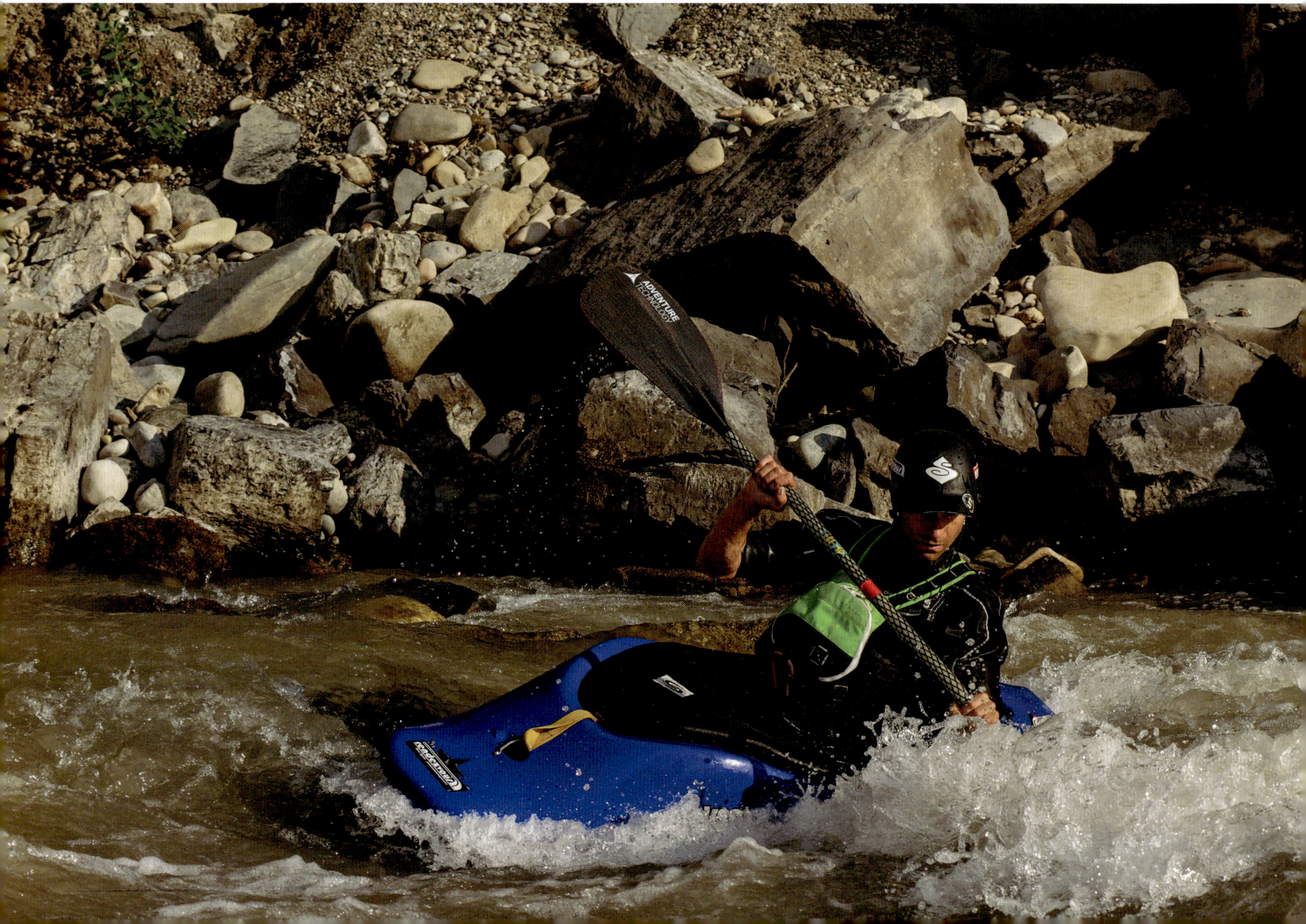

aura of prosperity. Obviously, there's nothing wrong with any of that, nor with the dream of parents wanting to create opportunities for their children.

However, serious ecological consequences can ensue when large landscapes are altered and wildlife habitat whittled away. Many corners of the West are littered with blighted frontier ghost towns that did not consider the costs of their booms nor how to avoid inevitable busts. In 2014 the Bakken Formation, found along the borders of North Dakota, Montana, and the Canadian province of Saskatchewan, passed the threshold of producing one billion barrels of crude oil, speaking to the enormity of the wellspring of fossil fuel beneath the ground. Some North Dakota boomtowns have been inundated with a transient labor force that dwells in itinerant "man camps," overwhelming local infrastructure, bringing social problems, and despoiling the land with pollution.

Development in the Bakken Formation is following a pattern. The Jonah Field in Wyoming, where natural-gas production started to ramp up in the late 1990s, was supposed to be a setting where industrial-strength natural-gas production would be proved compatible with greater sage-grouse, but grouse have disappeared. Many experts say this outcome was predictable. Back in 2006, Clait E. Braun, former avian research program manager for the Colorado Division of Wildlife, crafted a report titled "A Blueprint for Sage-Grouse Conservation and Recovery" that spelled out the problem. (Braun and wildlife researcher Jessica Young are credited as codiscoverers of the Gunnison sage-grouse as a new species.) Braun also has been lead author or contributor to more than fifty-four articles on sage-grouse and more than fifty technical abstracts. And he has testified before Congress as well as serving as editor of the *Wilson Journal of Ornithology*.

"The blueprint contains specific recommendations that, if implemented by land management agencies, are designed to increase sage-grouse abundance by 33 percent by 2015 and increase sage-grouse distribution by 20 percent by 2030," Braun wrote. Key recommendations included minimizing habitat fragmentation, avoiding drilling and road construction in wintering habitats and within 3.3 miles (5.3 kilometers) of lek sites to protect breeding and nesting habitat, avoiding large-scale (more than 50 acres, or 20 hectares) sagebrush burning or removal, eliminating juniper that has encroached upon sagebrush, and optimizing livestock grazing methods to maintain and enhance sage-grouse habitats.

All these recommendations sounded reasonable—and, in fact, what would have been good for grouse would also have benefitted many of the hundreds of other species in addition to water and air quality—but the recommendations were not systematically embraced due to political and social resistance. Braun's fears that not enough was being done to halt the loss of sage-grouse numbers in areas being overtaken by energy development were corroborated when the US Bureau of Land Management (BLM), the federal permitting agency for oil and gas development on public lands, released the findings of an independent review called the "Powder River Basin Sage-Grouse Viability Study" in 2012.

In the Powder River Basin, located in northeast Wyoming along the boundary with Montana, a modern epic boom of coal strip-mining and coal-bed methane gas extraction is exerting a huge impact. If it continues at its present pace, conservationists express real worries that continued habitat fragmentation will result in "functional extinction" of sage-grouse populations there. (Functional extinction is the concept that before a species is completely gone from Earth, their ecological role has been greatly or completely diminished and the "loss" of their role in the ecosystem affects other species.) Together, the expanding grid of land disturbance, coupled with the arrival of mosquito-carried West Nile virus, is eliminating habitat and killing birds. In fact, the BLM specifically noted the possibility of catastrophic impacts from West Nile.

When the Gunnison sage-grouse was listed as threatened, just as Braun predicted it would be, he felt no sense of vindication, only sorrow.

An ongoing modeling of existing greater sage-grouse populations and lek locations being carried out by researchers from the University of Montana and noted biologist David Naugle, principal author of the 2012 BLM report, predict that of 370 known active leks in and around the Powder River Basin, only 6 may survive the onslaught of energy extraction and West Nile virus, the latter exacerbated by mosquito populations actually reproducing in coal-bed methane wastewater ponds.

Dustin Bleizeffer, editor of the online news site Wyofile who has been covering energy issues in Wyoming for years, said that sage-grouse sit perched on the edge of functional extinction in northeast Wyoming. "The sage-grouse population here in northeast Wyoming has been slashed more than 82 percent since 2001 due to intense coal-bed methane gas development—more than thirty thousand wells drilled over a short period of time—along with thousands of newly constructed produced-water impoundments, thousands of miles of new roads and powerlines, as well as a handful of West Nile Virus outbreaks," he wrote after the BLM report was released in 2012.

As sage-grouse populations have continued to fall or only maintain their numbers, the clarion message is that remaining sage-grouse habitat must be secured.

"Every new study on sage-grouse in oil and gas country indicates that we must protect leks with increasingly larger buffers," said Mark Salvo of Defenders of Wildlife (DOW) in an official statement. "We've been saying it for years: sage-grouse are a landscape species; they need large expanses of undeveloped sagebrush to persist."

The boosters of free enterprise counter that enforced protections for sage-grouse will spell trouble for America's ranching, oil, gas, and coal industries in the West. But, in fact, policy experts say oil and gas development will continue apace. The majority of the most-promising prospects for energy drilling are already being developed. Millions of other acres already have been leased for future development. What advocates of the sage-grouse say is that in order for the bird to persist, industry must be required to minimize its development footprint. Indeed, sophisticated drilling techniques exist that enable companies to extract more oil and gas but through far less ground disturbance.

According to Braun, who has a true bird's-eye view, concern about habitat protections affecting the financial bottom lines of energy development companies is real, but the magnitude of alleged impacts on the industry is overblown. "Properly done, we can have oil and gas and sage-grouse, but we have to have some areas where there isn't huge landscape manipulation," he said.

Still, the rallying cry of the energy industry—and states like Wyoming that rely upon royalties filling their coffers (royalties are payments that energy companies make to federal and state governments based upon a percentage of the profits they make from drilling)—is that sage-grouse are equivalent to the northern spotted owls of the Pacific Northwest.

Many fearmongers claim that protecting the sage-grouse will be tantamount in its effects to what happened when the federal government intervened to rescue the spotted owl, bringing with it dramatic logging reforms enacted on US Forest Service lands in the 1990s aimed at saving owls and the related suite of forest-dwelling creatures relying on similar habitat.

What energy companies fail to mention is that those reforms in how national forests were being managed—unsustainably and at a huge ecological cost—were actually necessary. By and large, restrictions to better protect species that fell under the umbrella of the northern spotted owl have worked, but there have been myriad other landscape-level benefits. No longer are rivers coated

An aerial perspective of the Foote Creek Rim Wind Farm and Elk Mountain in Carbon County, Wyoming. Good siting of wind farms is critical to avoid sage-grouse habitat loss and wildlife mortalities of bats, raptors, and other birds.

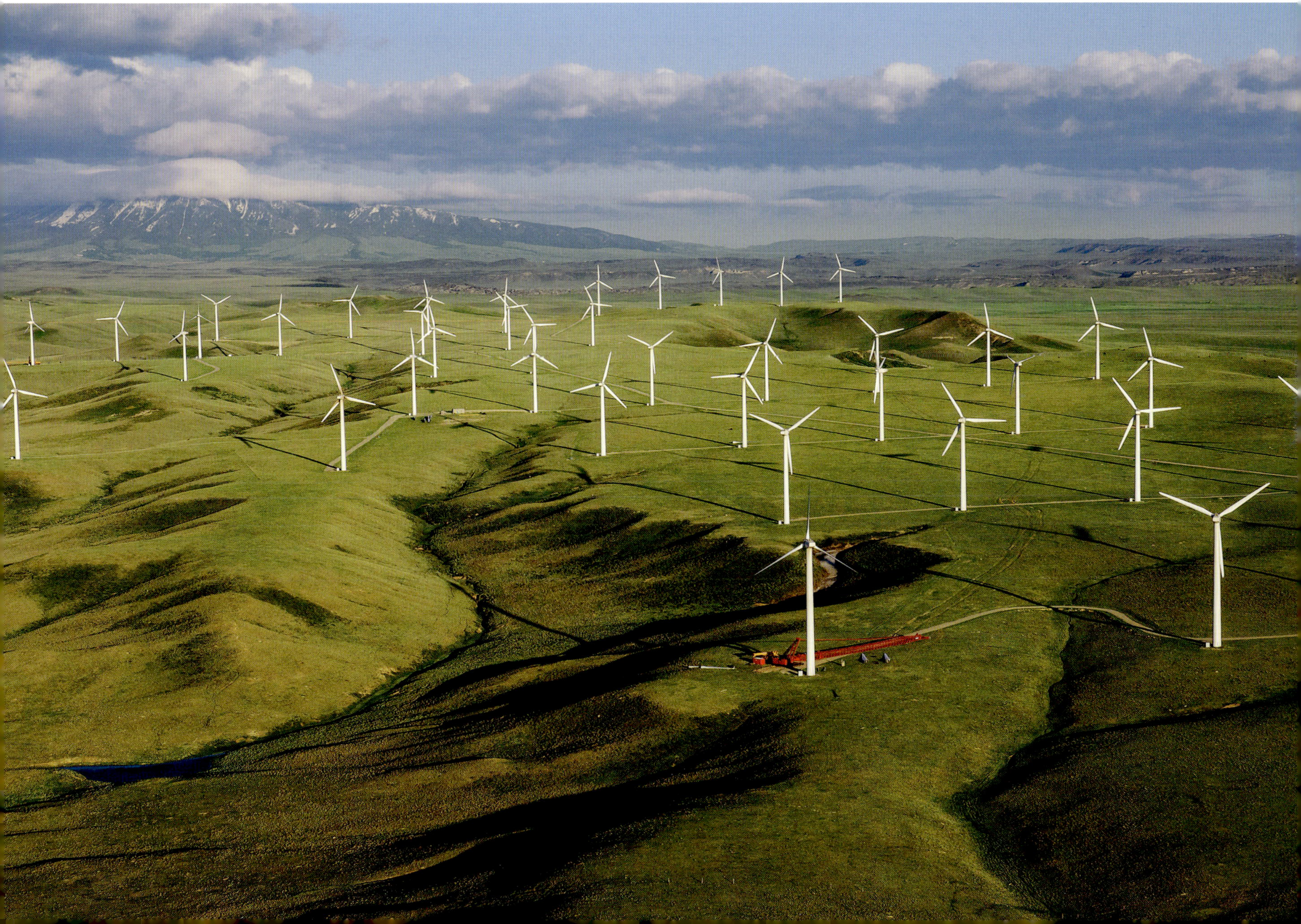

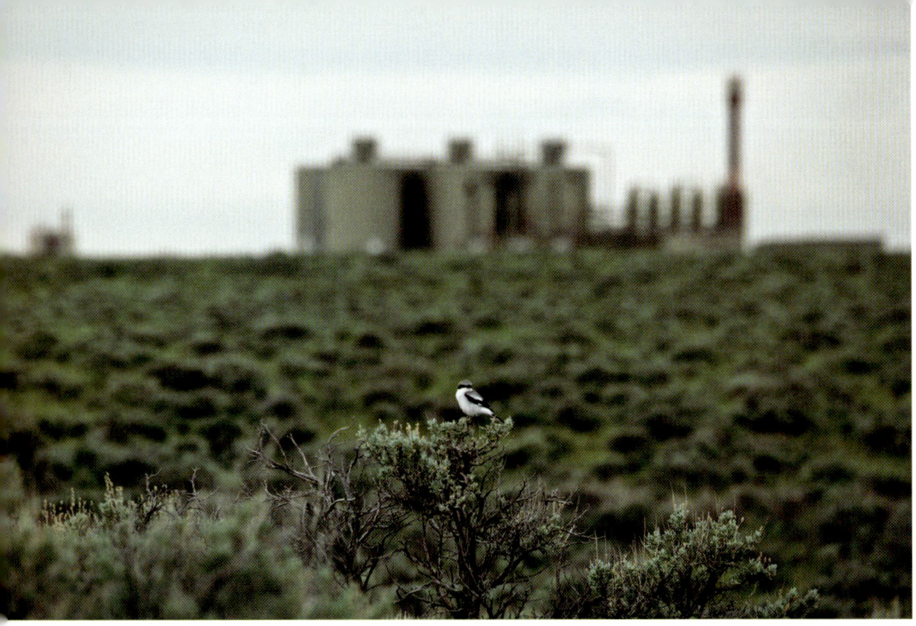

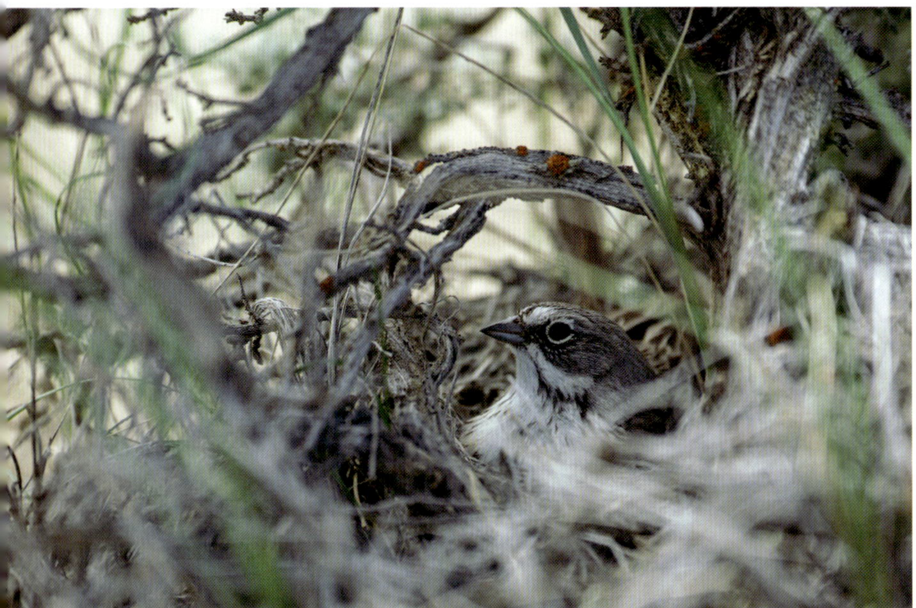

ABOVE, TOP A loggerhead shrike near a nest on a tall sagebrush stand along a sandy wash in the Jonah Field, Sublette County, Wyoming. Numbers of these masked songbirds have dropped significantly in the last century.

ABOVE, BOTTOM A brooding female sagebrush sparrow on her nest with nestlings underneath (Jonah Field, Sublette County, Wyoming)

in muck from clear-cut hillsides eroding into streams, which devastates water quality for people downstream and spawning terrain for salmon. That said, the negative fallout—from timber mills shutting down and jobs being lost—that lingers in some logging communities is still palpable.

ECOLOGY = ECONOMY

Fortunately, an age of new thinkers is showing that the economy-versus-ecology paradigm is a false dichotomy. The survival of the sage-grouse does not have to be sacrificed in order to have a healthy economy, and neither does saving the bird mean that ranching and energy development must come to a standstill, suggest biologists Naugle and Braun. Decisions being made need to be smart ones, based upon the best available science.

Not long ago, a think tank called Headwaters Economics pored over data compiled by the US Census Bureau and other federal agencies. The Headwaters researchers tried to draw a correlation between residents' incomes and the classification of surrounding landscapes. In particular, they wanted to know if protected landscapes make a difference, one way or another, in affecting peoples' bottom-line earnings. What they found is that protected landscapes yield a dividend.

"The notion used to be that if you weren't mining the landscape of its ore, or cutting down the forest for its trees, or covering the range with cattle, you were doing something wrong and your economy would stagnate," said Ray Rasker, cofounder of Headwaters. "But many communities in the West are evolving beyond that frontier mind-set. They're thriving, not in spite of being surrounded by protected public lands and putting certain kinds of development off-limits, but because of it."

According to a 2013 Headwaters study, prepared in cooperation with Stanford University, western nonmetropolitan counties with protected federal lands had faster employment growth and higher

Important Gunnison sage-grouse habitat lies at the base of Tomichi Dome. This area, where private and public lands interface, exemplifies the importance of collaborative conservation in the Gunnison Basin, Colorado.

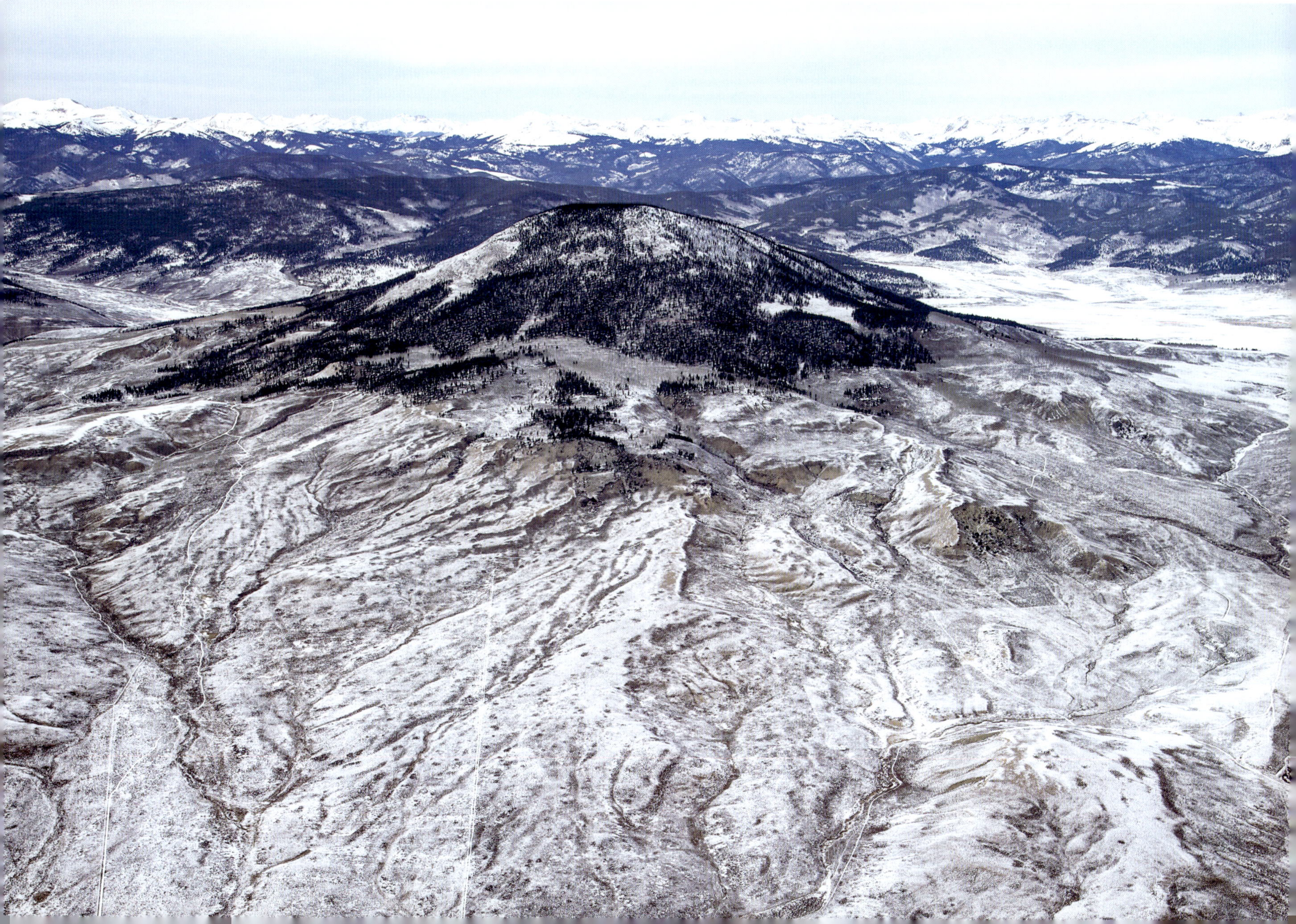

per capita income than counties without protected lands. Counties that had more than 30 percent of their land base in federal protected status increased jobs by 345 percent over the last forty years. By comparison, similar counties with no protected federal public lands increased employment by 83 percent. In addition, in 2010 per capita income in western nonmetropolitan counties with 100,000 acres (40,000 hectares) of protected public lands was on average $4360 higher than per capita income in similar counties with no protected public lands. Healthy landscapes lure mobile job creators, entrepreneurs and business owners who want to situate their companies in places that offer a higher quality of life to their families and workforces. Where there are healthy landscapes, there is, more often than not, cleaner air and water, more recreational opportunities, and healthy wildlife populations, too, Rasker says. This is valuable currency in the twenty-first century for employers trying to win the favor of top-flight talent.

Economic data provided by the National Park Service also shows that in the Greater Yellowstone Ecosystem, which has Yellowstone and Grand Teton national parks as its heart, nature tourism anchored by the national parks is credited with generating more than $1 billion annually in commerce for the region.

Ted Turner, the founder of CNN who became a "bison baron" in the West and is today the second-largest private property owner in the country, said that economic prosperity does not automatically translate into habitat destruction. "Capitalism isn't the problem," he told me when I interviewed him for my book *Last Stand: Ted Turner's Quest to Save a Troubled Planet*. "The problem is how we've been practicing capitalism as an enterprise that delivers short-term gain for a few at the long-term expense of the environment," he continued. "That has to change."

Turner has found a way to turn a profit and use his properties as "arks," providing imperiled species like grizzlies, gray wolves, sage-grouse, eastern indigo snakes, Bolson tortoises, Chiricahua

TED TURNER
THE NEW WEST
Todd Wilkinson

Ted Turner, the second-largest private property owner in the United States, said in my interview with him, "We have to move beyond the mind-set of owning land only to liquidate it of merchantable assets." He embraces a principle he calls "the triple bottom line," illustrated by three legs of a stool.

The first leg focuses on making money, and Turner noted that no matter how well intended private land conservation efforts are, they need to be economically self-sustaining. Turner said, "It's unrealistic to make conservation a burden for those who are trying to do the right thing. Society should want to keep (those people) on the land so that their stewardship can continue to be passed on to future generations."

Second, whenever Turner pursues an activity, he aims to minimize harm or carry the activity out in a way that actually leaves the land healthier. "You're more likely to achieve profitability over the long term if you respect the limits of the land," he said. "Mother Nature has her own economy of efficiency that has been refined by millions of years of fine-tuning. Many of the problems in farm and ranch country today stem from decisions to maximize profit in the short term at the expense of everything else."

The final leg of the stool, supporting local communities, involves treating the people working for you well and spending your money in local communities, where the ripple effect can be significant.

The good news? Turner's ethic is rubbing off on others, whether they own one acre or a hundred thousand. §

A bison cow and calf graze on spring green grasses in Lamar Valley, Yellowstone National Park, Wyoming. Ted Turner manages 51,000 bison on fourteen ranches, the largest private herd in the world. Some of the bison on his Montana Flying D Ranch, in the Greater Yellowstone Ecosystem, are descendants of the few hundred survivors from historic times in Yellowstone National Park.

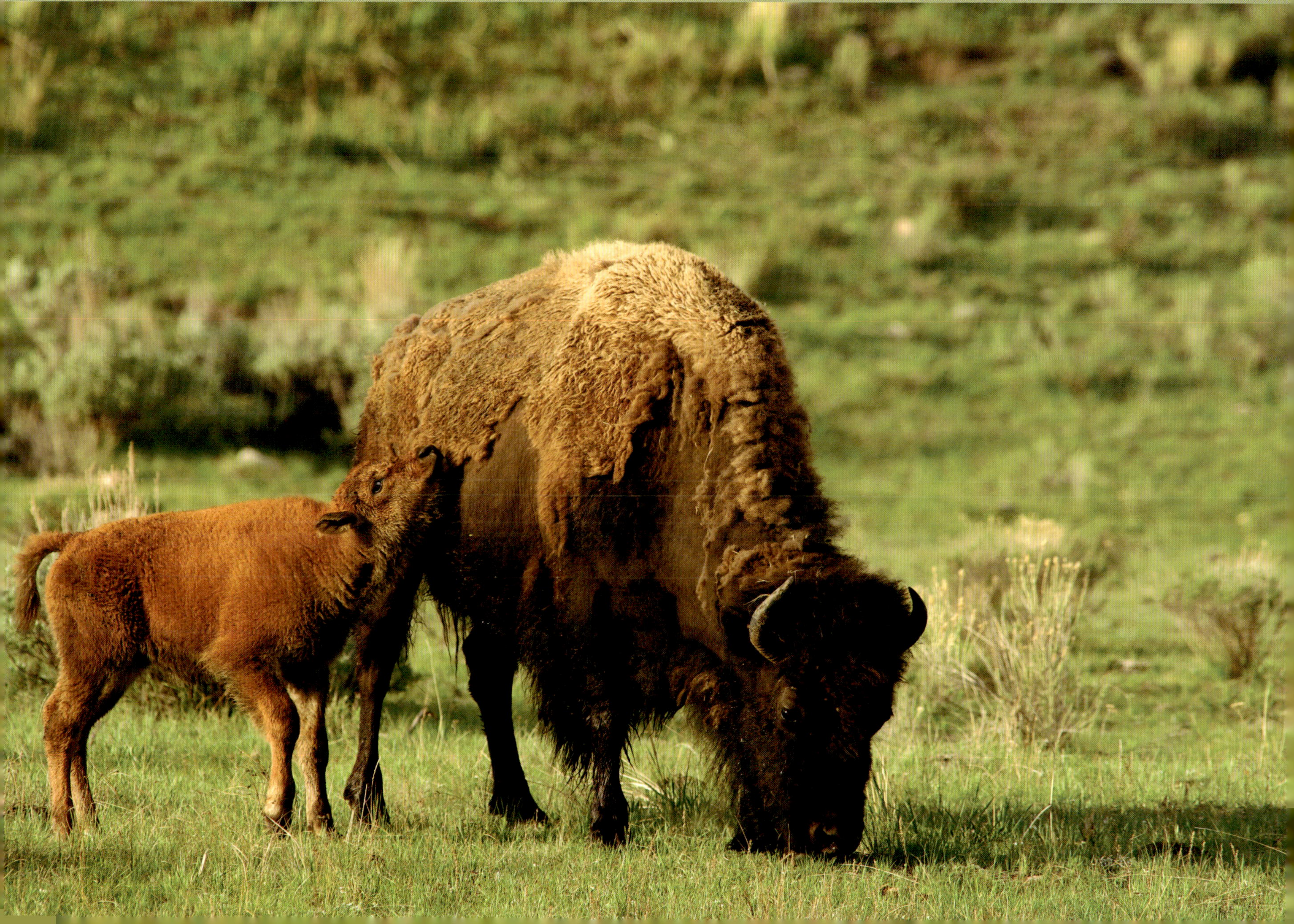

leopard frogs, and other wildlife a place to roam, by implementing what he calls "the triple bottom line": making money, minimizing harm, and supporting local communities.

One of Turner's ark properties in the southeastern United States, Avalon Plantation in Florida, is at the forefront of restoration efforts for federally endangered red-cockaded woodpeckers. The US Fish and Wildlife Service has made the plantation a centerpiece of its Safe Harbor program, showing that private property owners who help recover species are not, contrary to widespread public opinion, penalized.

"Those who claim that one can't hug a sheltering tree and simultaneously aspire to have economic prosperity and strive to give all people a decent quality of life are being disingenuous," Turner said. "On my lands, I have set out to prove that the polemic of environment versus economy is a false dichotomy, that you can be a tree hugger and still have your name appear in *Forbes.*"

Landowners of every political persuasion are emulating Turner, including some of the largest private property holders in the West who together hold deed to millions of acres. (For one example, see the profile of Heather Stirling in this book.)

WHITHER THE SAGE-GROUSE WITH CLIMATE CHANGE?

Looming as perhaps the greatest challenge facing modern civilization is climate change, caused and being exacerbated by the burning of carbon-rich fossil fuels. While natural gas is regarded as a "bridge fuel" between the old fossil-fuel economy and one that runs more cleanly on renewable, alternative energy, a raging debate is occurring over how much coal should be left in the ground and never exploited. Another aspect is that ecologists have a growing appreciation for the role of nature as a "carbon sink"—the ability of soil and water and living plants to absorb carbon dioxide out of the atmosphere.

Scientist Joe Fargione with the Nature Conservancy said there's a tremendous amount of carbon captured in the roots of healthy plants and even more in the soil. When ground is plowed, disturbed, and cleared of cover, about 40 percent of the carbon, stored in the top foot of soil, is released into the atmosphere. Many landowners have placed conservation easements on their property that prevent it from being converted into subdivisions covered

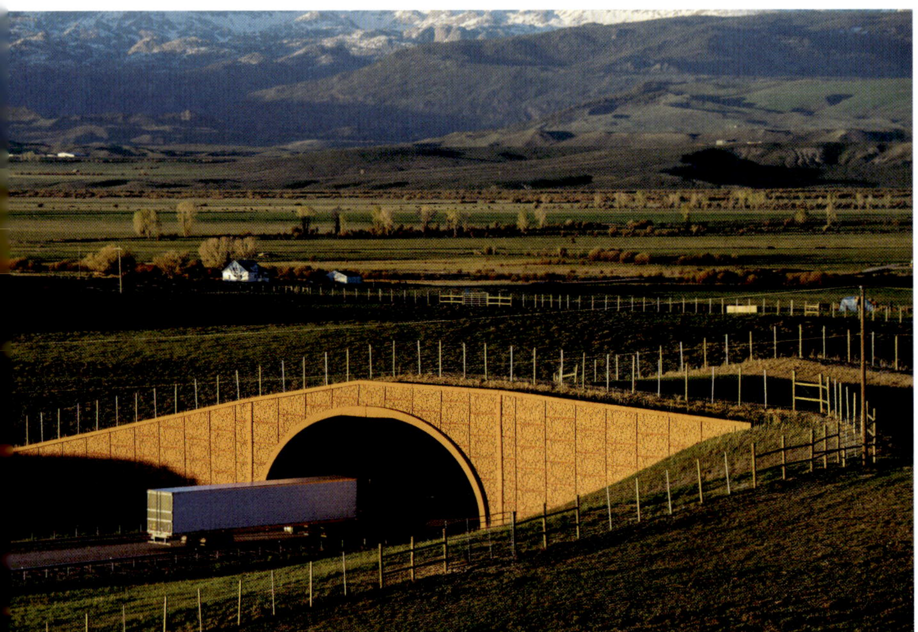

Trapper's Point, Wyoming, is a narrow spot on the seasonal migration corridor of pronghorn and mule deer that is now a paved highway. Mounting animal fatalities led to wildlife bridges being built in two locations in the Upper Green River Basin, allowing wildlife to roam freely while reducing highway fatalities.

in asphalt, concrete, and dense grids of energy-development drill pads. In exchange for tax advantages, which enable them to better pass on their ranch or farm to the next generation, these landowners agree to refrain from harming the natural integrity of the land. Conservation easements increasingly are being used as tools to preserve the sagebrush sea ecosystem and its inhabitants.

As for alternatives to fossil fuels, the promising potential of clean renewable energy generated by solar and wind remains, for now, a boutique economic enterprise—though some of the prime sage-grouse recovery states (Wyoming and Montana especially) have high capacity for wind power. Wind and solar farms are expected to increase in size and number as infrastructure improvements reduce costs of production. Researchers point to Wyoming, which holds the largest number of greater sage-grouse among all of the eleven sagebrush-sea states and has huge potential for generating wind power. An analysis shows that less than 25 percent of the areas in Wyoming with commercial-grade wind resources overlap with the best sage grouse habitat.

Of course, no form of power generation is absolutely benign, as biologist Naugle points out in his authoritative 2011 book *Energy Development and Wildlife Conservation in North America,* a compilation of his writings and those of two dozen other experts on a vision for landscape conservation that will protect wildlife while ensuring future energy supplies. The propellers on giant windmills can kill birds, and some of the corridors carrying the electricity to market from both solar and wind farms could slice through sage-grouse habitat.

"Unfortunately, the threats are many," wildlife researcher Braun noted. "Sage-grouse survived overgrazing by livestock, but now they are facing multiple whammies—energy development plus livestock grazing plus wind farms plus subdivisions and climate change; it's devastating. And it will be disastrous if efforts to protect the bird are uncoordinated and done piecemeal."

Pausing for a moment, Braun finally said, "I want to be optimistic, but it's difficult to be. There is no political will or foresight to make sure sage-grouse exist. I hope that policy makers prove me wrong."

THE DOMAIN OF POSSIBILITY

Doom-and-gloom scenarios—whether about prospects for sage-grouse or the livelihoods of people who live on the sagebrush sea—sell newspapers yet hold little power to rehabilitate the fortunes of sagebrush species. It isn't as if policy makers contemplating options don't know what to do. The sage-grouse is one of the most studied North American birds in modern times. The prescription for its survival is straightforward: protect the best and biggest sweeps of sagebrush that remain on both public and private land. Those sweeps, called core-area habitats, encompass the most important pieces of sage-grouse habitat ranging from where the birds mate and nest in warm months to finding cover during the winter.

According to the Public Lands Council and National Cattlemen's Beef Association, public-land ranchers who have access to hundreds of millions of acres of BLM, Forest Service, native reservation, and state lands also hold title to 120 million acres (50 million hectares) of private land, a sizable stretch of which is located in the sagebrush sea.

Where hope lies is in the stories of compatibility that defy either-or scenarios. Middle ground is the domain of possibility. There are numerous examples of good people out on the land doing what they can to make sure that sage-grouse have a place.

What do the visionaries searching for answers and trying to make a difference look like? They come in all forms. One of them is rancher Brian Rutledge, who runs a herd of cattle in Colorado just south of the Wyoming border. For ranchers like him, conservation has been part of his family legacy. Here's another line on his résumé that might surprise you: Rutledge is Central Flyway

Conservation Strategy and Policy Advisor for Audubon Rockies. It is among several noteworthy organizations at the local, regional, and national level trying to rally the public.

Audubon has been playing a pivotal grassroots coordination role in protecting sage-grouse habitat in the West, especially as part of its own Sagebrush Ecosystem Initiative, which began in 2007. It is also a vital contributing member of the Sage Grouse Initiative, launched in 2010 by the US Department of Agriculture's Natural Resources Conservation Service. By "harvesting the financial power of the Farm Bill," the latter has an ambitious goal of enlisting ranchers, conservationists, and other large private-property owners to focus on better stewardship of 78 million acres (32 million hectares) of sage-grouse range in the eleven states where the habitat is located. More than seven hundred ranchers have already enrolled.

"I'm a rancher, I'm a cattleman and a horseman, but before anything else, I'm a conservationist because my livelihood and the West I love depend on it," Rutledge told me. "Conserving sage-grouse doesn't have to be politically intractable. Wyoming is showing us a way forward."

If there's any state where one might expect a defiant posture to exist against preservation of the bird, it is Wyoming. But, in fact, Wyoming realized years ago that if it wanted to prevent federal listing of the sage-grouse under the Endangered Species Act, which would bring more stringent federal restrictions on public land use, then it had to step up to the plate, Rutledge said.

Two-thirds of Wyoming's economy is derived from a combination of fossil-fuel extraction (oil, natural gas, and coal), mining of minerals, and agriculture (ranching and farming). The state also is home to the best and biggest remaining sweeps of sage-grouse habitat. To reconcile these two facts, Wyoming has pioneered a plan that is arguably a model for the West. The question is, will it hold up?

Wyoming's plan is based on a strategy that was initially crafted by former Democratic governor Dave Freudenthal and implemented by his successor, Republican governor Matt Mead. Under the plan, 15 million acres (6 million hectares), or 24 percent of the state, is limited to disturbing no more than 5 percent of a 640-acre (259 hectares) section in sage-grouse core-area habitats, a departure from business as usual less than a decade ago. Some 85 percent of remaining occupied sage-grouse habitat will be protected against intensive disturbances like those occurring in the Powder River Basin in northern Wyoming and in Jonah Field.

"Incredible strides have been made in Wyoming," Rutledge said, "because we've figured out how to work together as stakeholders and, so far, not fighting and suing each other. Of course, an essential part that's hard for some to accept is that we don't have a choice. Either we craft a solution, or one is going to be foisted upon us (by the federal government)."

The energy industry wants at least one thing that ranchers, hunters, anglers, tourism officials, and, yes, even politicians also want, Rutledge said. "Everyone wants a set of rules that are predictable, binding, and enforceable—that you can plan your business and your life around," he explained. "Lawsuits filed to force sage-grouse listing (under the Endangered Species Act) or prevent it create chaos and turmoil. The courts are not settings that build goodwill. They cost time and money. Worst of all, they undermine public support for ecosystem protection."

He hears the skepticism from some environmental groups and biologists like Clait Braun. Braun noted in a 2014 guest essay printed in the *Casper Star-Tribune* that even the Wyoming Department of Game and Fish documented a decrease of about 60 percent in the number of male sage-grouse counted on leks since the state moved forward with its strategy of protecting core-area habitats.

Rutledge heeds the concerns of skeptics: namely, that the energy industry can't be trusted, that it is out only to maximize short-term profit, and that politicians, who receive hefty campaign contributions from energy companies, will seek to create loopholes for harmful development to invade sage-grouse core-area habitats.

Rutledge is a pragmatist. Under Wyoming's plan, "The metrics of compliance with habitat protection will be assessed annually, and if adjustments need to be made, they'll happen through a process we call 'adjustable impact management,'" Rutledge said, noting that Wyoming's Sage Grouse Implementation Team (comprised of state officials, landowners, and representatives of both the energy industry and conservation organizations) will work together to oversee the state plan. The Sage Grouse Implementation Team, separate from efforts playing out in the Sage Grouse Initiative, is being watched closely by the other ten sagebrush-sea states. "We're focused on not only safeguarding the core-area habitat, but mitigating the impacts on land that is already disturbed," Rutledge continued. "This is a new era for sage-grouse conservation, and Wyoming will be a proving ground."

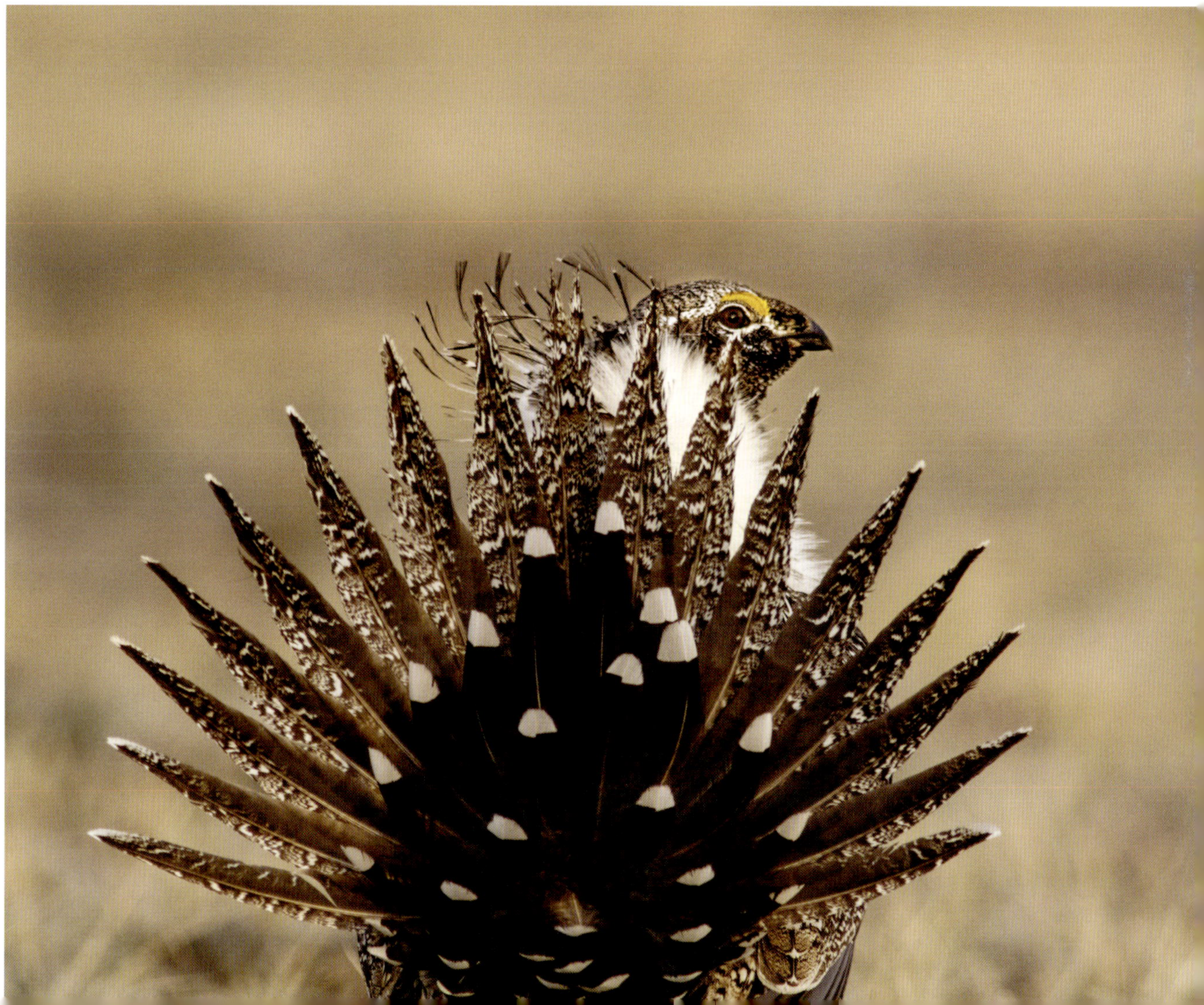

Greater sage-grouse males fan spiky tail feathers while displaying to impress females (Sublette County, Wyoming).

THE GREAT BASIN'S STRUGGLE

Habitat protection needs to happen fast because declines in sage-grouse continue. Out in the Great Basin, on the western front of grouse range between eastern California and Utah (where 20 percent of remaining sagebrush exists), the Sage Grouse Initiative is confronting a different set of issues. Pinyon, juniper, and cheatgrass are invading sagebrush habitat from opposite ends of the altitudinal spectrum: Pinyon and juniper, native shrub trees, have, in the absence of fire, been marching down higher-elevation slopes and outcompeting sagebrush. Cheatgrass, an exotic forage introduced to the range by ranchers a century ago to fill the bellies of cattle, is spreading upward from valleys on the back of soil disturbance, such as overgrazed pastures and, ironically, the proliferation of wildfires.

According to experts with the Sage Grouse Initiative, conifers such as pinyon and juniper have expanded their range in the Great Basin by 600 percent and now encompass 14 million acres (5.6 million hectares)—that's equal in size to seven Yellowstone National Parks. And sage-grouse are notoriously finicky: just 4 percent tree cover can prompt breeding grouse to abandon a lek. Incursion by these trees not only takes out sagebrush but profoundly influences the territorial habits of grouse and can have consequences for reproduction. As an example, researchers with the US Forest Service estimate that in Oregon alone there are 875,000 acres (354,000 hectares) of pinyon and juniper trees within 3 miles (4.8 kilometers) of leks.

Meantime, researchers point out that 50 percent of existing Great Basin sagebrush has been invaded by cheatgrass, which not only leaves sagebrush more vulnerable to burning but also usurps native wild plant communities in the "canopies" of sagebrush, leading to loss of associated wildlife right down to the insect level. And it can alter, too, the amount of water held in the soil.

In response, the Sage Grouse Initiative has enlisted ranchers to beat back cheatgrass, help facilitate range restoration with the seeds of native plants, and modify how cattle graze. Simultaneously, aggressive efforts are under way to clear pinyon and juniper. Not only is this slowing the loss of sage-grouse habitat, but both tactical measures actually stand to improve grass conditions for grazing cattle on private as well as public land.

But again, the clock is ticking. Botanists say that if a substantial reduction of pinyon-juniper is not undertaken in the Great Basin within the next few decades, 75 percent of the interloping conifers will have grown into large trees that completely overtake the native bunchgrasses and sagebrush, eliminating them as native seed sources, reports the US Forest Service. The good news is that tree removal and a regimen of controlled burns, where appropriate, has resulted in greater sage-grouse returning to their old haunts.

WILL SMALL STEPS FOR SAGE-GROUSE BE ENOUGH?

Along with these macro issues, myriad threats contributing to sage-grouse losses are exceedingly nuanced in their local impacts. For example, every spring as birds come home to breed on their leks, cocks and hens alike are flying into barbed wire fences. Thousands likely die in such collisions annually. A fairly simple antidote is marking fences with highly visible tags to make birds aware of the deadly strands. University of Idaho researcher Bryan Stevens conducted a study funded by the Sage Grouse Initiative that showed fence tagging and relocation of more than 530 miles (853 kilometers) of fences saved an estimated 2800 sage-grouse.

Still, in terms of effectively mitigating sage-grouse losses, there exists a certain climate of denial regarding the magnitude of the response that is necessary. While the states of Montana, North Dakota, and South Dakota have moved to reduce or curtail their long-established hunting seasons on sage-grouse in response to bird losses, ecologists note that the major declines of birds are not owed to hunters but loss of habitat.

Meanwhile, in Utah, Nevada, and other states, the US Department of Agriculture's Wildlife Services agency has been enlisted to spend millions of dollars to kill thousands of corvids (ravens, mostly) that eat sage-grouse eggs and young sage-grouse chicks. The absurdity is that, just as with hunters, predation by other avian predators ranks low on the list of actual significant threats.

Critics say some of those trying to stave off federal listing under the Endangered Species Act of greater sage-grouse and Gunnison sage-grouse are targeting the little things and not grasping the big picture.

A group that does get the big picture is the Wilderness Society, which sees the ecological integrity of the sagebrush sea as being as important as the highest and most remote points of mountain wilderness. Because it is so close, sometimes we lose perspective of the tranquil majesty that is right in front of our face. It's like the ditty calling us home to the range, says Nada Culver, the Wilderness Society's senior director of agency policy and planning. "The greater sage-grouse actually inhabits some of the wildest public land in the American West—truly a real larger-than-life landscape where the (mule) deer and the (pronghorn) antelope play," she says. "We see the sage-grouse and its homeland as an integral part of the fabric of the West; land that holds all together, supporting the livelihood and spiritual enjoyment of many. We hope that all Americans appreciate this amazing bird as part of their heritage."

THE SAGEBRUSH SEA AS A BELLWETHER FOR PLANET EARTH

Indeed, the greater and Gunnison sage-grouse are analogous to canaries in the coal mine. As prominent scientists warn, humankind is confronting the leading edge of the sixth major species extinction episode on the planet. It will be exacerbated by climate change and rising human population—and, in America, inward growth to the West (the fastest-growing region in the United States) and coming shortages of freshwater. How we deal with sage-grouse augurs the future.

"This is the most ambitious conservation challenge the US government has ever undertaken," rancher and Audubon executive Rutledge said. "It's portrayed as being about only the sagebrush ecosystem. But it's really about every ecosystem on the planet that we ignore or take for granted and how we can bring ourselves back in line to live more compatibly with nature. It's an opportunity to repair the damaged legacy we're about to leave our children."

Terrell D. Rich, an avian ecologist with the US Fish and Wildlife Service, and a team of scientists posed a question in an assessment they prepared for the US Forest Service: "If we cannot successfully conserve sage-grouse and the sagebrush ecosystem in the US given our theory, our knowledge, and our large blocks of public land, then one wonders how we can succeed for other species, for other land ownerships, and for other regions of the world?"

I believe we can succeed, but it will require a sea change in how we think. It requires embracing self-sacrifice to achieve a larger, nobler good. Sage-grouse serve as a mirror that we hold up to ourselves, reflecting back the character that we possess as stewards of creation.

I like to imagine, as I inhale the sagebrush aroma of the West, that our descendants fifty years from now will know the timeless thrill of total immersion in the sagebrush sea. Celebrating the survival of an umbrella-species bird isn't just a logistical exercise or a scientific imperative. I often muse back upon my first encounter with a sage-grouse in human form. Native people don't venerate or value the sage-grouse because of what it does for them; their ceremony reminds us that we are its sacred keeper, and its survival is linked to our own. §

SAGE MIGRATIONS
TIMELESS FLIGHT OF SANDHILL CRANES
Dave Showalter

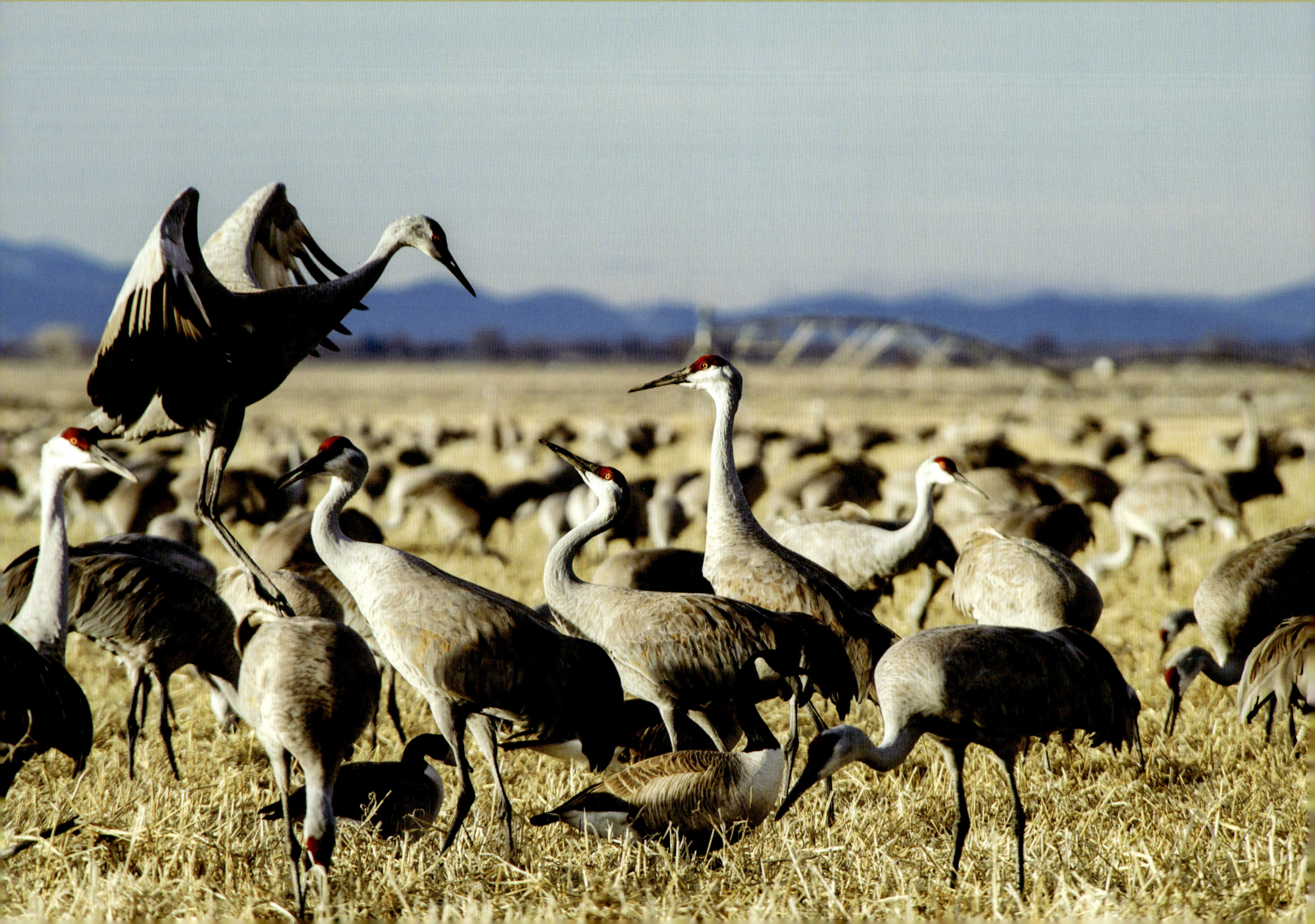

FOR SANDHILL CRANES, SAGEBRUSH IS THE FILAMENT that holds their western habitat together. Though they are primarily perceived as wetland birds, sandhills are adaptable and use a wide range of habitats. They rely on wetlands for the protection of the surrounding waters, for food, and for nesting, yet the Rocky Mountain population of sandhill cranes also depends on the sage for cover, a variety of food, and the riparian habitats contained within it. Though many people think "dry" when they think of sagebrush habitat, it does contain wet meadows, springs, and seeps that provide crucial rest stops for migrating sandhill cranes.

Near winter's end, one day in February, more than twenty thousand sandhill cranes take flight on a south wind. They've spent the winter along the Rio Grande at Bosque del Apache National Wildlife Refuge in southern New Mexico. Here in the northern Chihuahuan Desert, they've enjoyed an easy winter in manmade wetlands and grain fields planted for them, as well as for snow, Ross's, and Canada geese. Now the spring migration begins, as the obligation and timing required to mate pull the cranes north en masse. Each morning, tens of thousands of waterfowl burst from the wetlands at dawn, completely filling the sky with a chaos of sound and motion in warming light.

In one of North America's great animal migrations, Rocky Mountain sandhill cranes leave their cozy winter environs to disperse to their summer breeding grounds in the central and northern Rocky Mountains. Many cranes nest in the sagebrush sea's Greater Yellowstone Ecosystem (GYE), roughly 1000 miles (1600 kilometers) north of their winter quarters. This determination to move between seasonal homes, to head north while it's still winter in Yellowstone, to rest and refuel at key stopover areas along the Central Flyway's western edge in the sagebrush ecosystem, has been hard-wired in Rocky Mountain sandhill cranes for eons. They nest, breed, and raise young, called colts, in wet meadows of the GYE, then make the return trip to Bosque del Apache in autumn, though the fall migration isn't as urgent as in spring.

Sandhill cranes feed in planted fields at Monte Vista National Wildlife Refuge, an important migration stopover in the San Luis Valley, Colorado.

I migrated with sandhill cranes on my own journey to view birds of the same flock over their entire migration route. From Bosque del Apache, the Rocky Mountain greater sandhill cranes follow the winding Rio Grande Rift Valley north, the first few dropping into the desert scrub of southcentral Colorado's San Luis Valley in early February. By April they've dispersed into the sagebrush habitat of northwest Colorado's Yampa River Valley and then northwest to the GYE, where they'll spend breeding season.

Although it's still winter in the San Luis Valley in February, the cranes are winged messengers of spring. News of their arrival travels rapidly through the communities of Monte Vista and Alamosa, to birders anxiously looking forward to crane festivals, a new beginning, another season to be inspired.

Sandhills continue to arrive, dropping into wet meadows and grain fields across the massive high-desert valley. In blue-black light

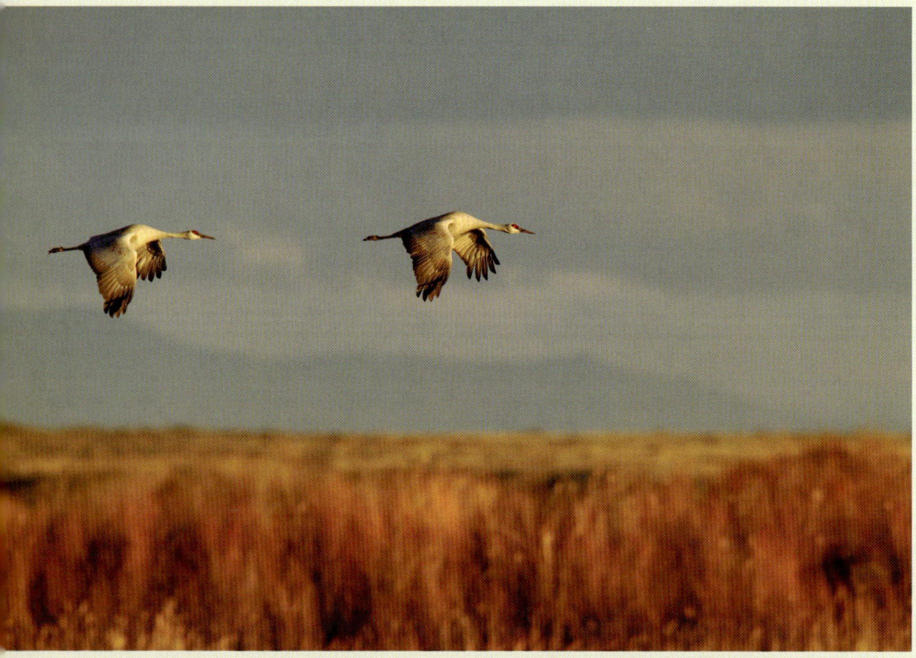

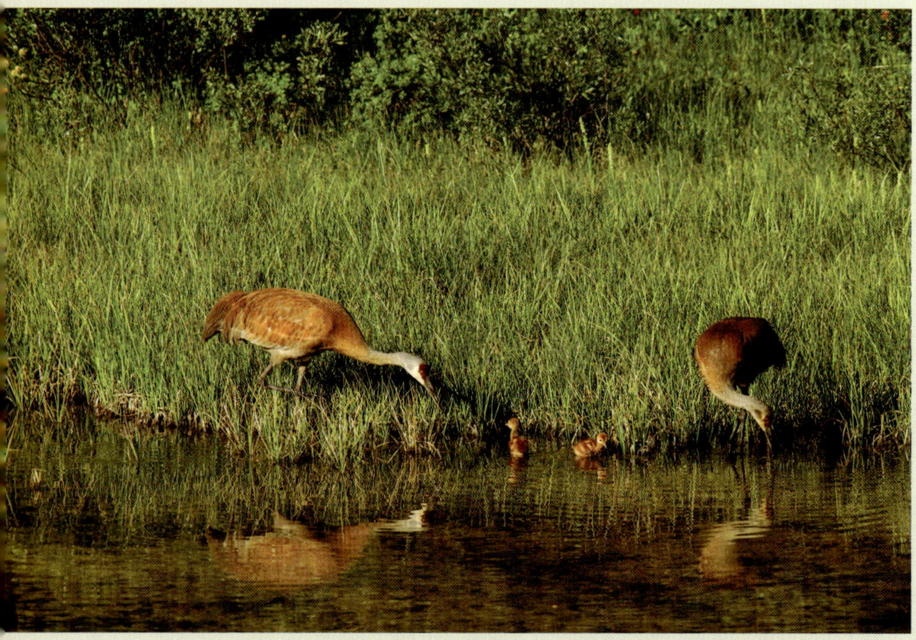

each morning, roosting cranes chatter a primeval, throaty *ga-rooo, ga-rooo*, and before sun rises on the south flanks of 14,345-foot (4372-meter) Blanca Peak, stringers of twelve, twenty, thirty cranes trace the snow-covered tops of the Sangre de Cristo Range. They keep coming, silhouettes like arrows against a gold backdrop, and suddenly they're circling and dropping from the sky, feet dangling like landing gear, wings outstretched, the tips like fingers, into fields of barley and alfalfa at Monte Vista National Wildlife Refuge.

Observing a few hundred of these massive gray birds with the patch of brilliant red skin on top of their heads, I am drawn in by their charisma and grace. Small family groups peck and scratch for grain. Momentary territorial squabbles establish dominant and submissive roles in an eye-blink. A pair of sandhills extend their long necks, beaks pointed to the sky, and call in perfectly synchronized harmony, their lifelong pair bond pronounced. And they dance, levitating on a breeze in suspended elegance.

"Sometimes cranes dance because they're nervous," said conservation photographer Michael Forsberg, "sometimes because they're in love, and sometimes cranes dance because they just like to dance."

On the western edge of the sagebrush sea, another crane migration stretches from the American River near Sacramento, California, north to Alaska along the Pacific Flyway, with stopovers in eastern Washington's sage country on the Columbia Plateau. In Colorado the San Luis Valley, 300 miles (480 kilometers) north of Bosque

TOP LEFT Sandhill crane pair returning to evening roost (Bosque del Apache National Wildlife Refuge, New Mexico)
BOTTOM LEFT A pair of sandhill cranes with colts, as their young are called, in Wyoming's Hoback River Valley. Sandhill cranes of the Rocky Mountain population nest in marshy areas throughout the Greater Yellowstone Ecosystem.

Sandhill cranes wade from an evening roost in Bosque del Apache National Wildlife Refuge, New Mexico, before flying to feed in grain fields. Roosting in shallow water helps protect sandhills from predators.

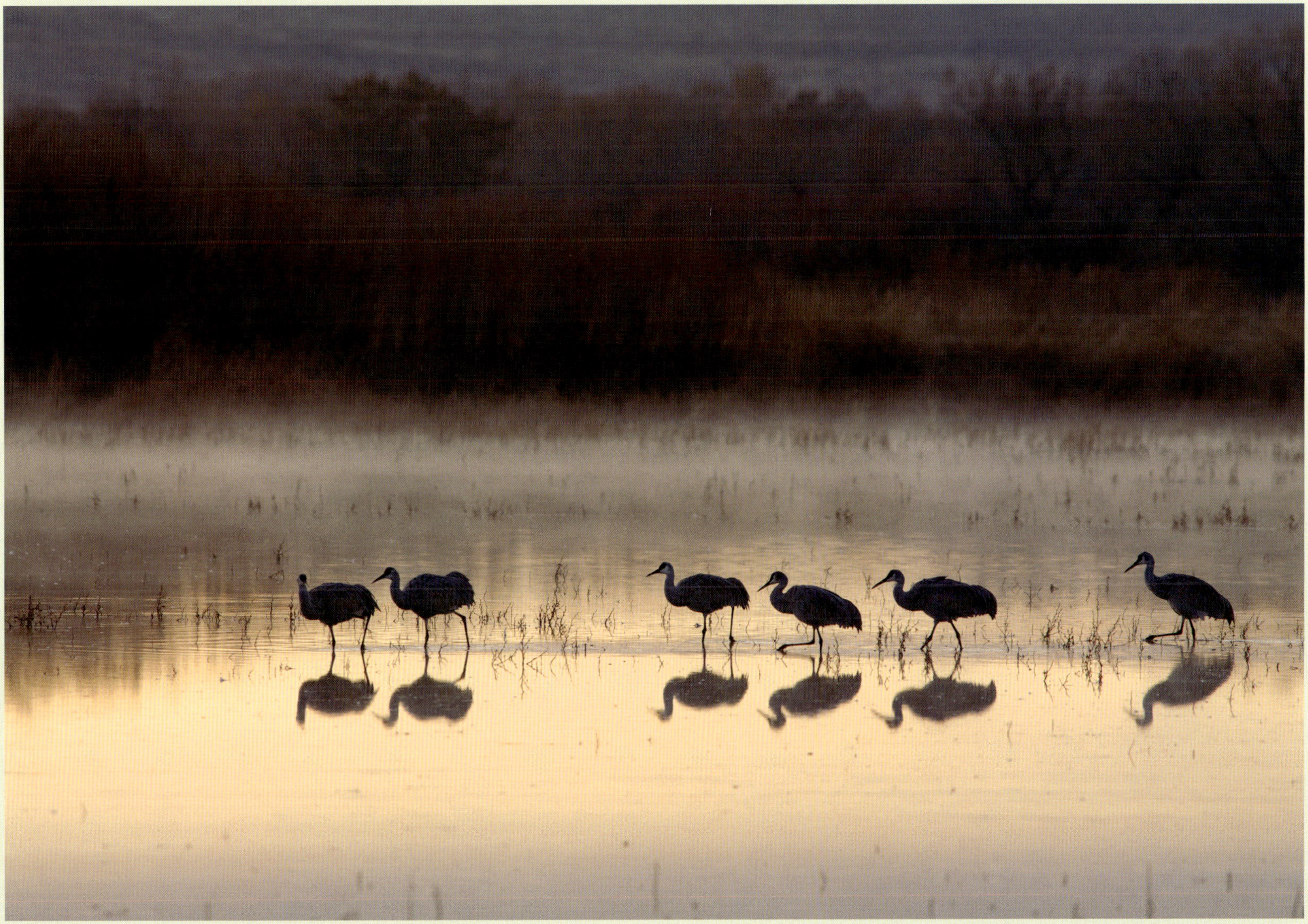

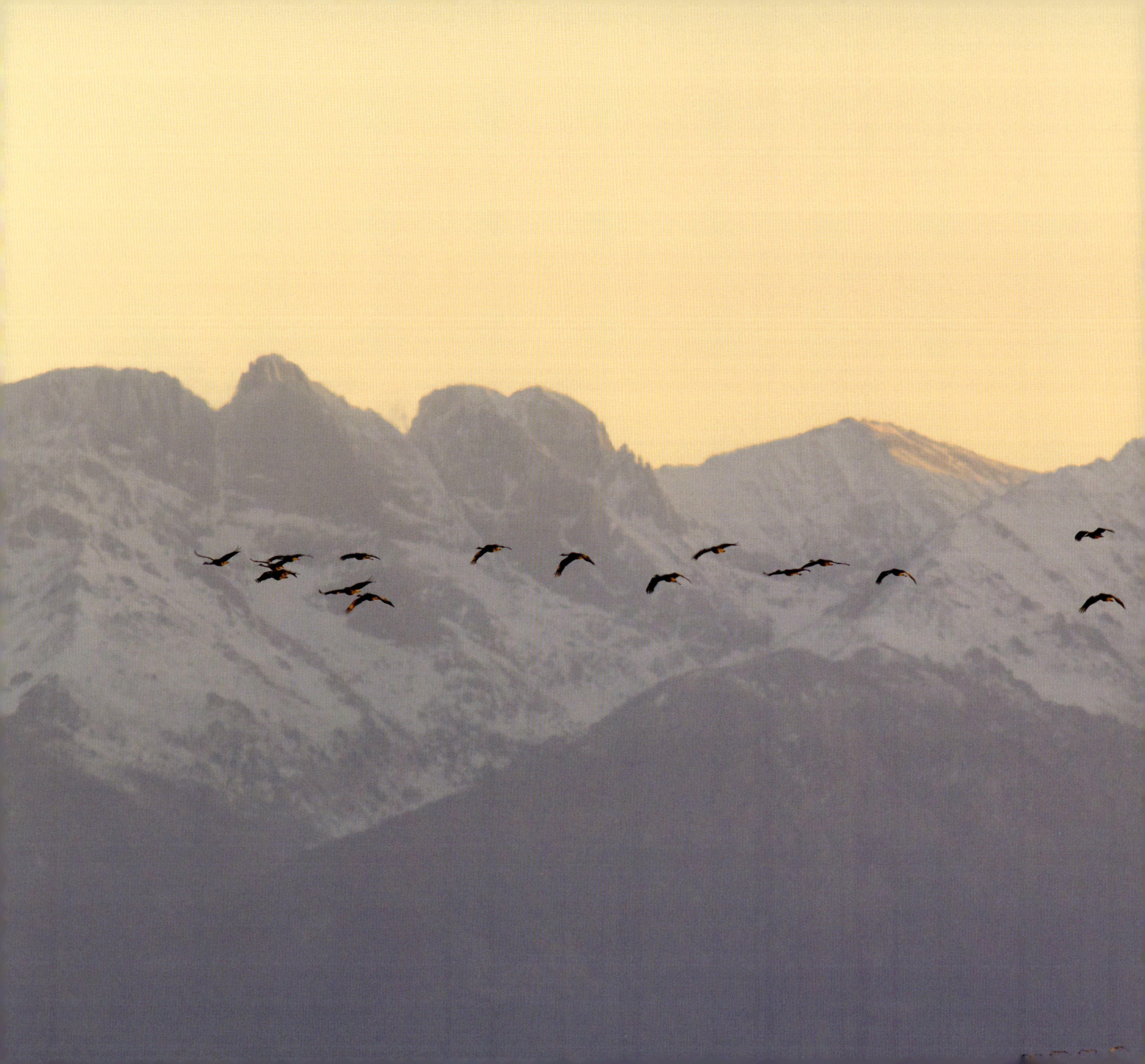

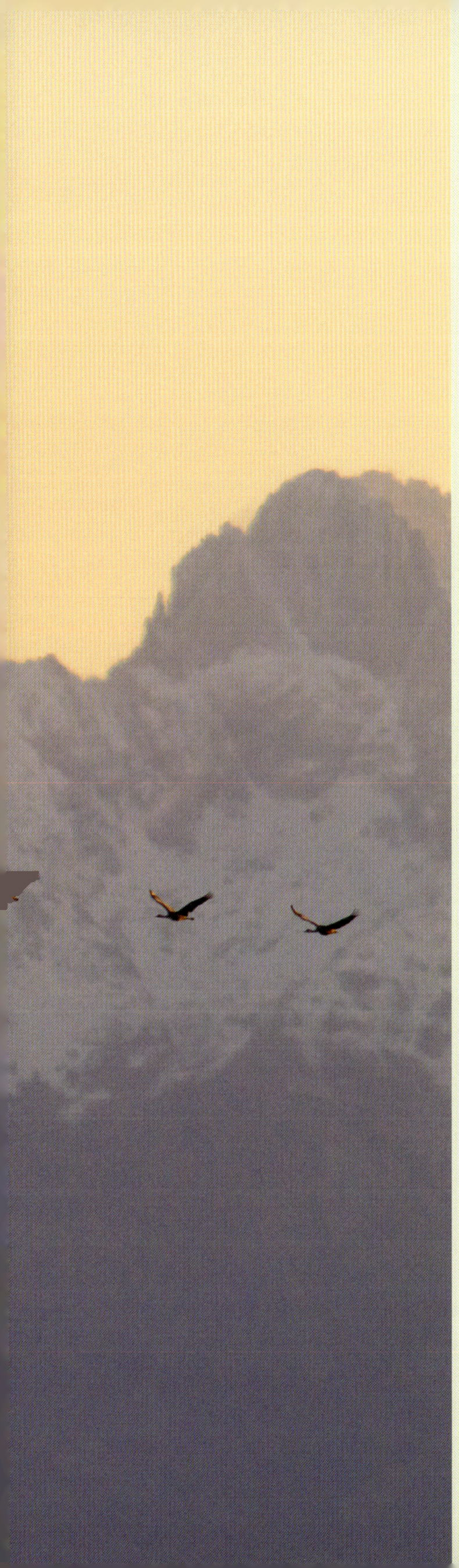

Sandhill cranes fly across the face of Kit Carson Mountain (14,165 feet/4317 meters) and Crestone Peak (14,294 feet/4357 meters), two giants in the Sangre de Cristo Mountains, as they leave their evening roost in the San Luis Valley, Colorado, in early morning.

del Apache National Wildlife Refuge, has been a perfectly situated Rocky Mountain sandhill migration stopover for millennia. A mosaic of expansive, shallow wetlands and grass once covered the San Luis Valley floor, the Rio Grande meandering in broad, sweeping oxbows.

While sandhill cranes can find refuges and protected areas for their winter and summer grounds, the critical stopovers along their migration path have steadily been disappearing. The San Luis Valley is now carpeted with potato farms, crop circles, and center-pivot irrigation machines. Irrigation pulls from the shallow water table, drying up wetlands used by sandhills and waterfowl. Irrigation water is growing scarcer, a climate change reality, so local farmers have formed a water sharing cooperative to sustain one of the top-producing potato areas in the country. Cranes, too, are part of the water economy; the Monte Vista National Wildlife Refuge floods fields and plants grain to provide suitable crane stopover habitat. As water shares are divvied up, what wetlands will remain for migratory birds?

John Rawinski, a local birder and crane advocate, has already seen wetlands diminish. "San Luis Lake near the town of Mosca has always been productive for migrating birds, but it has been empty for two seasons now; the wind routinely swirls the alkaline-rich dust into the air from the dry lake bottom. The cranes and other water birds have lost an important resting place. Who speaks for the cranes today?" Loss of wetlands not only reduces important resting and refueling habitat along the cranes' perilous journey, but overcrowding on wetlands also leaves birds vulnerable to predation and disease.

Globally, fourteen of the fifteen crane species are declining. Habitat loss, agricultural conversion, powerline collisions, and illegal hunting are common threats to cranes. The endangered whooping crane reached a low of twenty-one individuals before a remarkable recovery effort brought it back from the brink of extinction. Sandhill cranes are the only crane species that still have a stable population. We have an opportunity now to prevent them from reaching this brink by conserving stopover habitat on their migration flyways. §

A PHOTOGRAPHER'S JOURNEY:
THE VITAL CORE OF THE SAGEBRUSH SEA
Dave Showalter

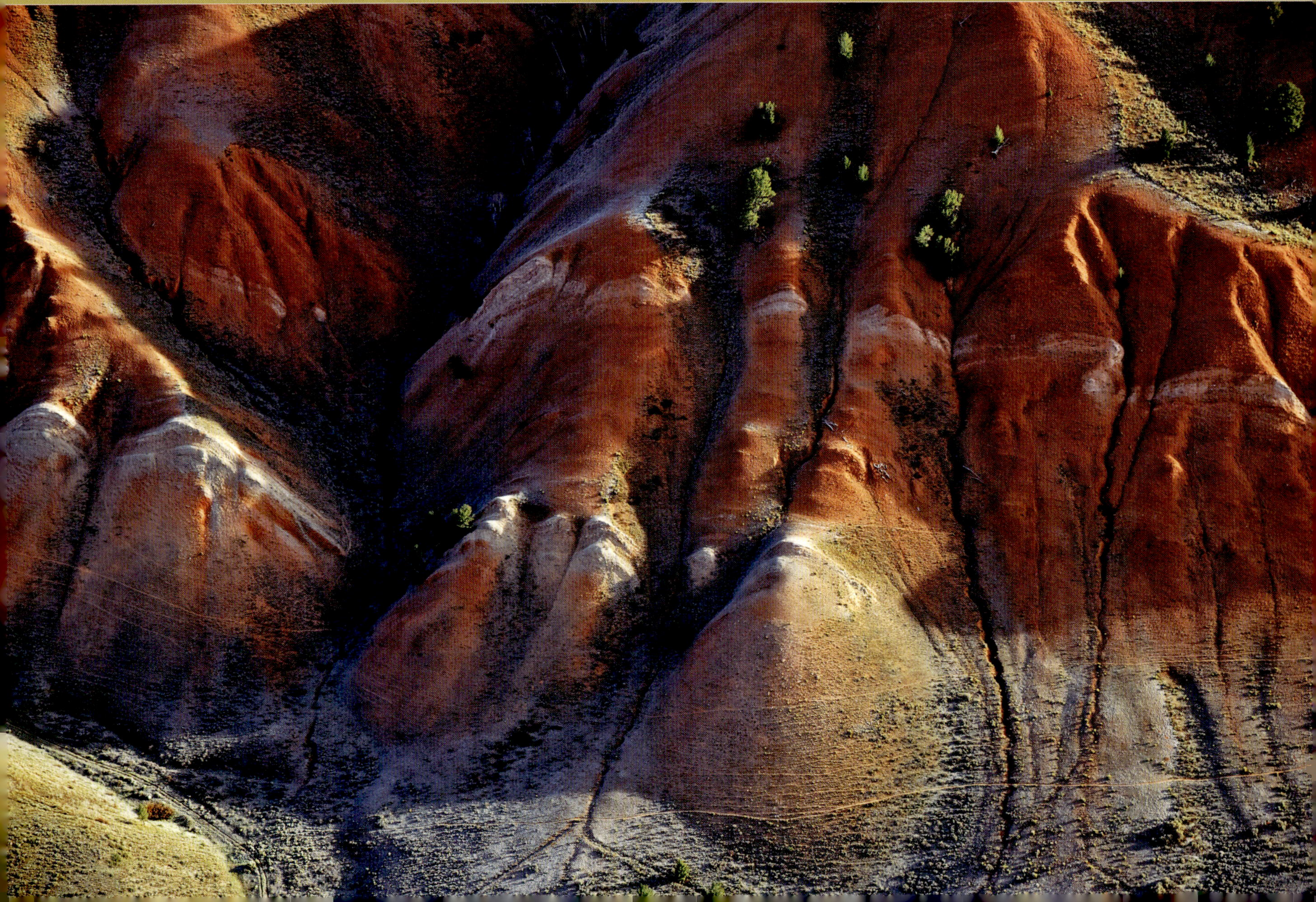

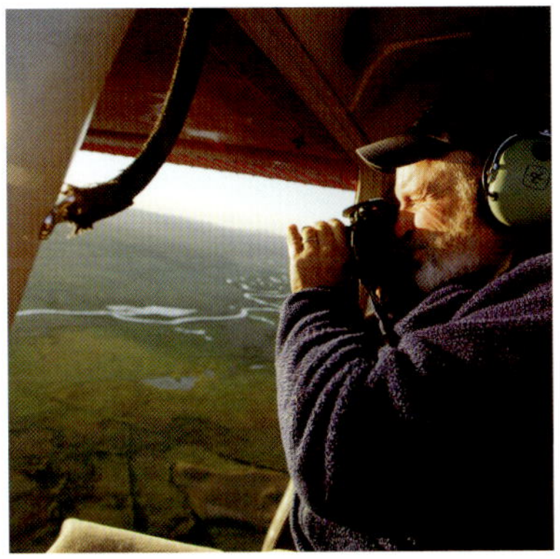

AS VAST AND COMPLEX as the West itself, the sagebrush sea can be difficult to comprehend. Not only do its landscapes vary dramatically but the issues facing this ecosystem are diverse as well. One region in particular, though, is at a crossroads, where human activity and big-energy development squarely collide with critical habitat for the sage-grouse, and thus for much of the wildlife of this region. From the Greater Yellowstone Ecosystem of Wyoming and Montana to the Gunnison Basin and southwest Colorado, this area is the beating heart of the eastern sagebrush ecosystem. I set out to make an intensive study of this region, to view it on a landscape scale, to explore it with my camera, and to seek out people who are contributing their voices and their efforts to finding a new way forward—a way that allows for balance between human needs and the conservation of this ecosystem on the brink. This is my journey.

ABOVE Aerial mission over the Green River Basin (Sublette County, Wyoming)

LEFT Pronghorn migration trails across the Red Hills, a bottleneck in the "path of the pronghorn" migration route (Bridger-Teton National Forest, Wyoming).

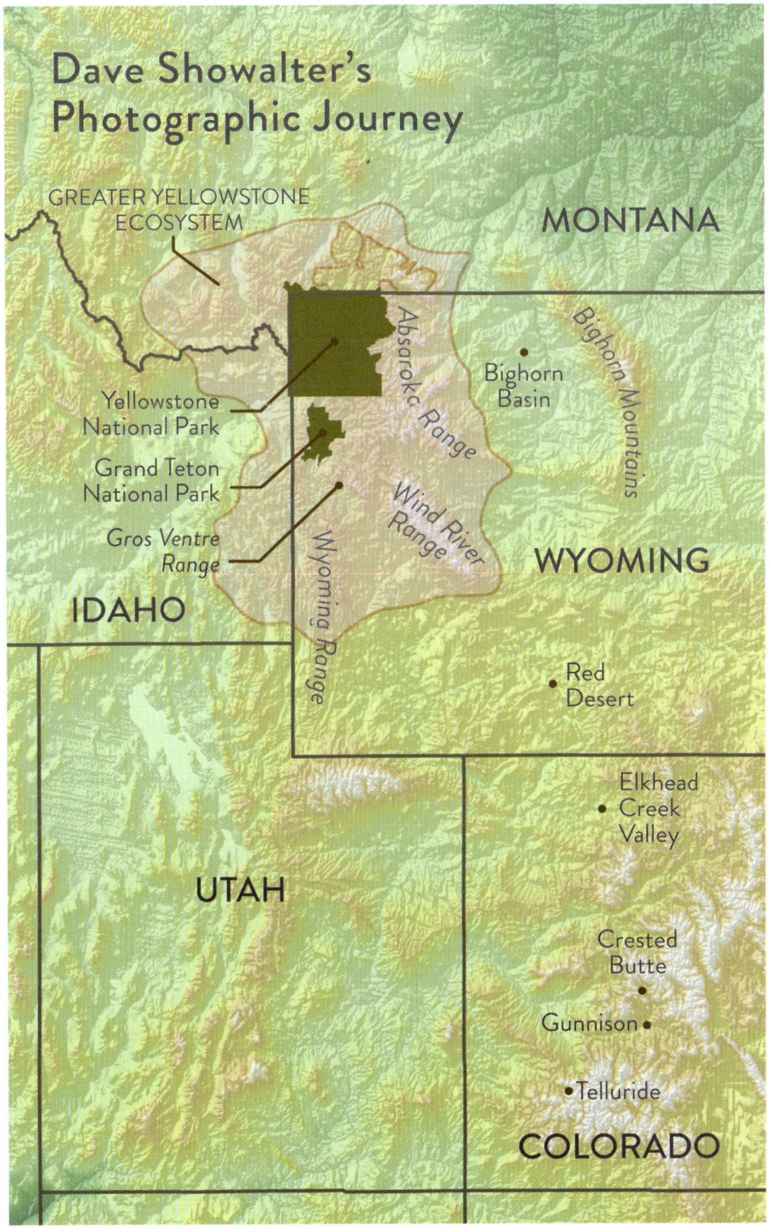

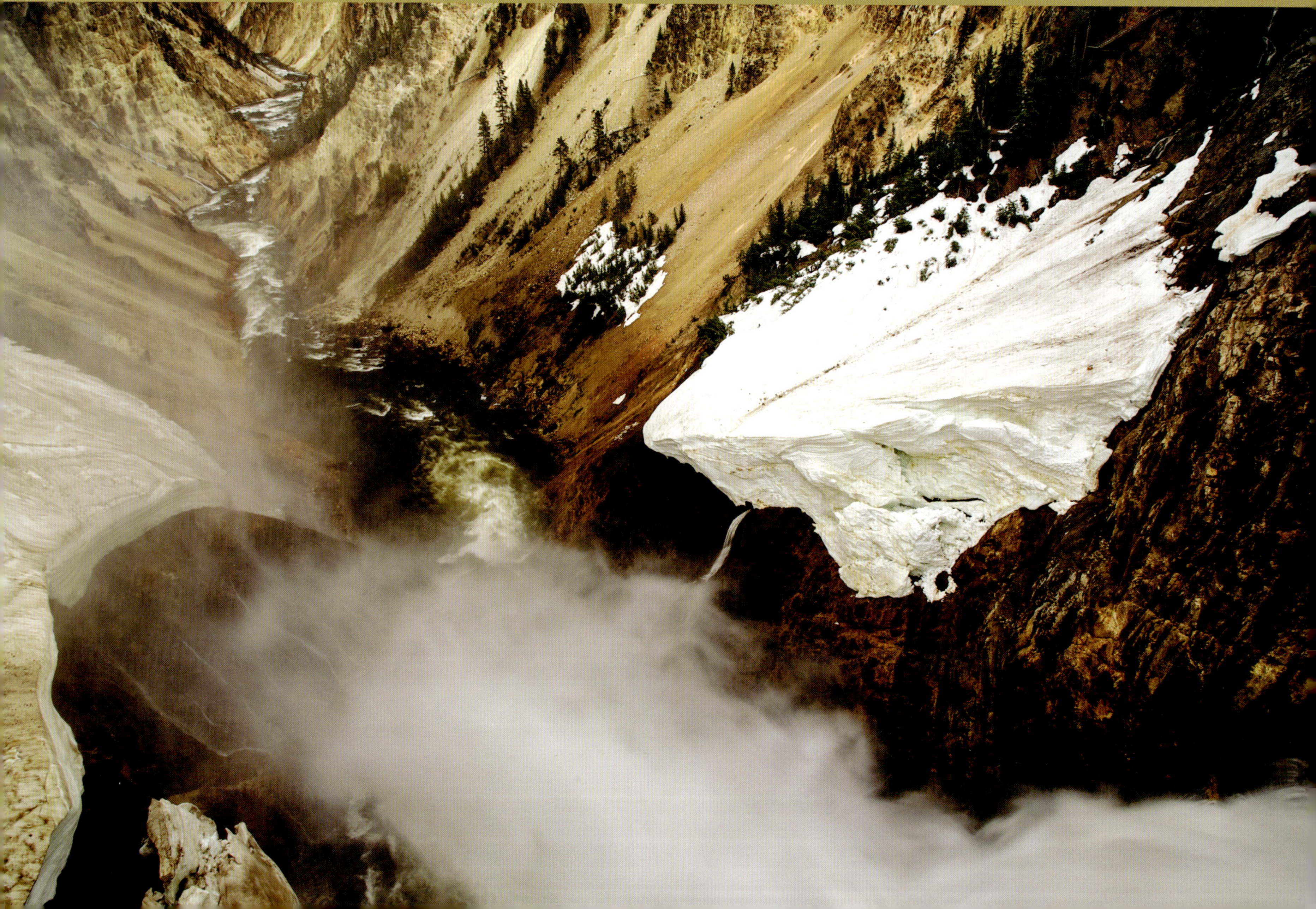

GREATER YELLOWSTONE ECOSYSTEM
MONTANA AND WYOMING

OPPOSITE Lower Falls of the Yellowstone River in Grand Canyon of the Yellowstone, Yellowstone National Park, Wyoming
RIGHT In Yellowstone National Park's northeast corner in Wyoming, this grizzly bear foraged for roots and turned over rocks for insects as light snow fell on a May morning.

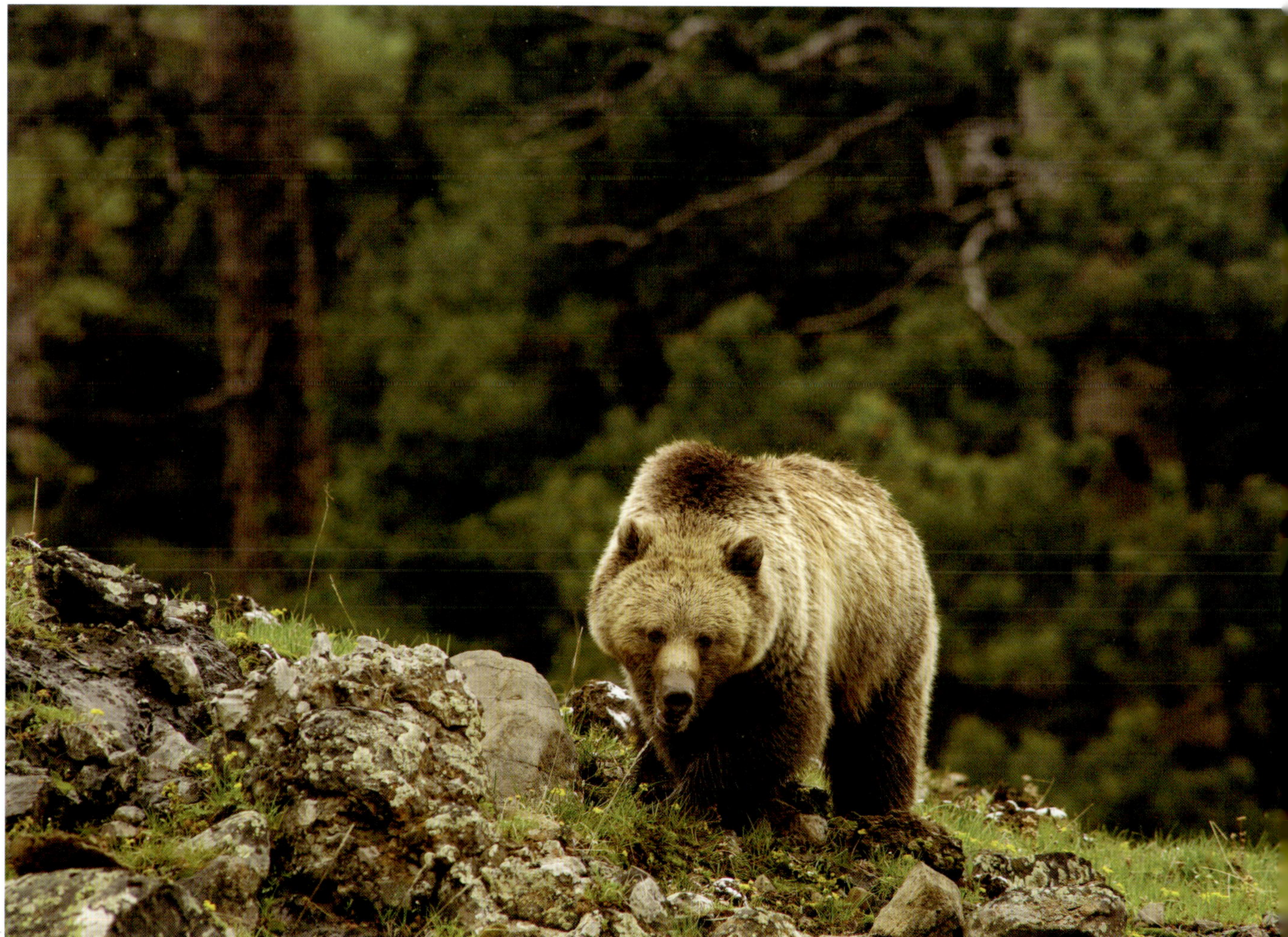

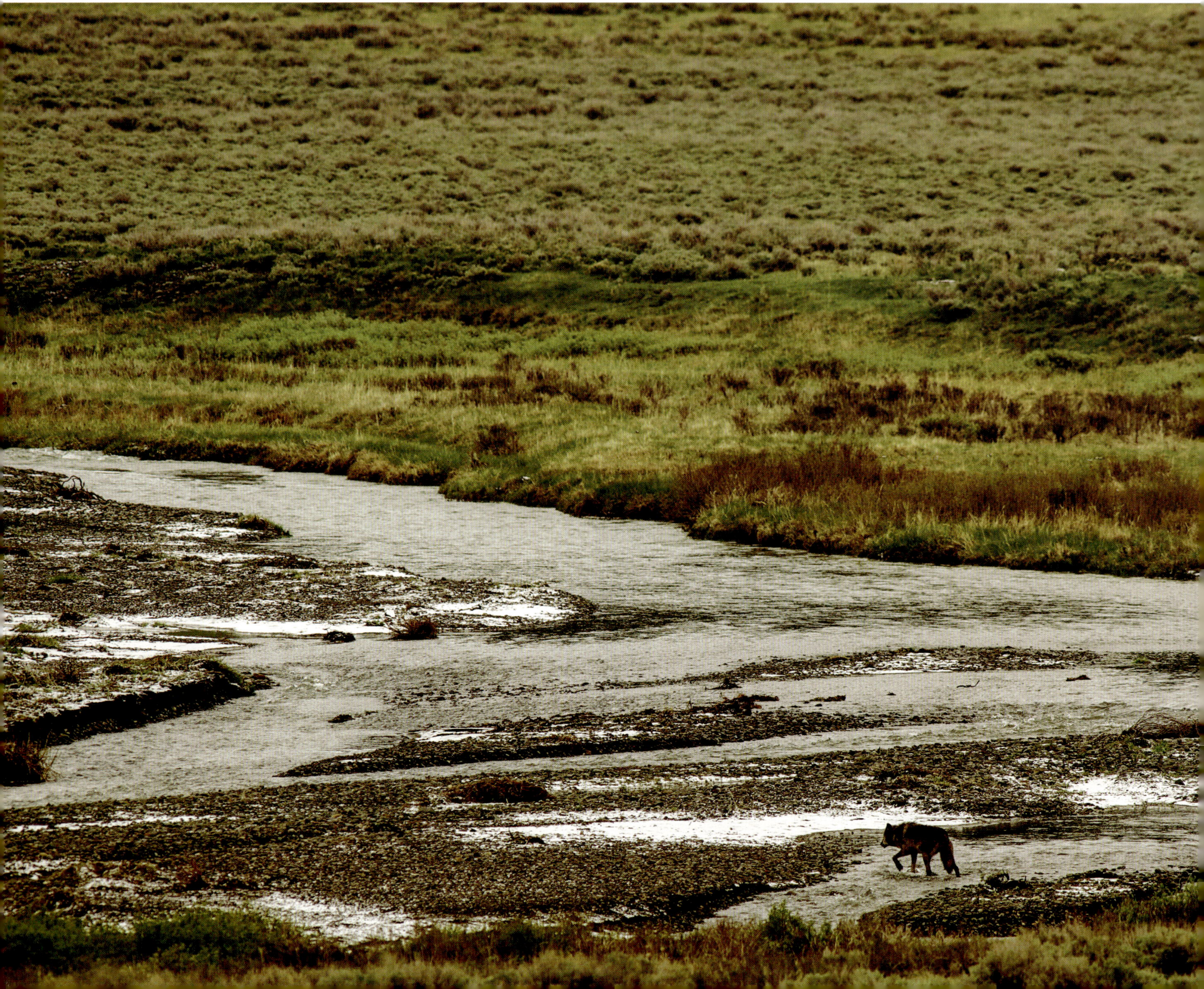

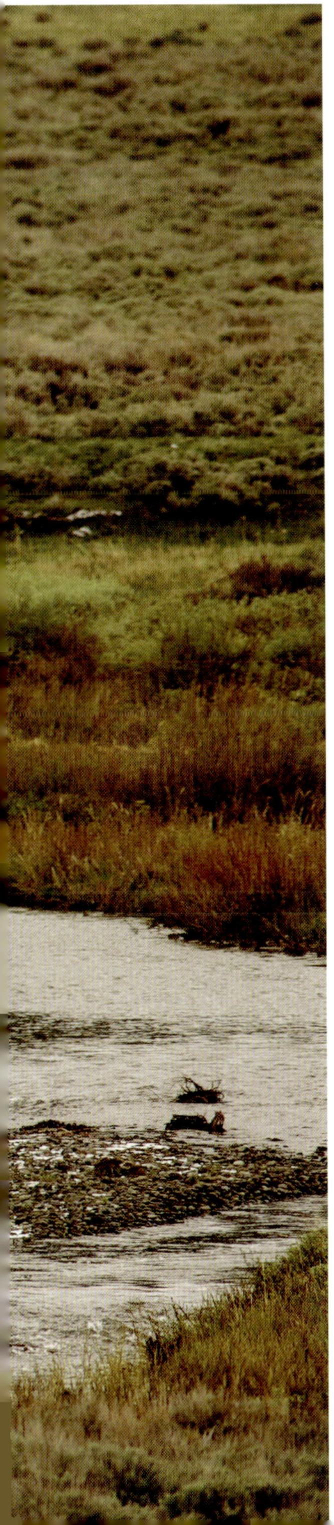
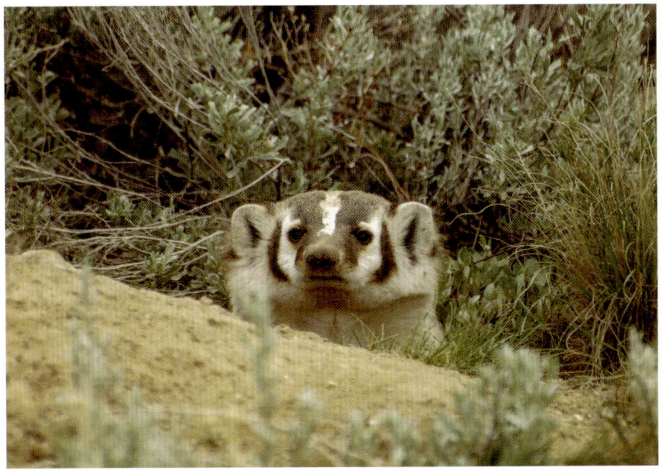
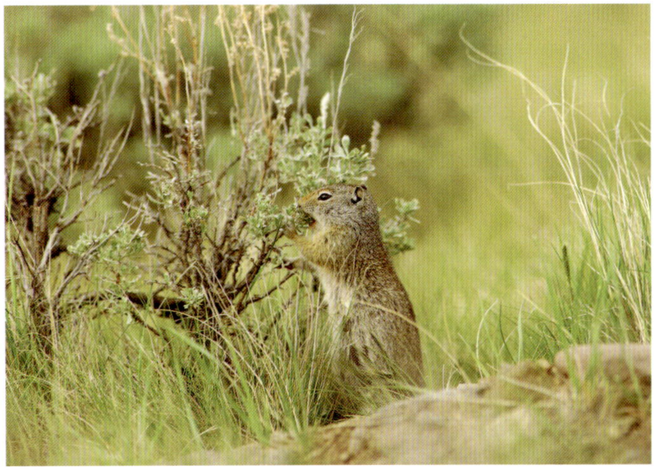
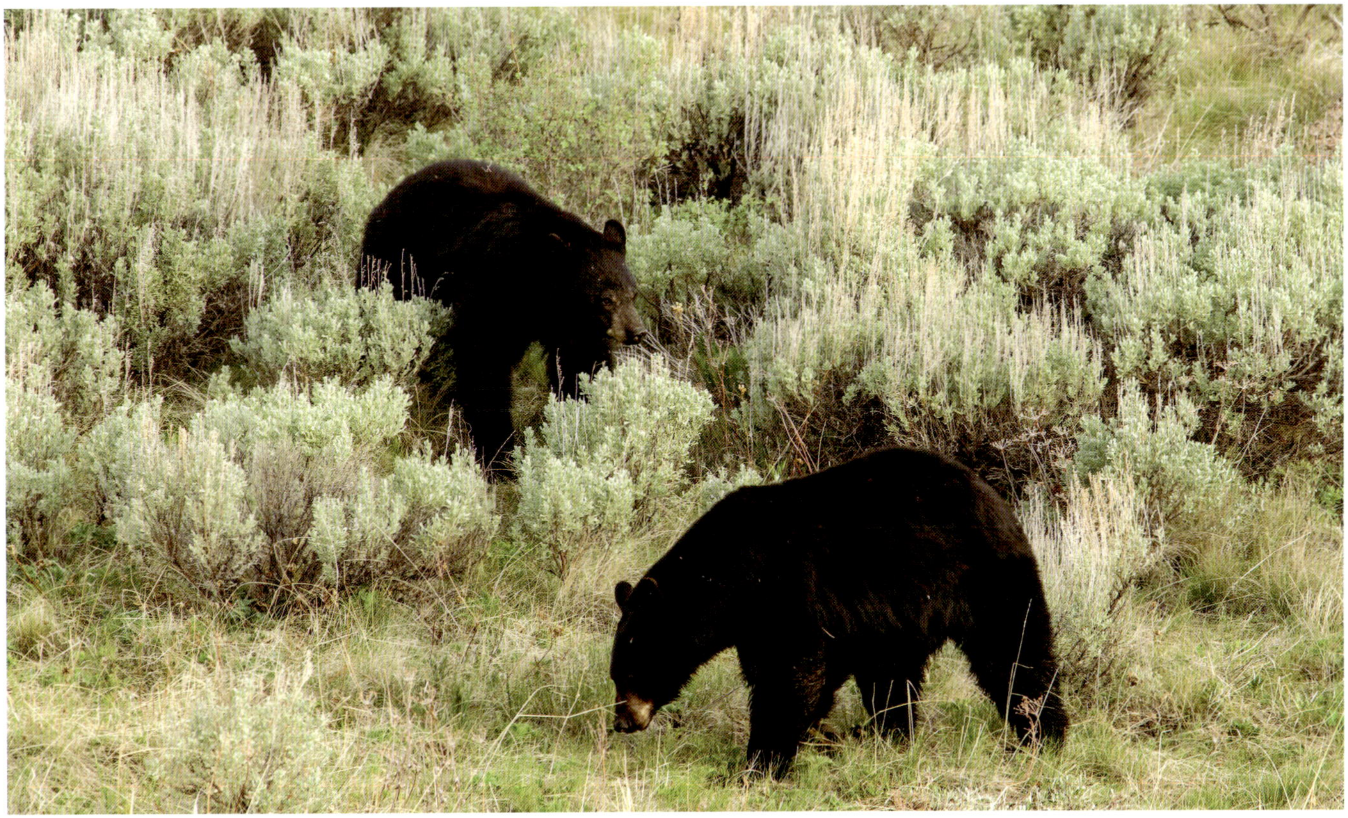

PREVIOUS PAGES, CLOCKWISE FROM LEFT A gray wolf crosses the Lamar River in spring in the Lamar Valley, one of the best places in the world to view wild wolves. | American badgers, such as this one in spring, are common throughout the sagebrush ecosystem, using their superior digging skills to feed on burrowing rodents. | A Uinta ground squirrel feeds on sagebrush in spring in Yellowstone National Park, Wyoming. These rodents hibernate in their burrows in fall and winter, spending part of spring and summer above ground. | Black bears graze in a mixed sagebrush and grass meadow (Yellowstone National Park, Wyoming).

Protecting her cubs from a boar grizzly in the area, this grizzly bear sow with subadult cubs runs from Swan Lake Flats in near darkness (Yellowstone National Park, Wyoming).

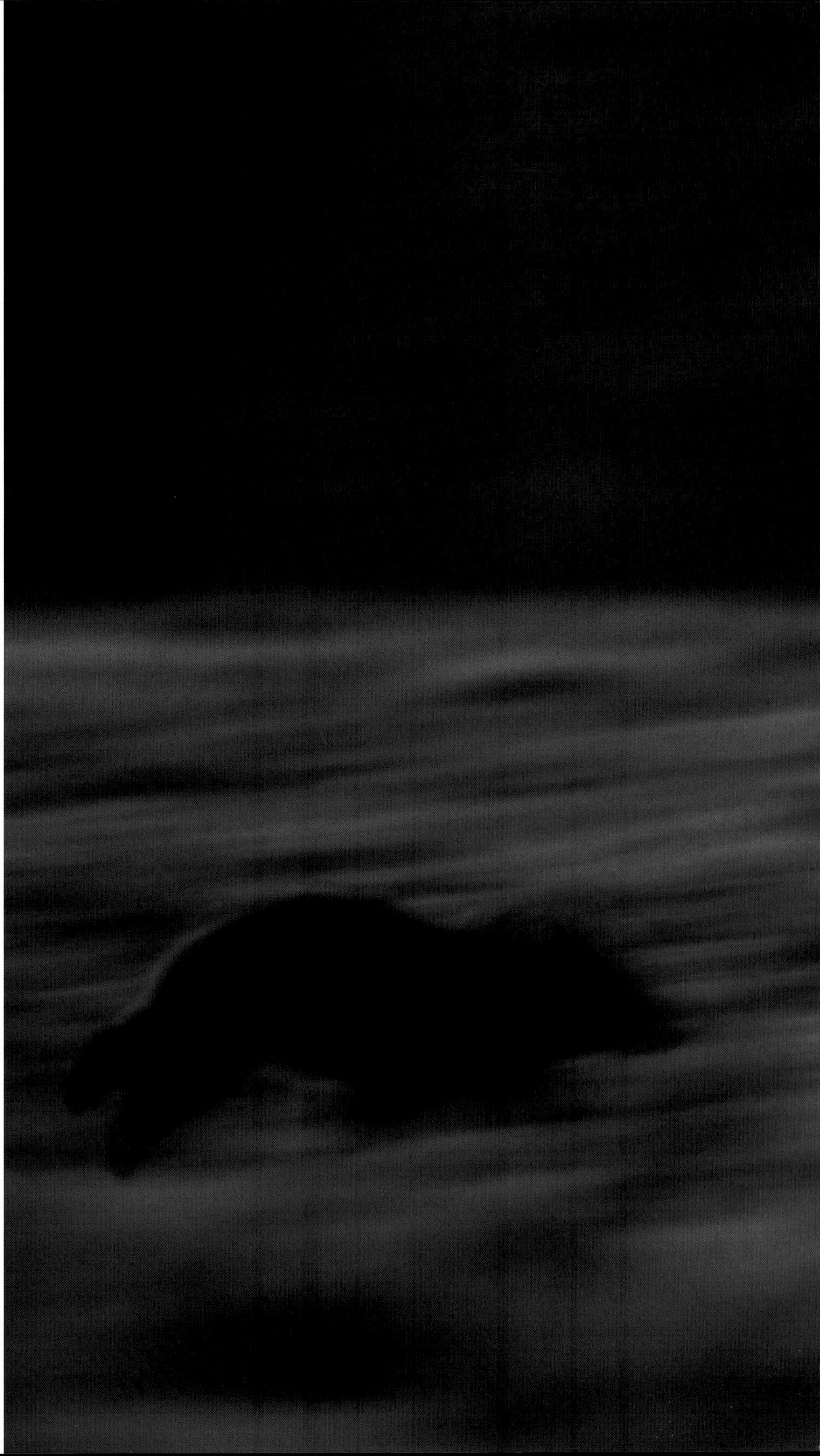

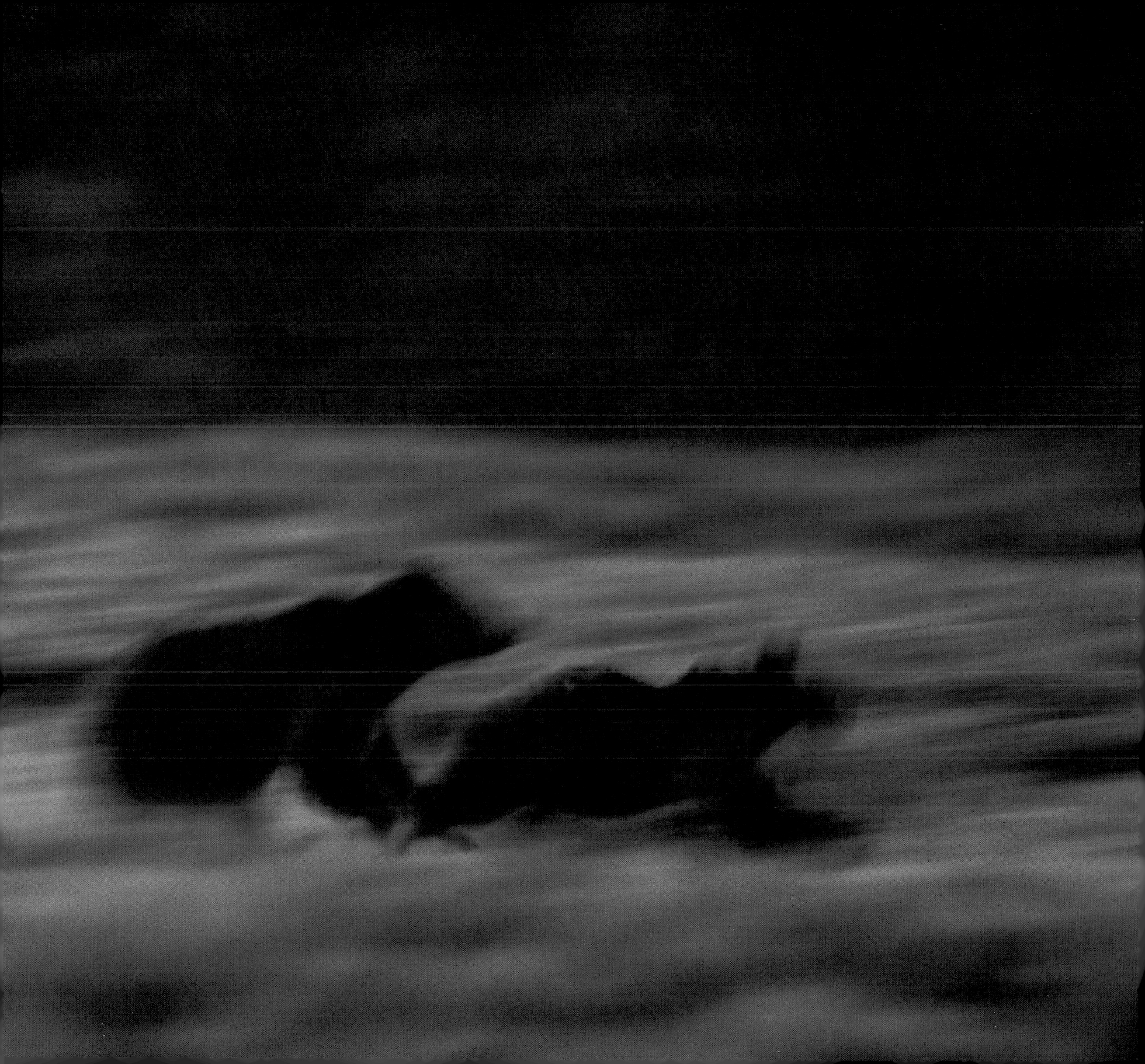

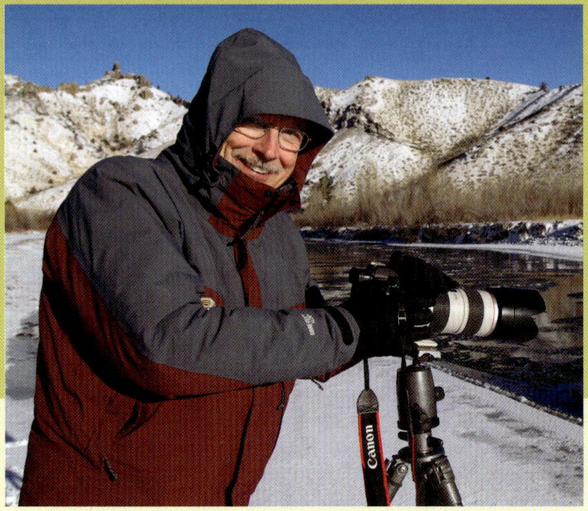

DAVE BURKE
GUARDIAN OF YELLOWSTONE'S WILD SIDE

DAVE BURKE CAN SEE an enormous swath of the land he protects from his living-room window. Heart Mountain, a blocky chunk of rock rising from the Bighorn Basin sagebrush, fills the skyline of a stunningly rugged western scene. Deer browse in the yard that Dave and his wife, Nena, landscaped to attract wildlife on the outskirts of Cody, Wyoming. Burke has studied Heart Mountain throughout a lifetime of exploring and photographing wildlife, sagebrush badlands, rivers, caves, and high mountains in Wyoming's northwest corner known as "Yellowstone's wild side."

Burke was an elected Park County commissioner from 2009 to 2012 who will tell you "it's all about the biodiversity" in the Bighorn Basin, where the rock faces of the Absaroka-Beartooth Front give way to a vibrant sagebrush-steppe. The Bighorn Basin is a vast sea of sagebrush, 100 miles (160 kilometers) across, bookended by the mighty Absaroka Mountains to the west and the distant Bighorn Mountains to the east. The variety of western wildlife east of Yellowstone National Park where Burke lives is the same as inside Yellowstone National Park, with people living, working, recreating, hunting, and fishing where grizzlies roam and big land mammals migrate. Growing up in nearby Powell, Wyoming, "always going on family hunting and fishing trips," and a lifetime outdoors had a lot to do with Burke becoming a steward for this part of the West, a steady hand in a maelstrom of development.

Lean and rangy in jeans, Burke has made the burly Heart Mountain hike dozens of times. He guided me one summer morning, where we took in the basin's breadth from a high perch on Heart's summit. During a volcanic event fifty million years ago, the three-hundred-million-year-old limestone rock beneath our feet broke off from the Beartooth Mountains in a mammoth rock slide and traveled at least 25 miles (40 kilometers) to its present location. It seemed that all of northwest Wyoming was in view from 8123-foot (2475-meter) Heart Mountain—from pointy Pilot Peak in the Beartooth Range, south to the hazy Owl Creek Mountains,

"The Absaroka–Beartooth Front is one of the last large, truly wild places that we have."

and east to the Bighorn Range. Two thousand feet (610 meters) directly below, pine forest gives way to crumpled gold and green sagebrush lands, then farther south toward Cody the irrigation pivot-sprinkler circles begin.

Facing intense pressure to drill and develop the Bighorn Basin stretched out before us, Burke has supported enhanced oil recovery in existing fields and a long-term sustainable plan to protect the Absaroka-Beartooth Front's remarkable natural resources. In four years as a county commissioner, Burke advocated for the Francs Peak Wilderness Area high in the Absarokas. He urged the Shoshone National Forest—a complex landscape of sagebrush, forest, and alpine ecosystems—to restrict all-terrain vehicle use while remaining off-limits to drilling. He fought for his beloved McCullough Peaks Wilderness Study Area—rugged, candy-striped sagebrush badlands with a resident wild horse herd that Burke loves to photograph. Burke's motivation runs deep: "The Absaroka-Beartooth Front is one of the last large, truly wild places that we have. It's still a wild forest that hasn't changed from the incursion of man. We have a special opportunity to keep it that way for future generations."

The recommendations of Park County commissioners helped shape the twenty-year plans for regional Bureau of Land Management lands and the Shoshone National Forest, big chunks of the West's natural heritage in a complex regional picture. Of his role, Burke said, "Commissioners have a unique opportunity to sit at the table with federal and state agencies to help mold future use of public lands. They represent and speak for the people when providing recommendations for lands and wildlife. It's so important that commissioners listen to residents and relay their desires. "

Back on Heart Mountain, between shutter clicks while standing on a rock that hasn't moved for 50 million years, Burke returned to the subject of our natural heritage: "It's not about how I use this mountain or you use this mountain. It's about how our grandchildren and great-grandchildren will use this mountain." §

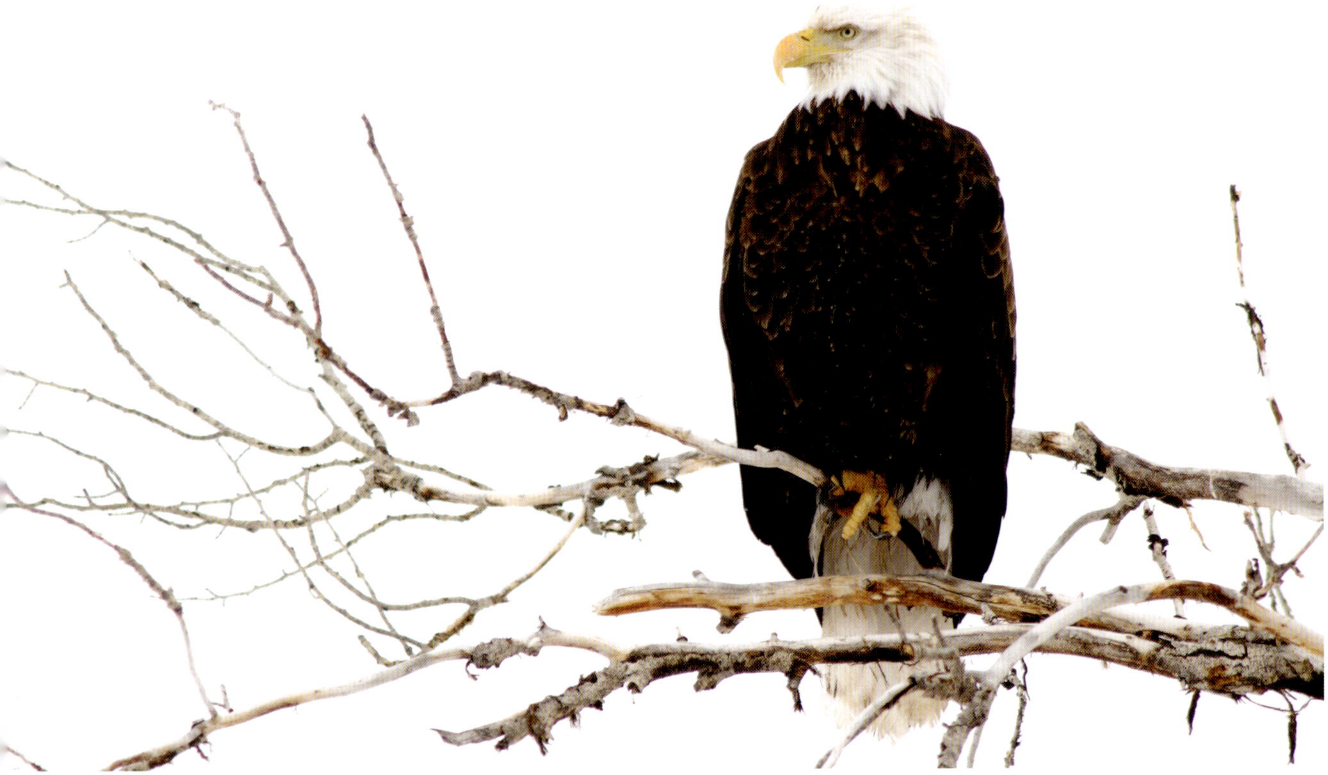

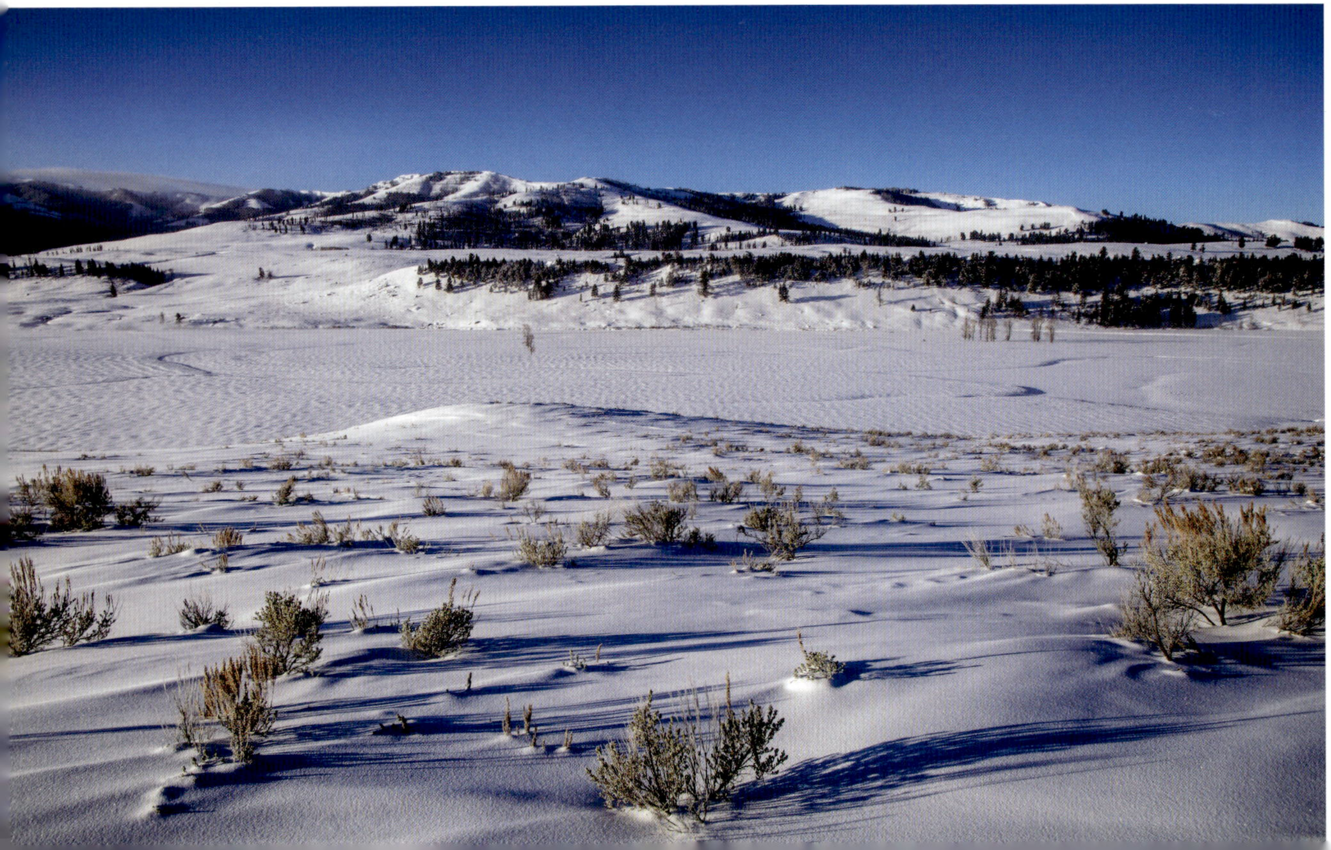

TOP LEFT Adult bald eagle perched near the Gros Ventre River in winter (Grand Teton National Park, Wyoming)

BOTTOM LEFT Sun and clear skies give the illusion of warmth on a -50 degrees Fahrenheit (-46 degrees Celsius) morning in Lamar Valley (Yellowstone National Park, Wyoming).

OPPOSITE A great gray owl dives at dusk to catch a small mammal near Jackson Hole, Wyoming. America's largest owl species, great grays can reach a height of 33 inches (84 centimeters) with a wingspan up to 5 feet (1.5 meters). Considered a boreal forest and western mountains species, great gray owls also hunt the forest edges of the sagebrush ecosystem.

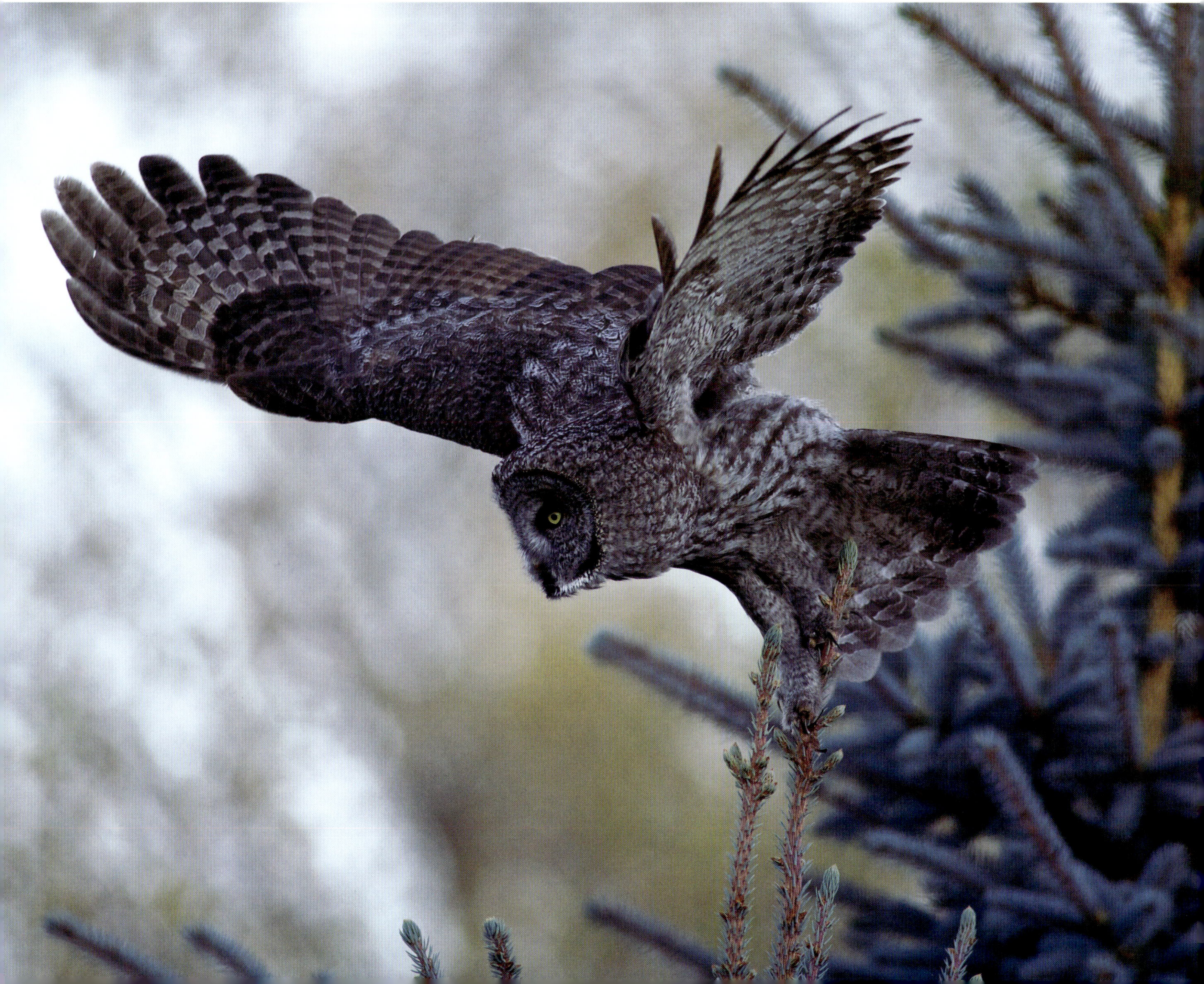

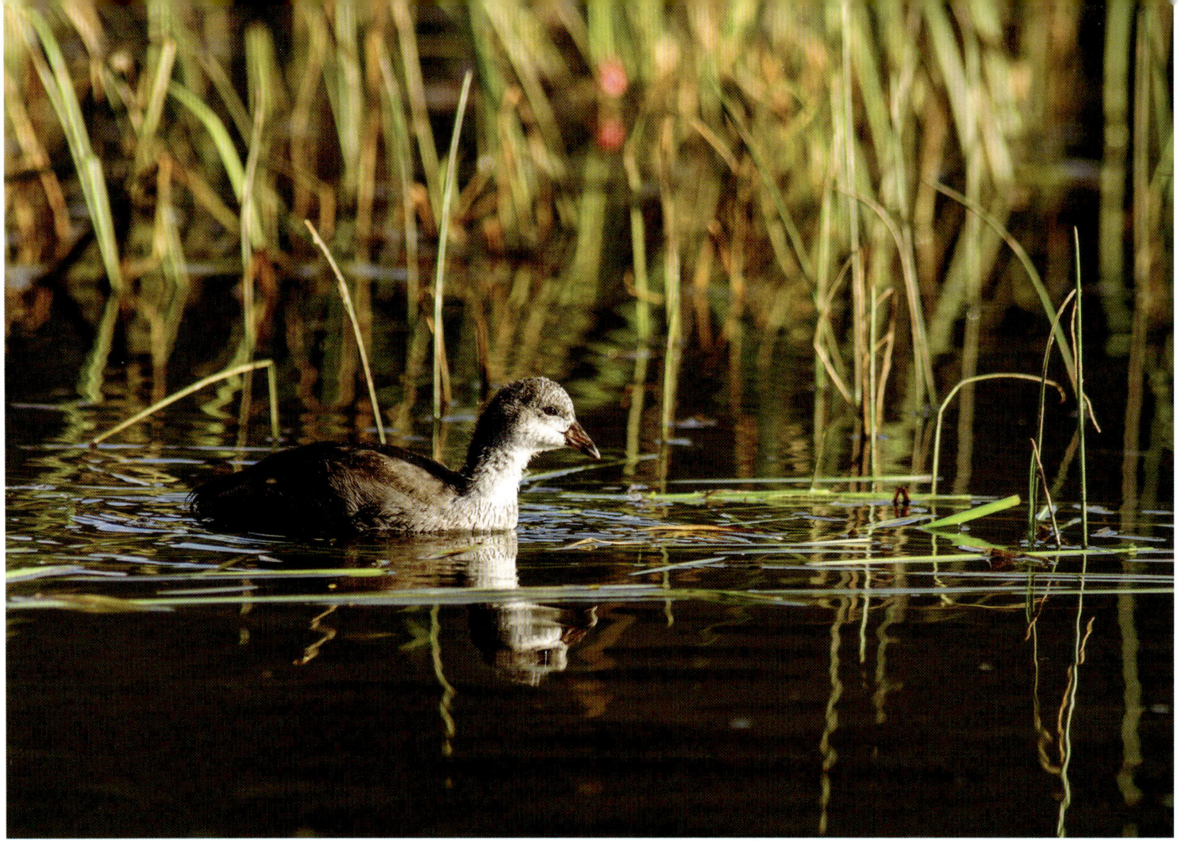

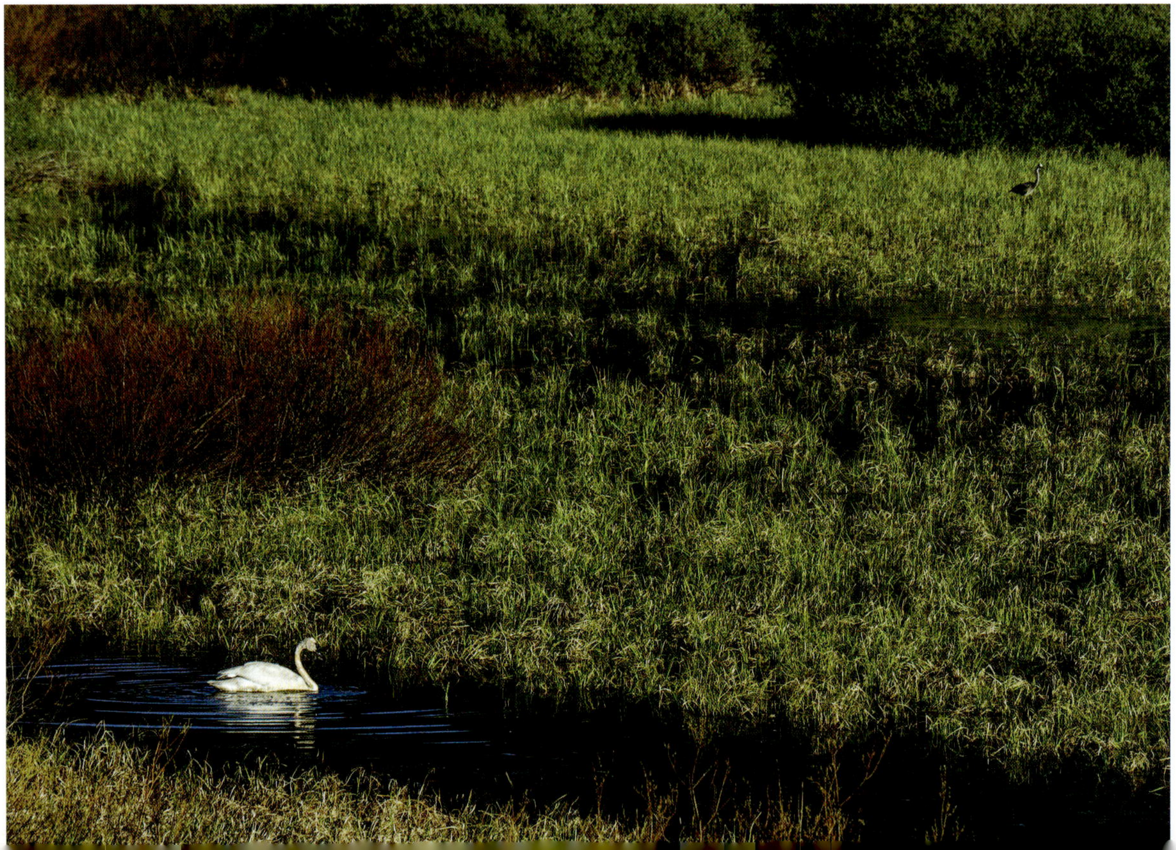

TOP LEFT A juvenile American coot is reflected in a wetland in the Gros Ventre River Valley in Bridger-Teton National Forest, Wyoming. Small ponds like this provide important riparian habitat for a variety of wildlife.

BOTTOM LEFT A trumpeter swan and a great blue heron in spring in the Snake River backwaters of Grand Teton National Park, Wyoming. Largest of North American waterfowl, the trumpeter swan has rebounded from near extinction in the early 1900s to become fairly common today.

A yellow warbler in willow along the Snake River in Grand Teton National Park, Wyoming

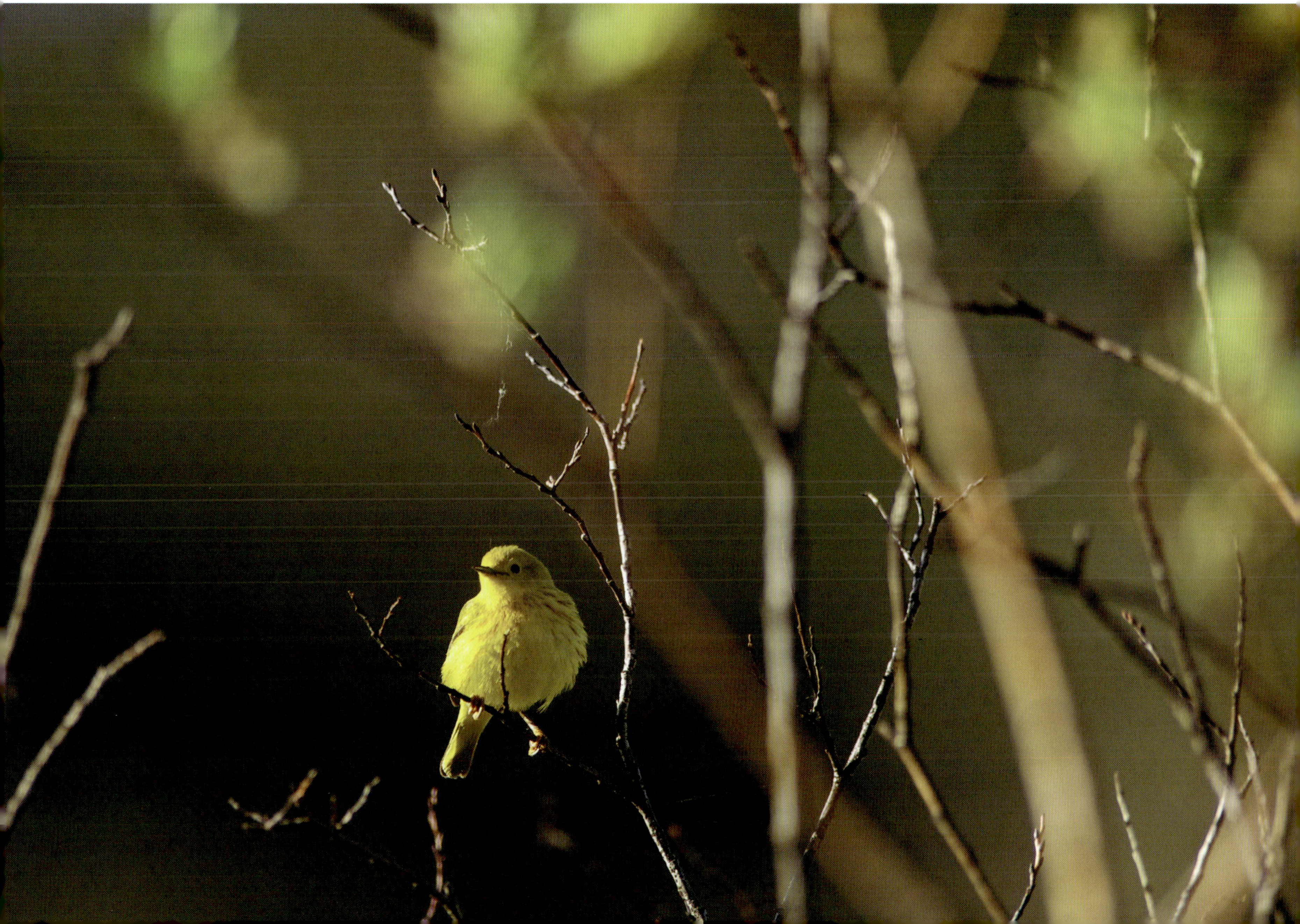

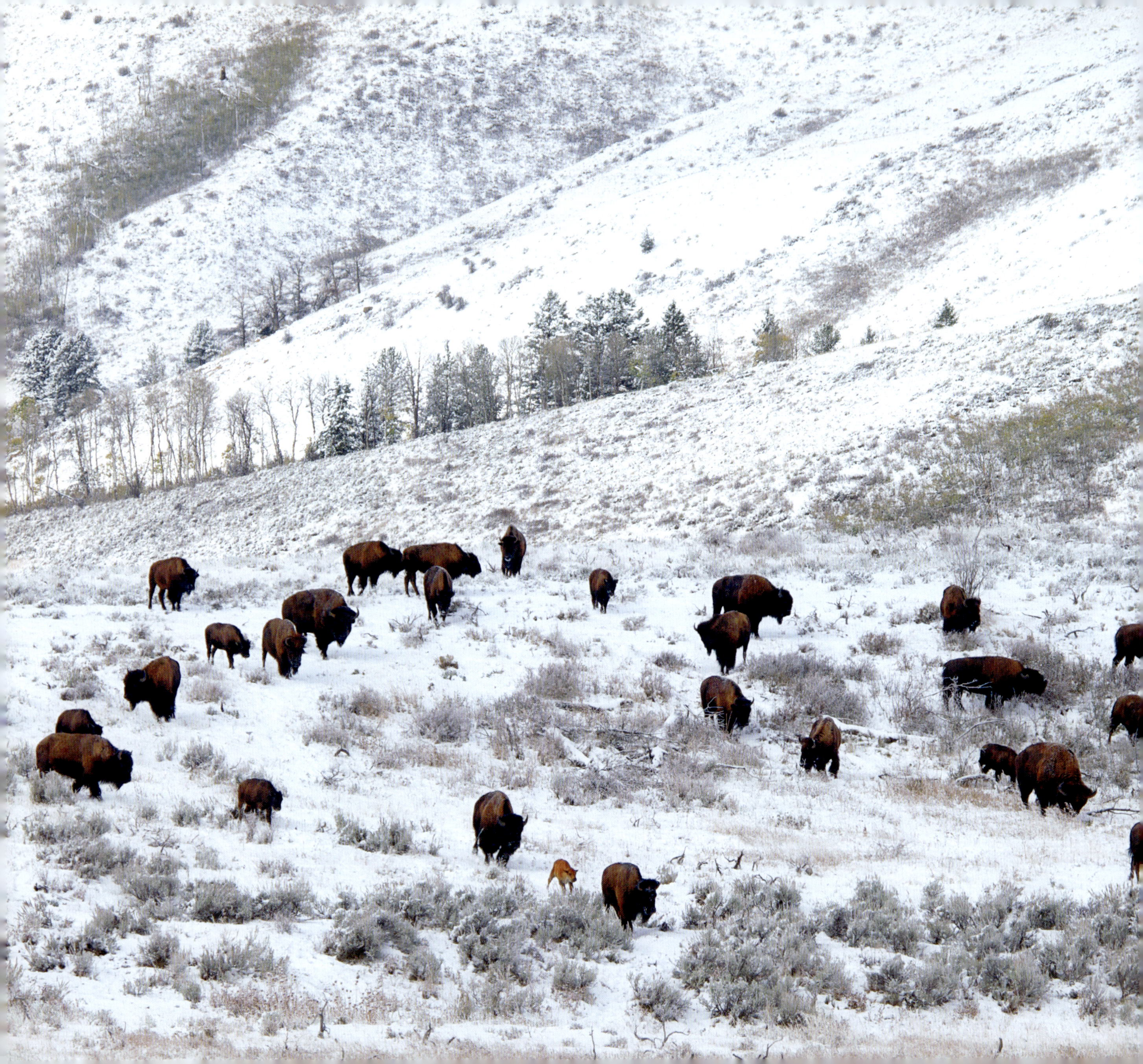

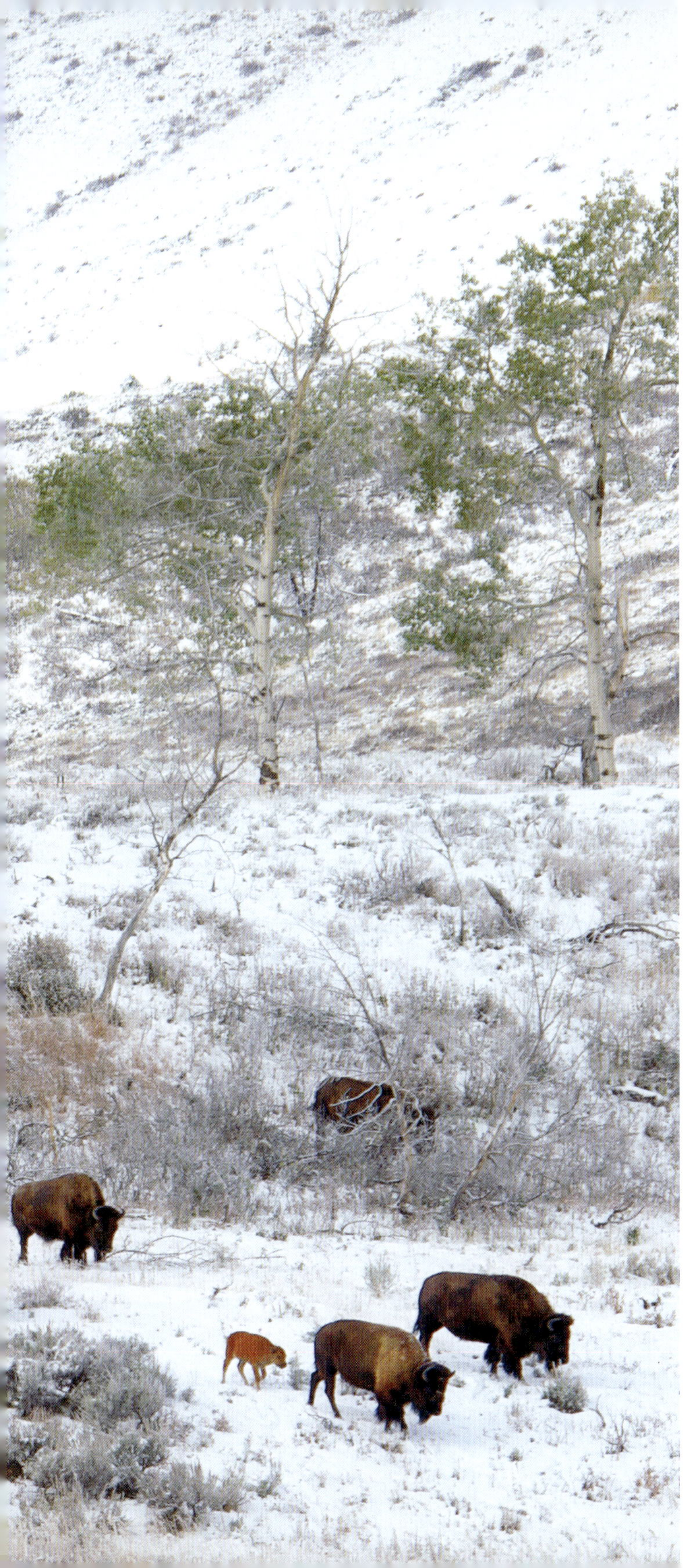

LEFT Bison with young calves move across the flanks of Blacktail Plateau in Grand Teton National Park, Wyoming.

BELOW A bull moose paws through snow while feeding on bitterbrush in Grand Teton National Park, Wyoming. Moose spend part of winter in sage to feed on antelope bitterbrush.

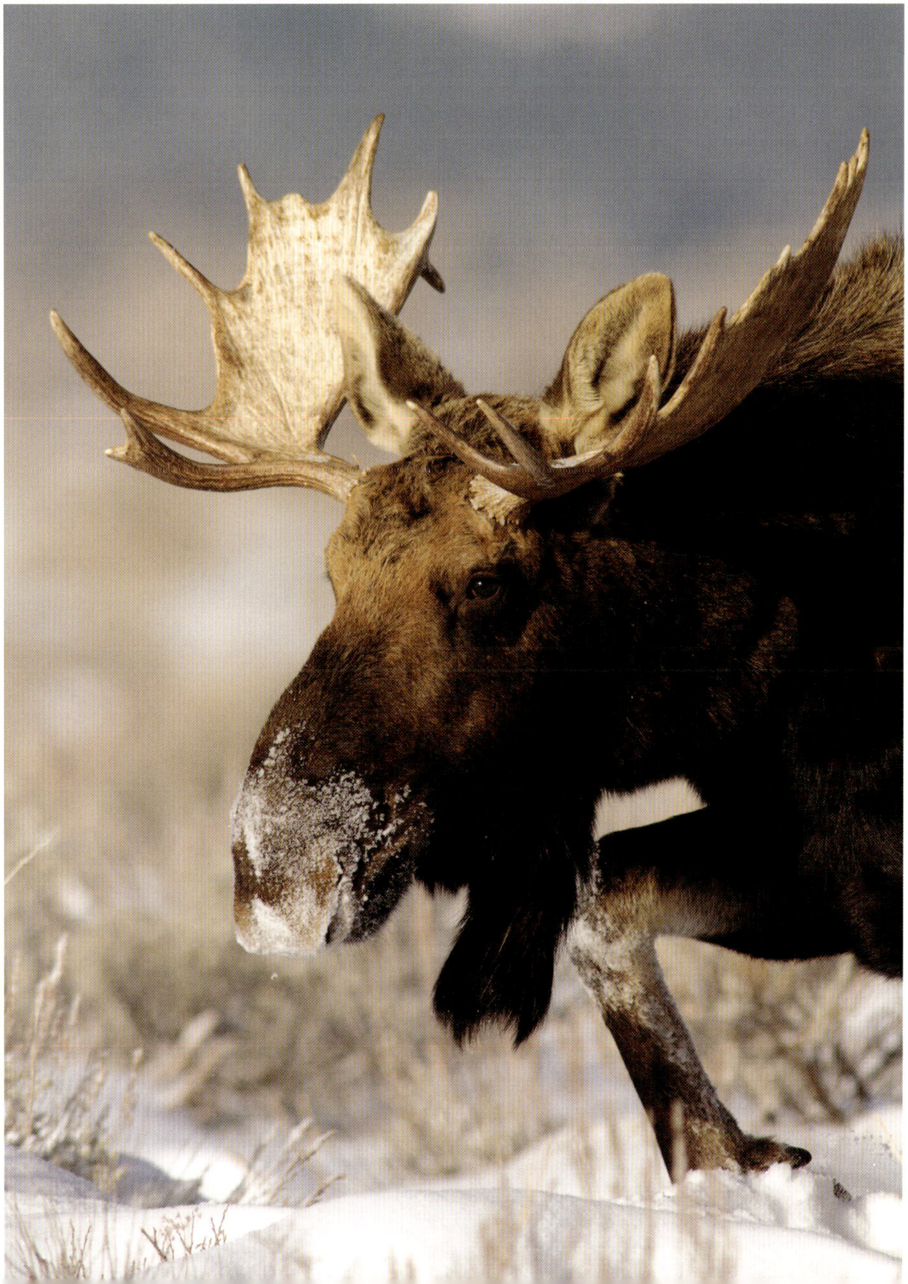

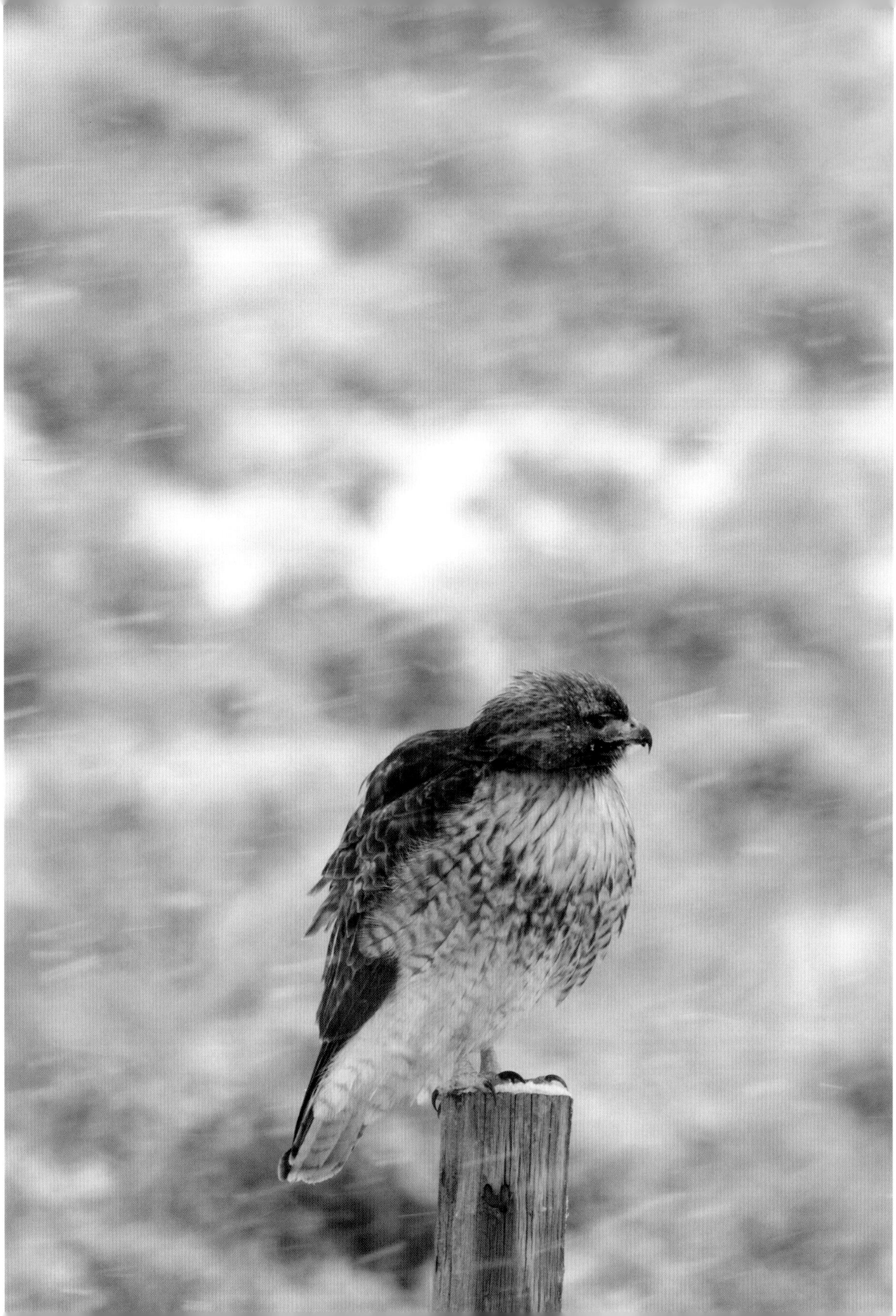

This red-tailed hawk faced a stiff wind during a spring snowstorm near Pinedale, Wyoming.

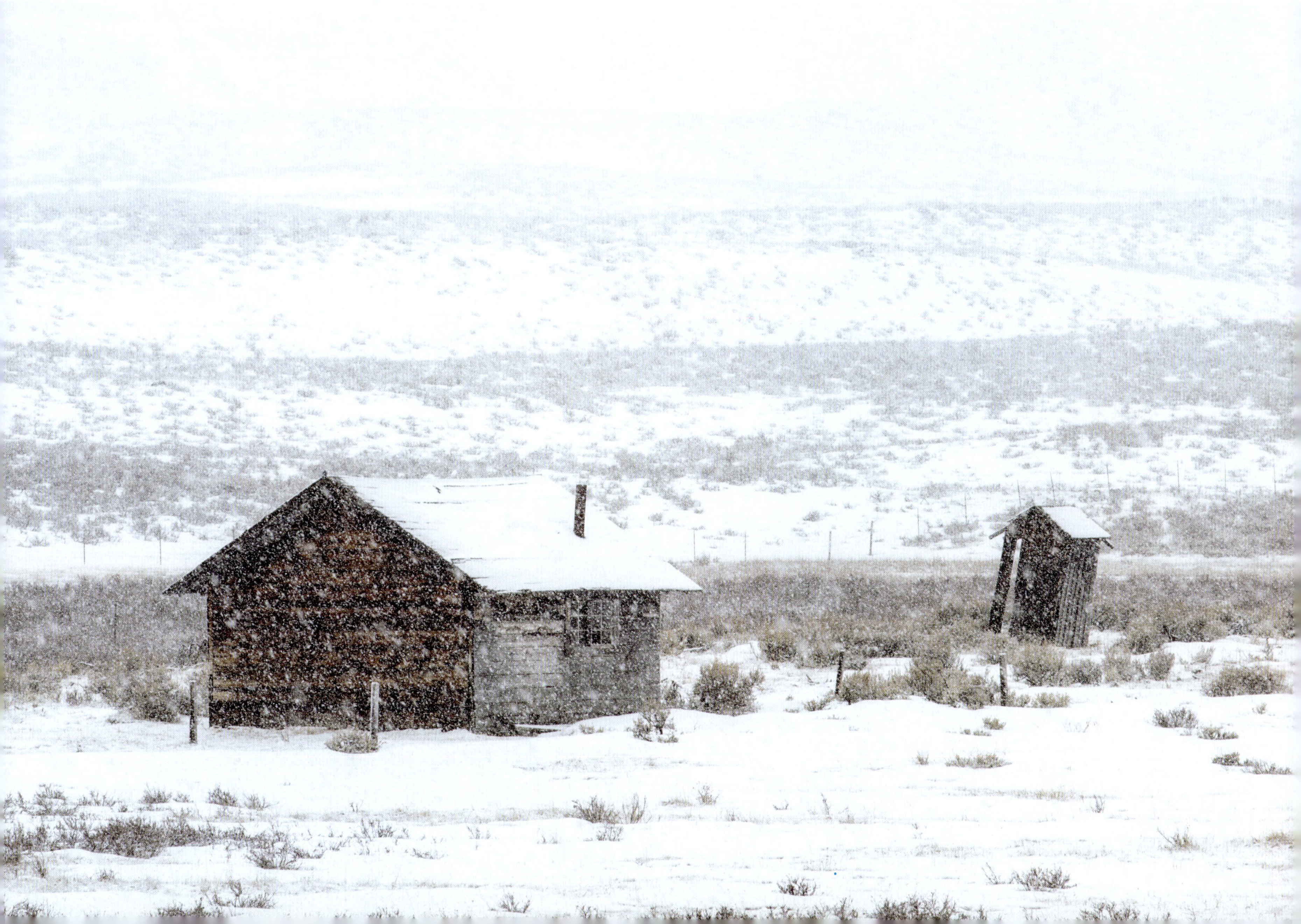

Snowy cabin in spring in the Green River Basin
(Sublette County, Wyoming)

A Swainson's hawk perches on a glacial erratic in Soda Lake Wildlife Habitat Management Area, Sublette County, Wyoming, with the Wyoming Range in the background.

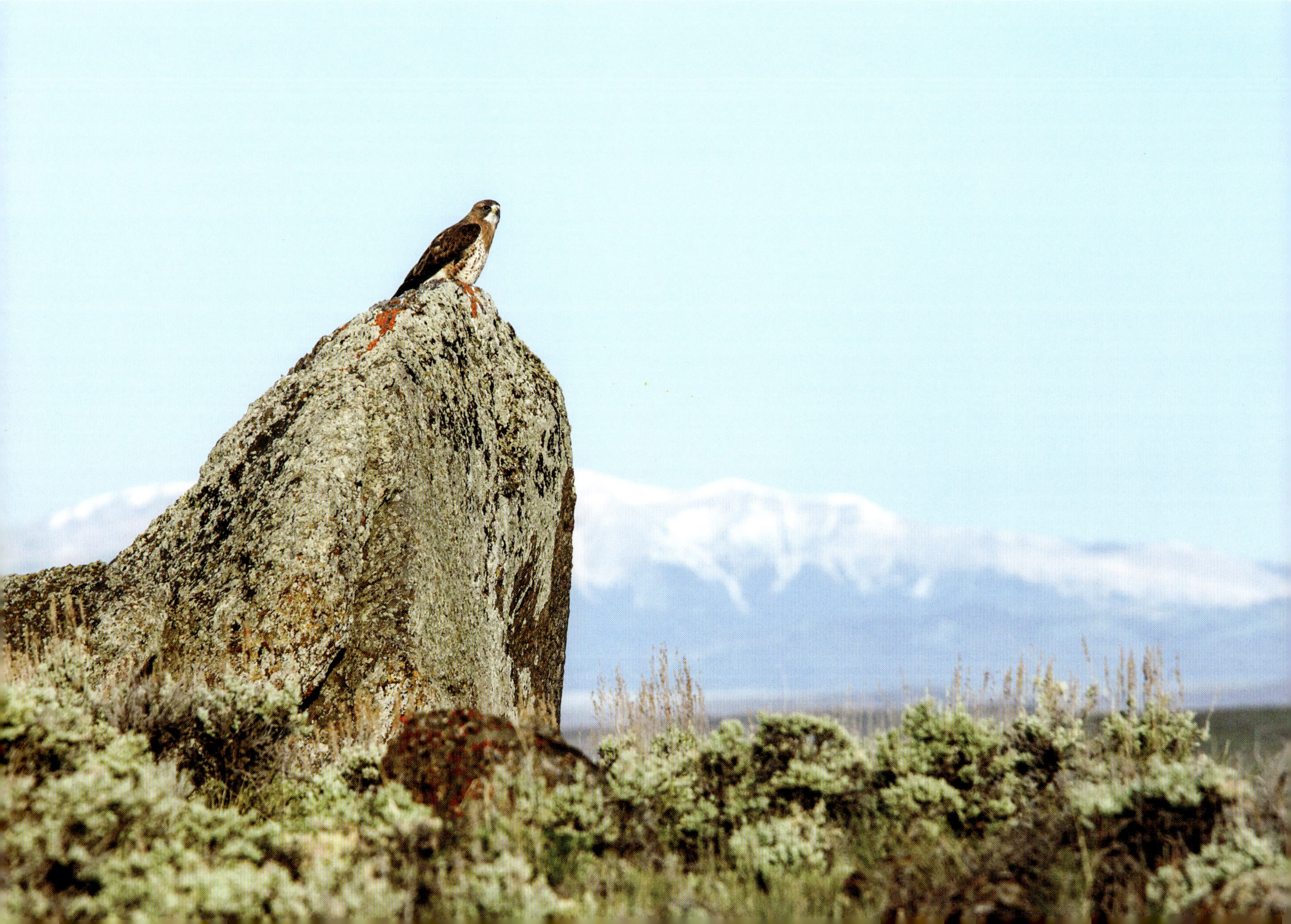

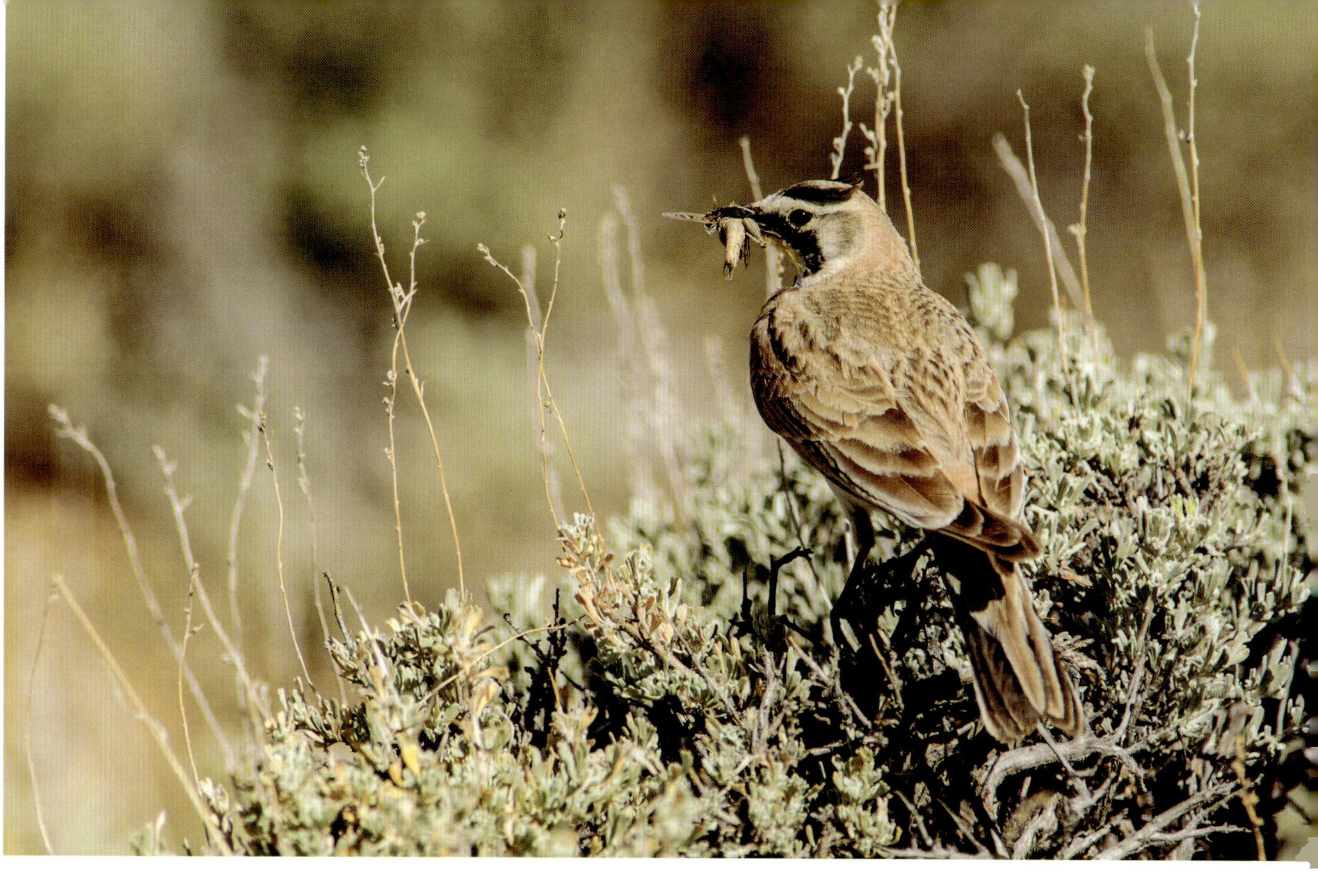

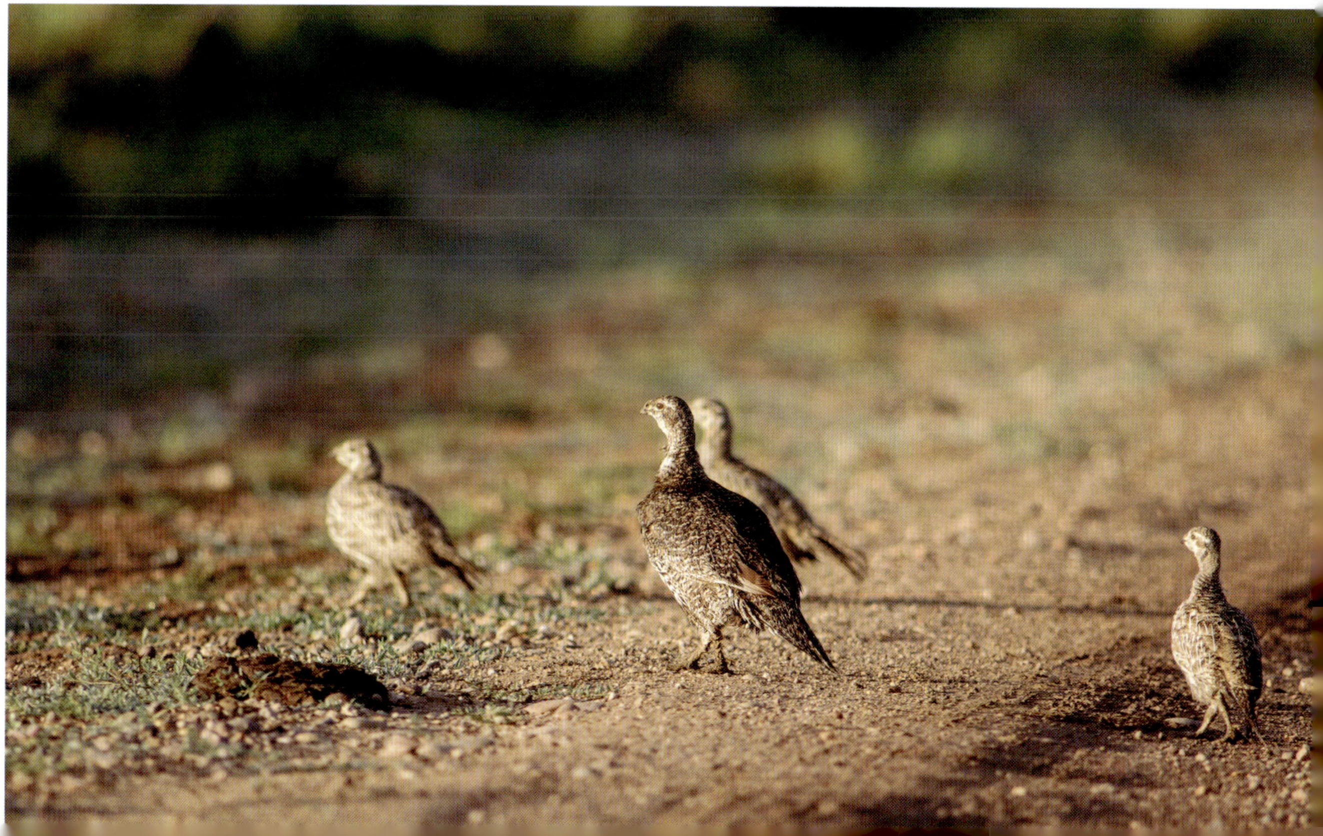

TOP RIGHT Horned lark with grasshopper (Sublette County, Wyoming)
BOTTOM RIGHT Greater sage-grouse persist near natural-gas development on the Pinedale Anticline near Pinedale, Wyoming. Here, a female crosses a dirt road with her brood of growing chicks.

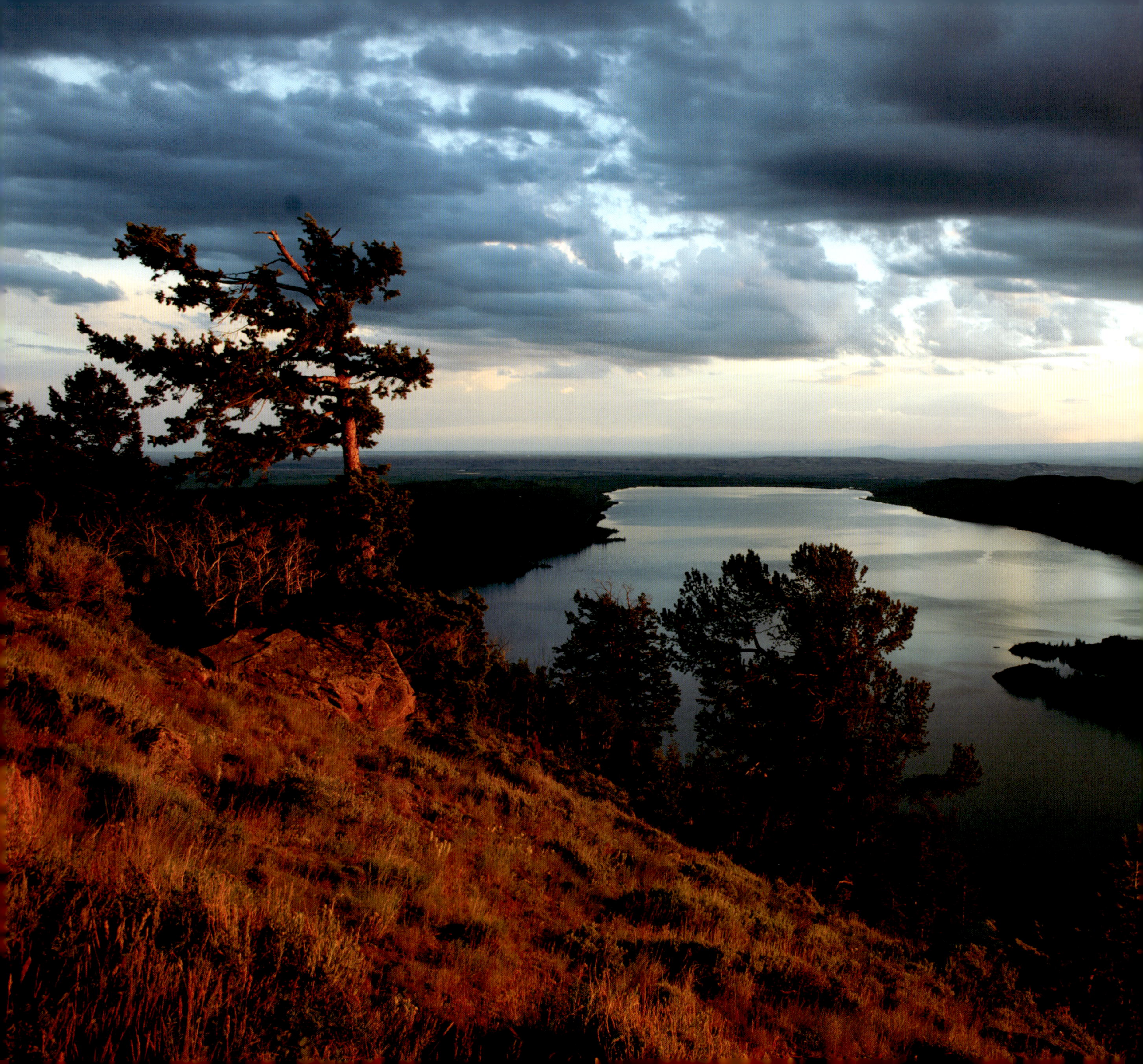

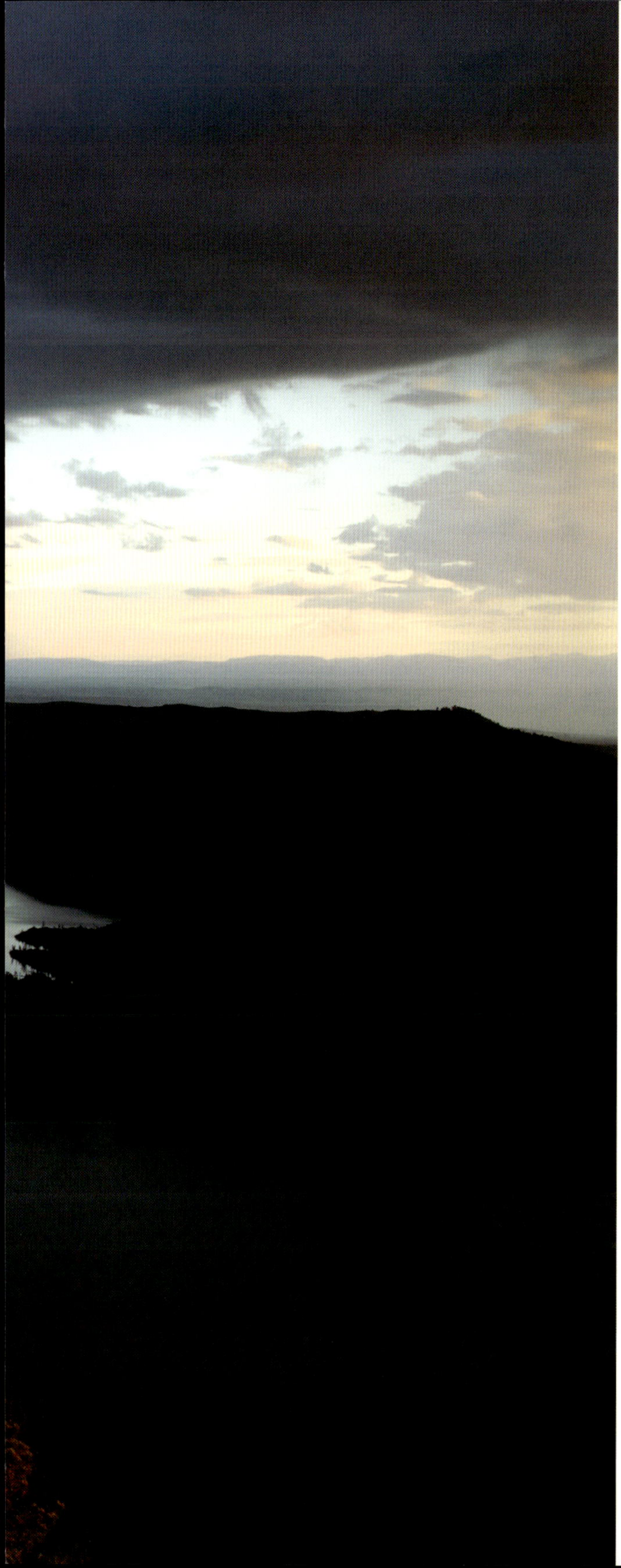

LEFT Summer storms pass over Fremont Lake near Pinedale, Wyoming. The lake, as well as Fremont Peak in the Wind River Range and other notable landmarks, are named for explorer and military officer John C. Frémont, who surveyed the area in 1842 while mapping the Oregon Trail (Bridger-Teton National Forest, Wyoming).
BELOW Golden eagles are the top avian predators in the sagebrush ecosystem, typically using cliff faces for nesting. They are fast, agile hunters throughout the western states, preferring jackrabbits and small mammals (Park County, Wyoming).

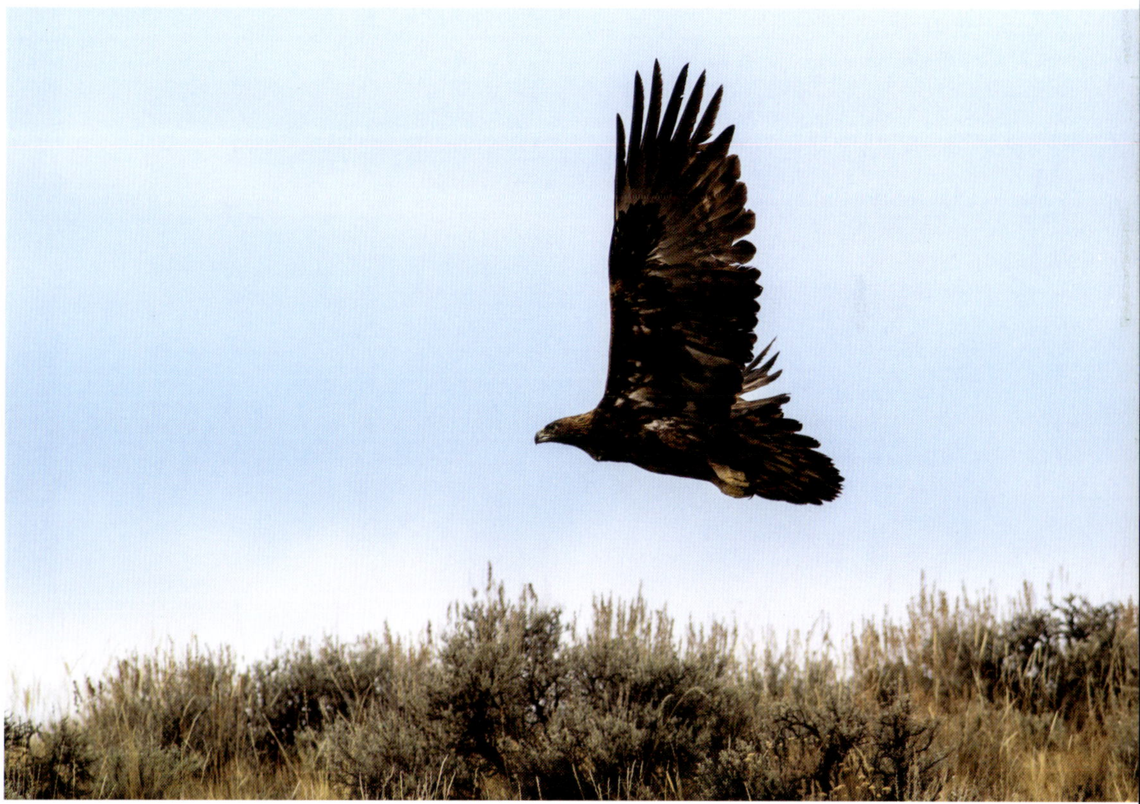

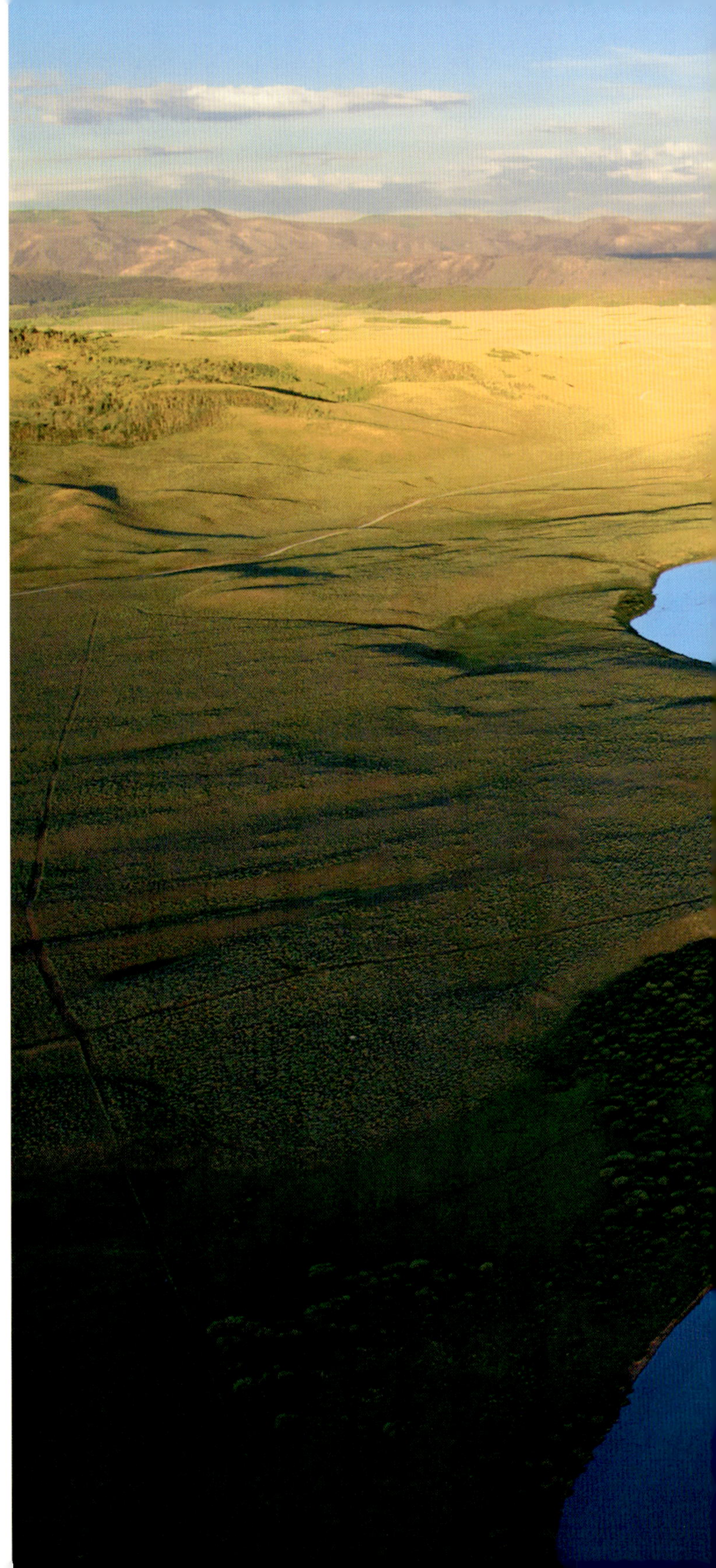

The Upper Green River curls back on itself before making a sweeping turn in Sublette County, Wyoming. Evening light paints the surrounding hillsides in warm hues.

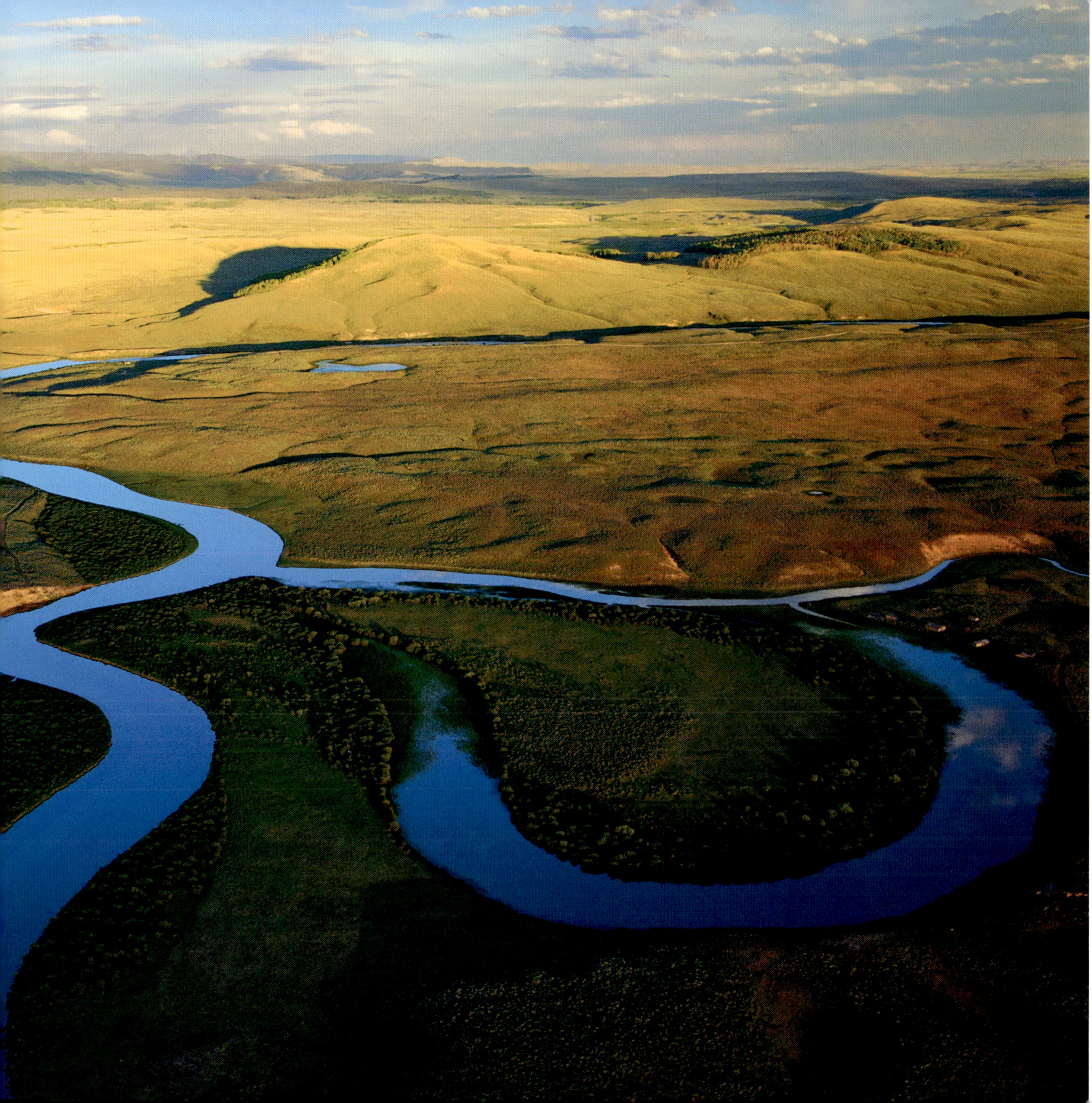

From Lookout Peak in the Wyoming Range, storm clouds over the Wind River Range catch warm sunset light. Most of the Wyoming Range, situated in Bridger-Teton National Forest in the southern Greater Yellowstone Ecosystem in Wyoming, is protected under the Wyoming Range Legacy Act.

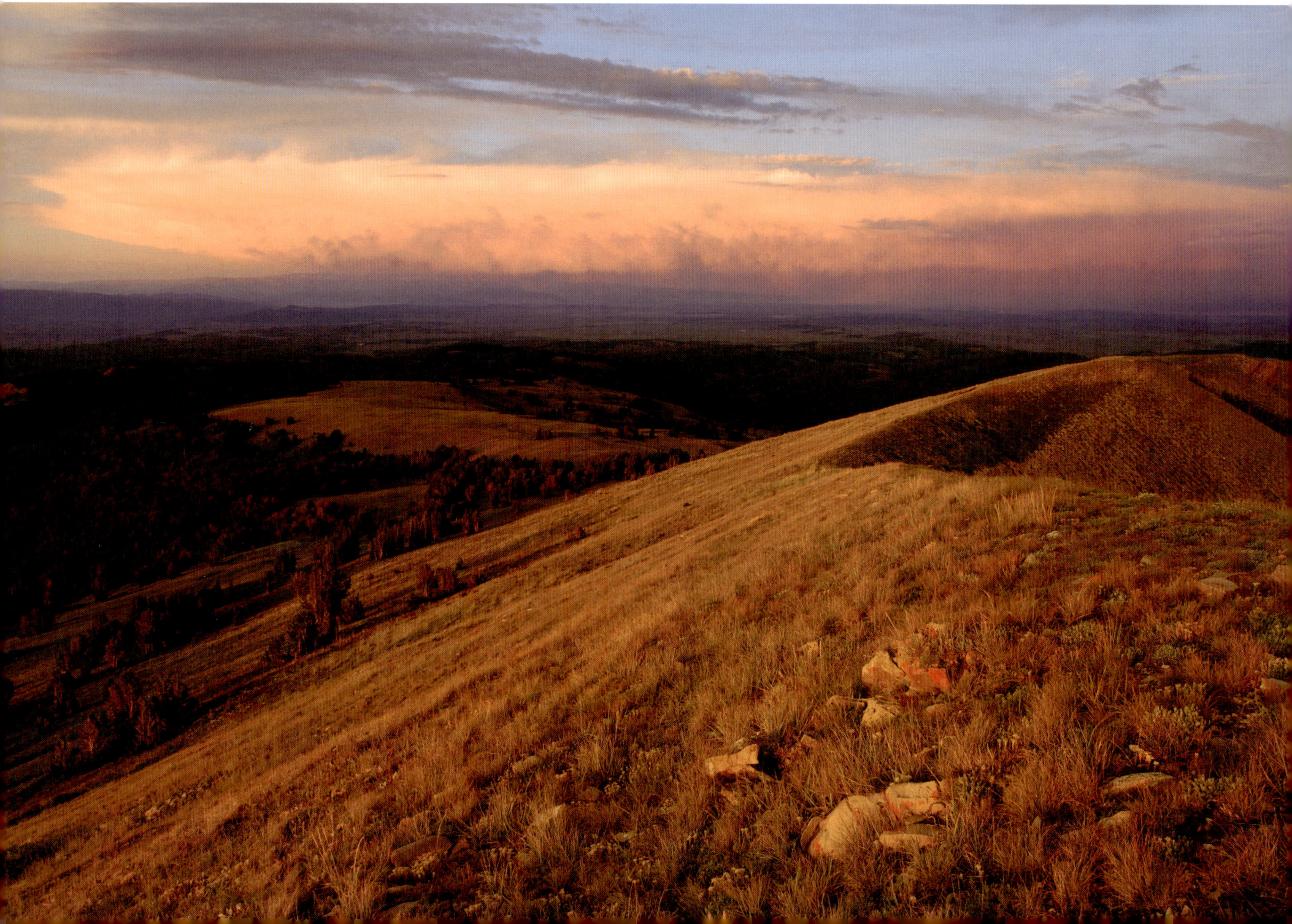

Stonecrop wildflowers in bloom in Soda Lake
Wildlife Habitat Management Area, Wyoming

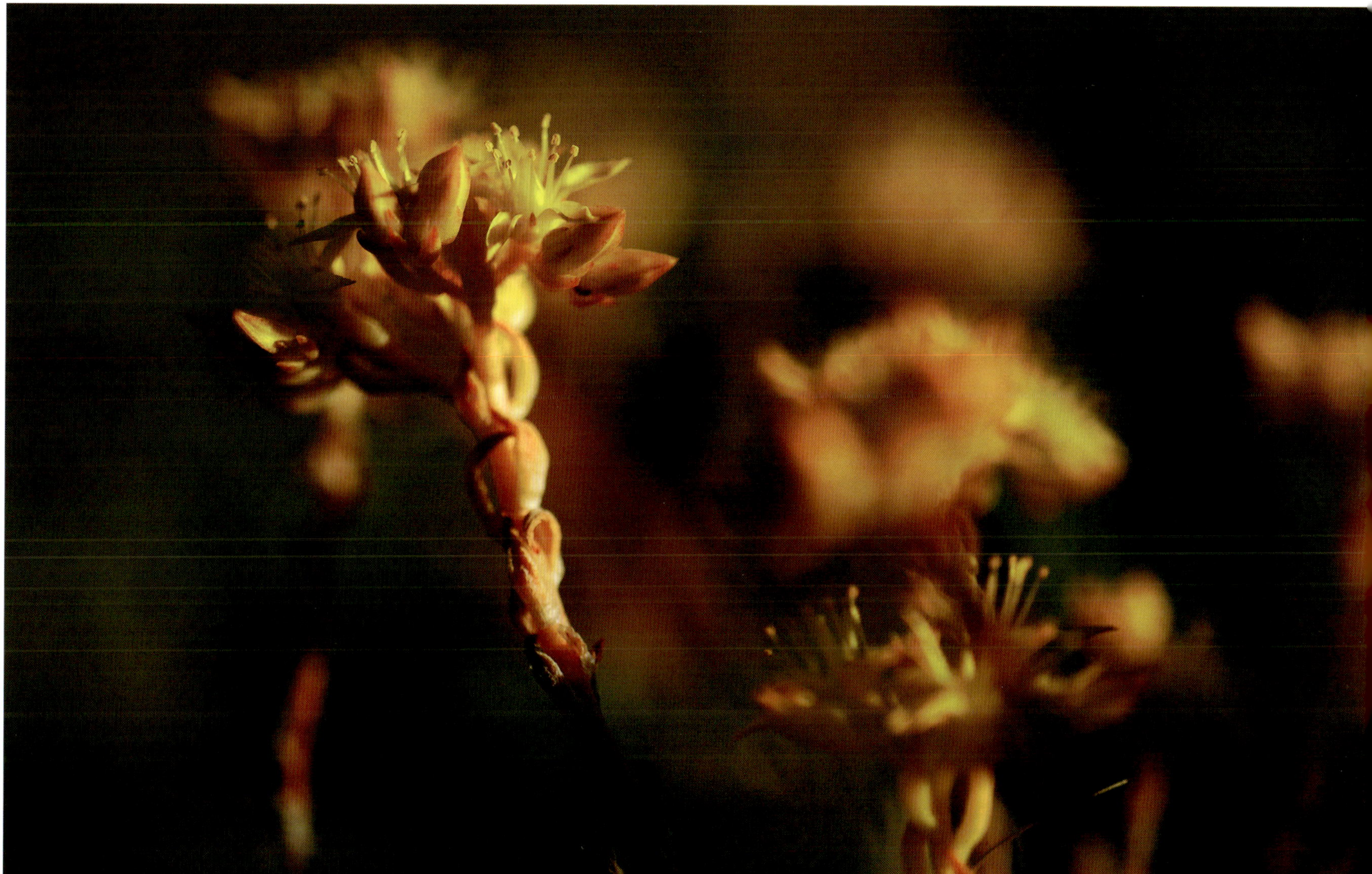

DAN SMITHERMAN
STANDING UP FOR THE UPPER HOBACK

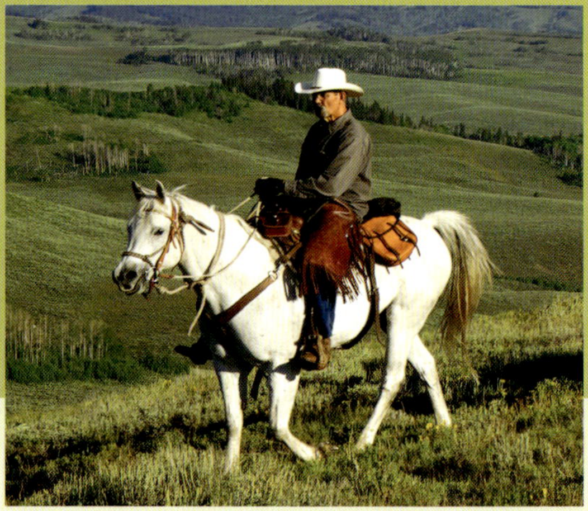

NEWS OF THE UPPER HOBACK energy leasing decision in 2006 tore through northwest Wyoming like a rifle blast. A Houston-based company announced plans to drill 136 natural-gas wells, with a network of roads to be traveled thousands of times by heavy trucks in the heart of the Bridger-Teton National Forest. If it could happen here, in the southern Greater Yellowstone Ecosystem, it could happen anywhere.

A grassroots coalition to save the Upper Hoback River Valley was led by outfitter Dan Smitherman, who had moved his family to Bondurant, Wyoming, in 1984 to carve out a quiet western way of life in a stunning mountain valley. Of the attraction to Bondurant, Smitherman related, "Everybody has something that connects to their soul. This is where I want to be and where I want to die." Forty minutes south of cowboy-chic Jackson, Wyoming, and Grand Teton National Park, Bondurant is a collection of ranches and ranchettes along the Hoback River cradled by the Gros Ventre Mountains and Wyoming Range, with the 8500-foot (2590-meter) South Rim connecting the mountain ranges.

Just west of town, below the Hoback River headwaters in the Wyoming Range sits a mountain basin with beaver ponds and creeks, mixed sagebrush and aspen, and conifer forest rimming the edges. It's a wildlife superhighway: a stronghold for moose; summer range for calving and fawning elk, deer, and pronghorn; home to all the toothy predators of Greater Yellowstone—grizzly and black bears, gray wolves, mountain lions. The 150-mile Red Desert mule deer migration travels to the Hoback, just the Hoback. On North America's longest mule deer migration, these deer cross highways, fences, lakes, rivers, and energy developments to reach the verdant slopes beneath the Wyoming Range, where they fawn and graze through the summer months. It's tough to get to, part of the attraction of this world-class resource. When the Upper Hoback was leased for drilling, folks from all walks of life came together to protest industrialization that would destroy this remarkable piece of Greater Yellowstone.

The Wyoming Range Legacy Act of 2009 that protected most of the range with the fundamental premise that it's too wild to drill couldn't protect leased inholdings, and the Upper Hoback drilling project was moving quickly. In 2010 Smitherman met with a fellow outfitter who told him, "You're wasting your time; they're going to drill there." Smitherman replied, "I'm going to

"When I asked why he decided to fight, Smitherman said simply, 'The word *can't* doesn't exist.'"

fight this" and walked out.

From that moment, Smitherman, who's been described as an "accidental activist," led an effort that grew in strength, aligning conservation groups with hunters, fishermen, wildlife watchers, hikers—seemingly everyone who loves this part of the West. Citizens for the Wyoming Range became a coalition to be reckoned with. A twenty-three-year Marine veteran who's traveled the world, Smitherman stands tall in a Stetson hat and speaks softly. When I asked why he decided to fight, Smitherman said simply, "The word 'can't' doesn't exist."

The coalition developed a website and a social media presence, packed public meetings, wrote letters to the Bridger-Teton National Forest, and grew in numbers. During the 2011 public-comment period, a reported sixty thousand comments—the vast majority of which were opposed to drilling the Hoback—were submitted. By then it was clear that the people, however diverse their backgrounds, were united in their opposition. In addition, new developments—such as endangered grizzly bears moving into the Upper Hoback and questions about environmental impact—required a new environmental review. There would be more time.

In October 2012, the Trust for Public Land (TPL), a nonprofit that creates parks and protects lands, announced it would purchase the gas leases and retire them forever, returning the Upper Hoback to the American people. In just a few months, the TPL successfully raised the $8.75 million to buy back the leases, and the transaction was completed in early 2013.

"This lady in Oklahoma sent me a five-dollar check with a nice note that said 'I really believe in what you're doing,'" recalled Smitherman, who was touched by the gesture. Talking about the coalition, he said, "I don't know if the number (of letters) was the most significant factor. What was important was the diversity across all backgrounds, socioeconomic groups, political beliefs, and ages. Everybody had some connection to the Upper Hoback."

Smitherman, now the Wilderness Society's Wyoming representative in the northern Rockies region, credits the grassroots coalition for the Upper Hoback watershed conservation victory, modestly downplaying his role: "Someone had to do it." He added, "We should transition to a more landscape-scale, collaborative, place-based decision-making process. Everyone gets a piece of the pie, and we still protect the most important places." §

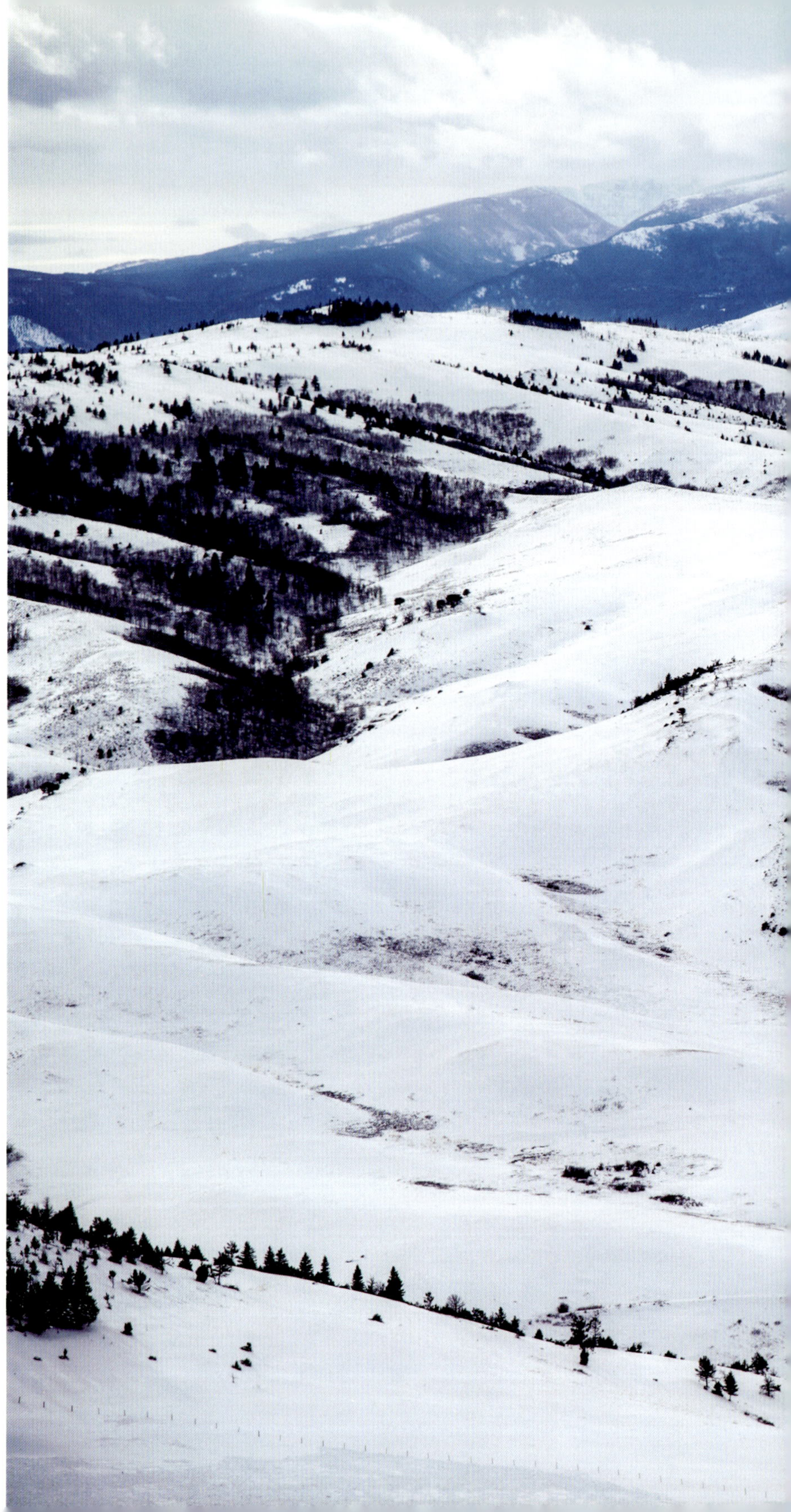

Ted Turner's Flying D Ranch
nestled against the
Gallatin Range, Montana

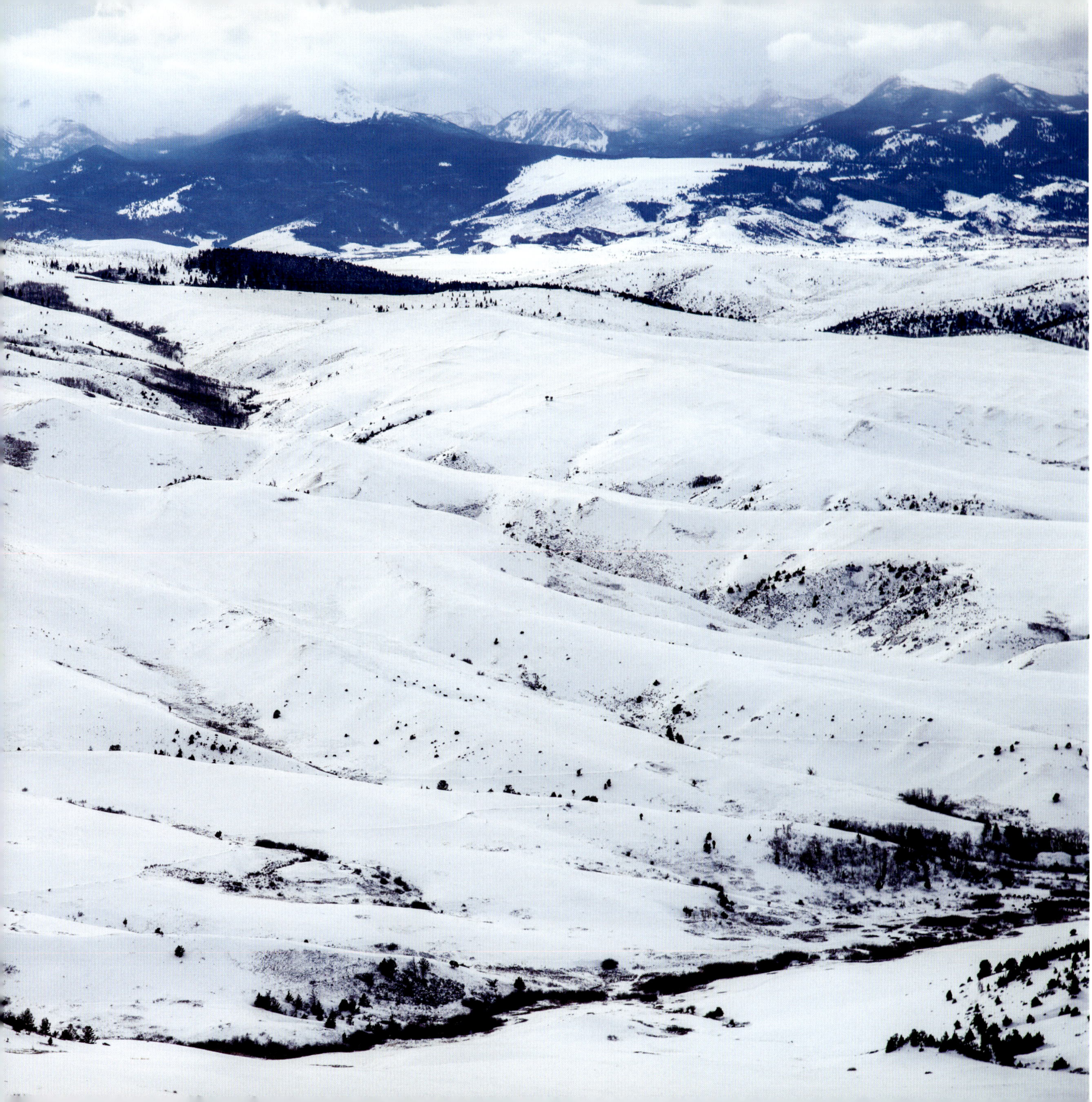

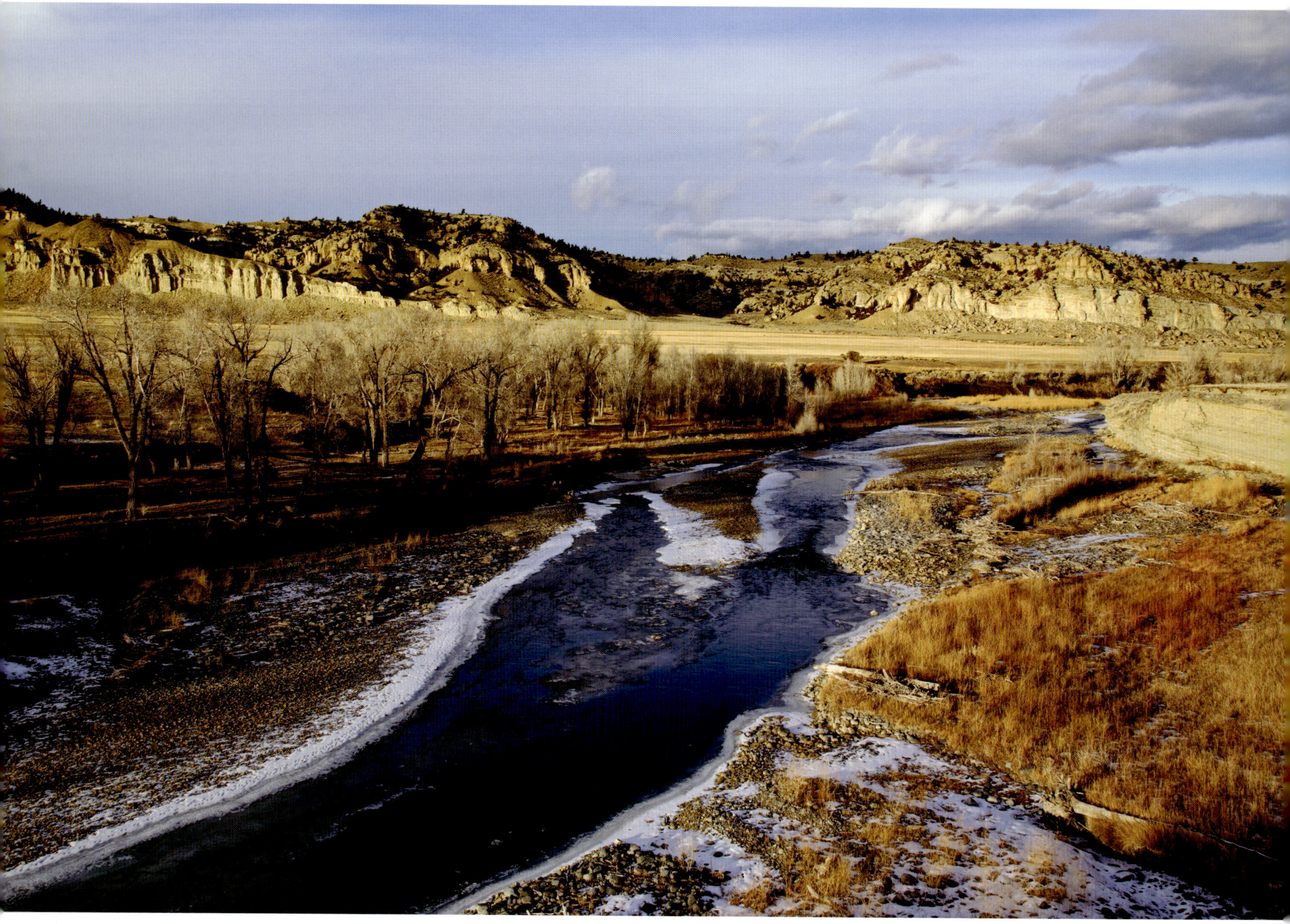

OPPOSITE The Greybull River in autumn transition to winter near Meeteetse, in Park County, Wyoming TOP RIGHT Scott Christensen of the Greater Yellowstone Coalition casts his fly in the Greybull River for Yellowstone cutthroat trout (Park County, Wyoming). The Greybull is an important fishery for the imperiled keystone species, which has suffered steep declines in nearby Yellowstone Lake, where a lake trout infestation has reduced the native trout population by 98 percent. BOTTOM RIGHT A beautiful Yellowstone cutthroat trout, ready for release back to the Greybull River (Park County, Wyoming)

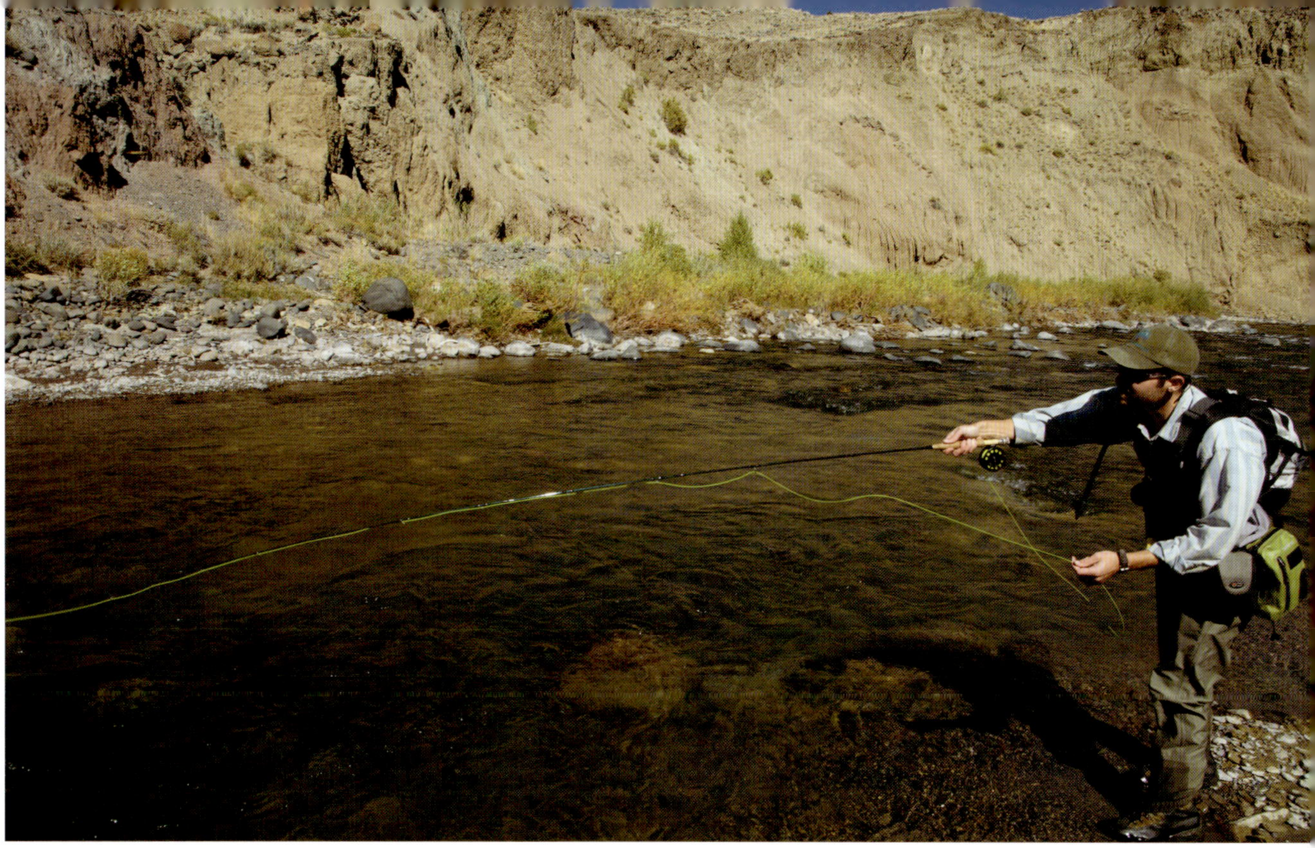

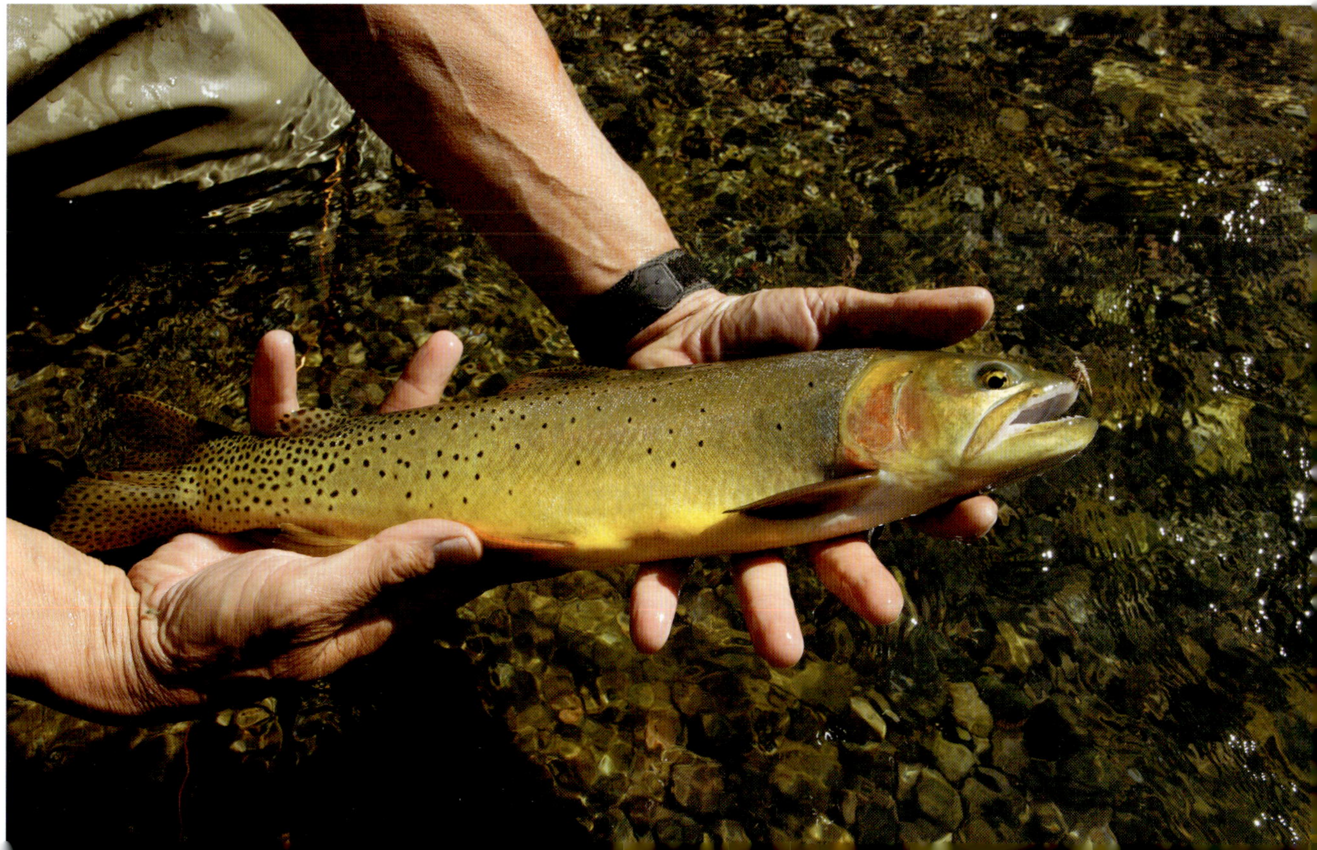

Condensate tanks, with a golden sunset backdrop behind the Wind River Range, are part of the Pavillion Gas Field, which lies in farmland near the town of Pavillion, Wyoming, where residents breathe toxic air and are unable to drink their municipal water.

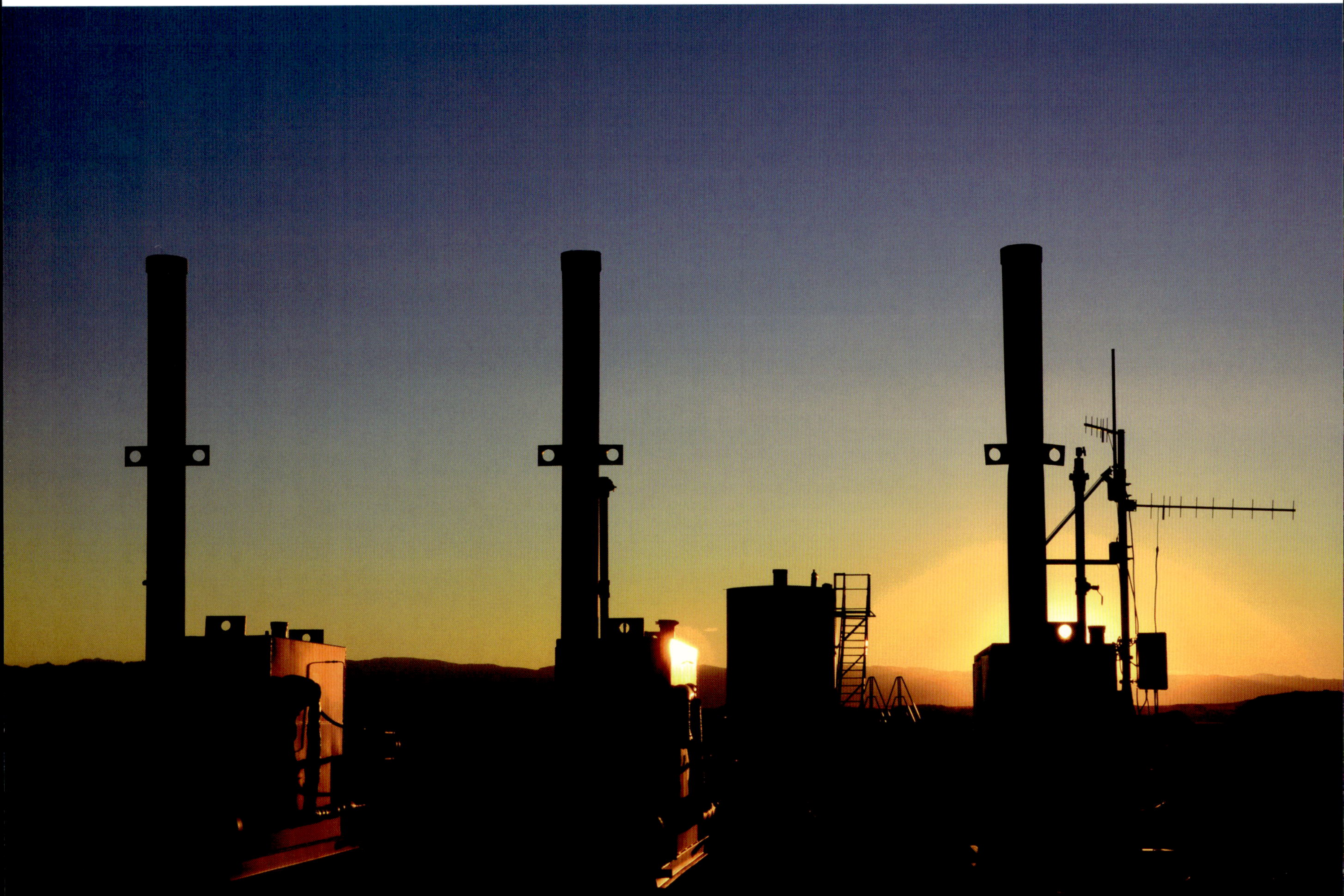

Legend Rock State Archeological Site, Wyoming, considered sacred by numerous tribes, has been visited by Native Americans over hundreds of years. Pecked in stone, the anthromorphs with horned headdresses are typically considered symbols of medicine or power among North American Indians. Many of these figures originate or end at cracks in the cliff, which researchers believe indicates a transition from one world to another dimension.

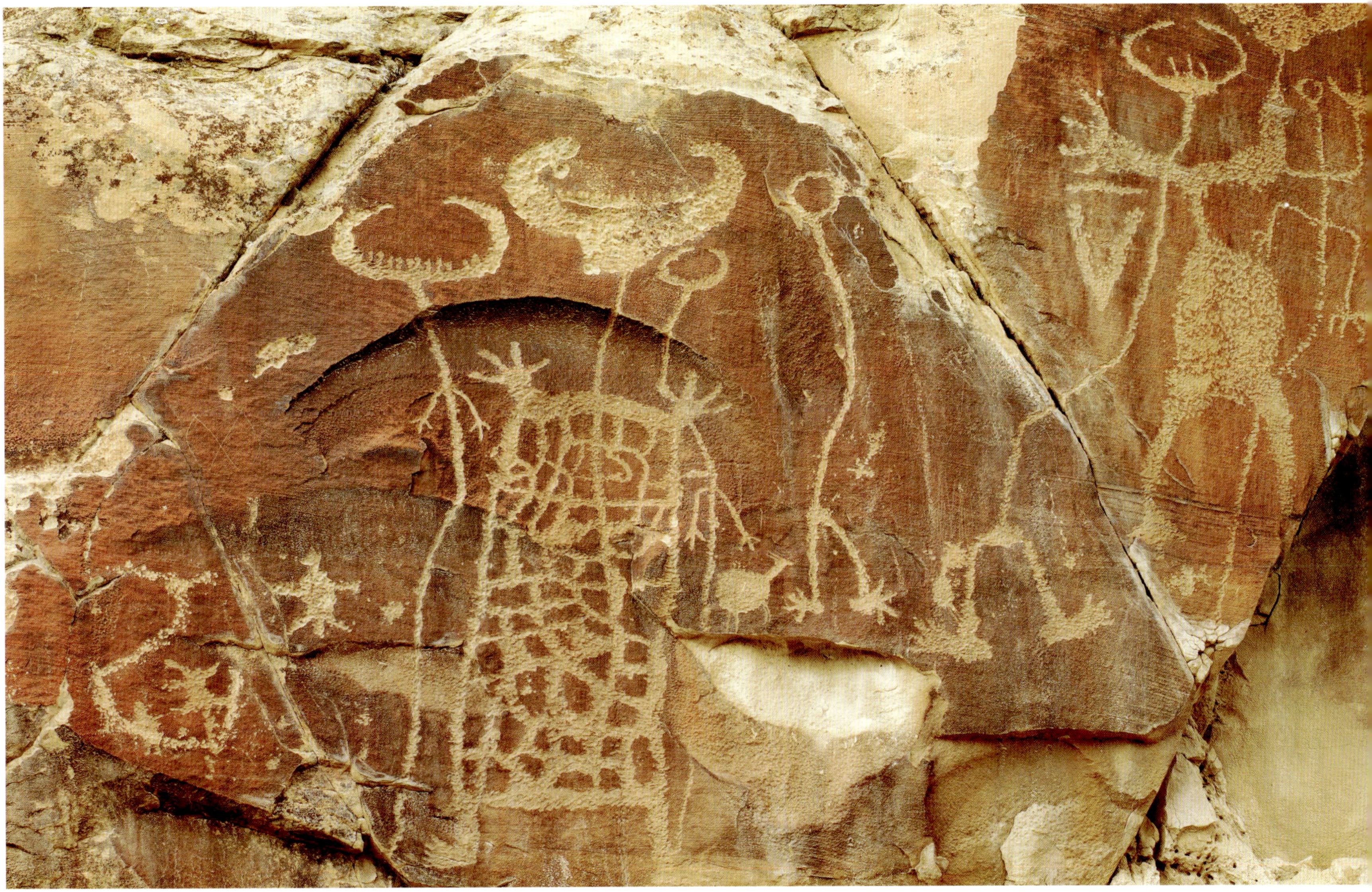

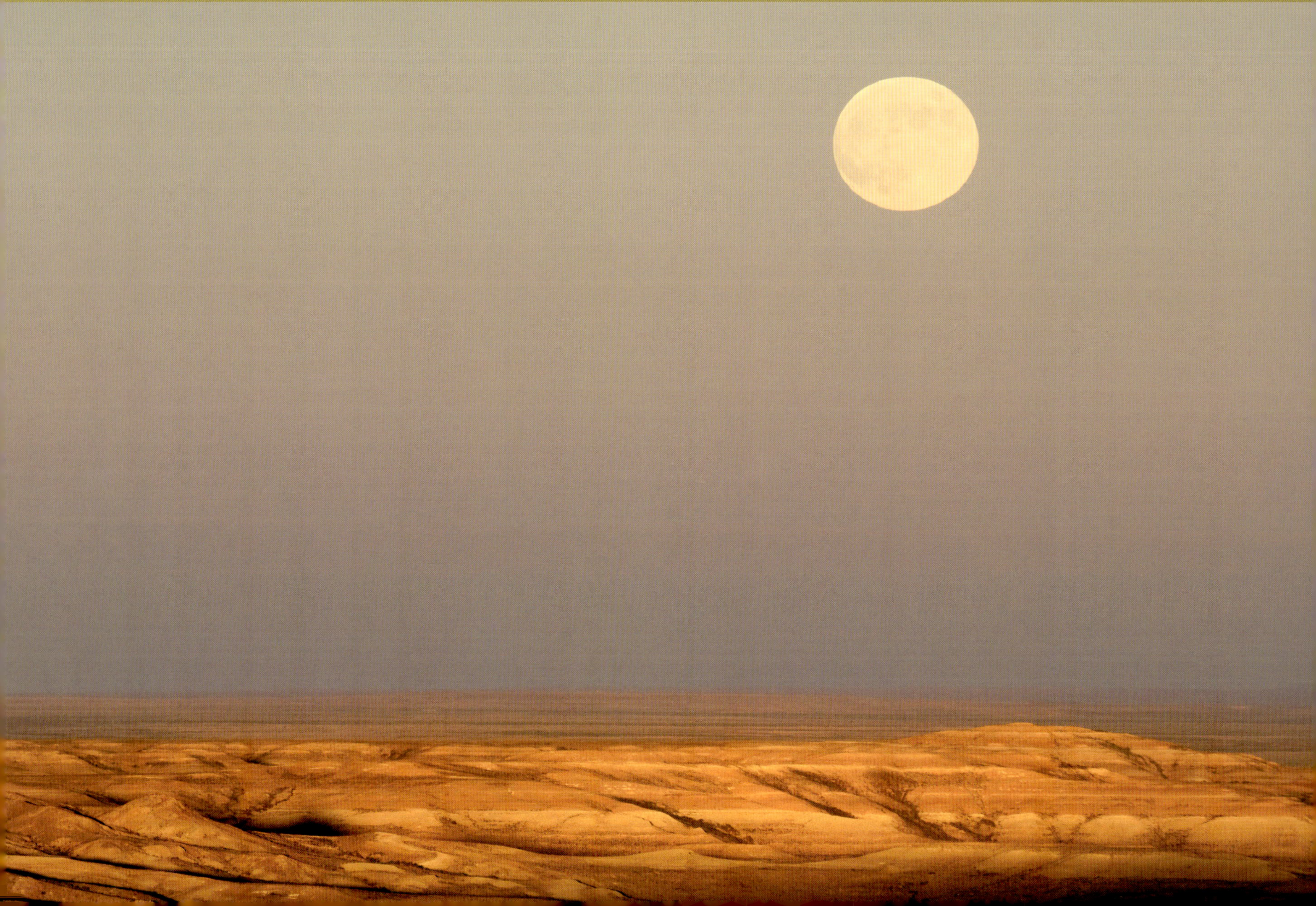

RED DESERT
WYOMING

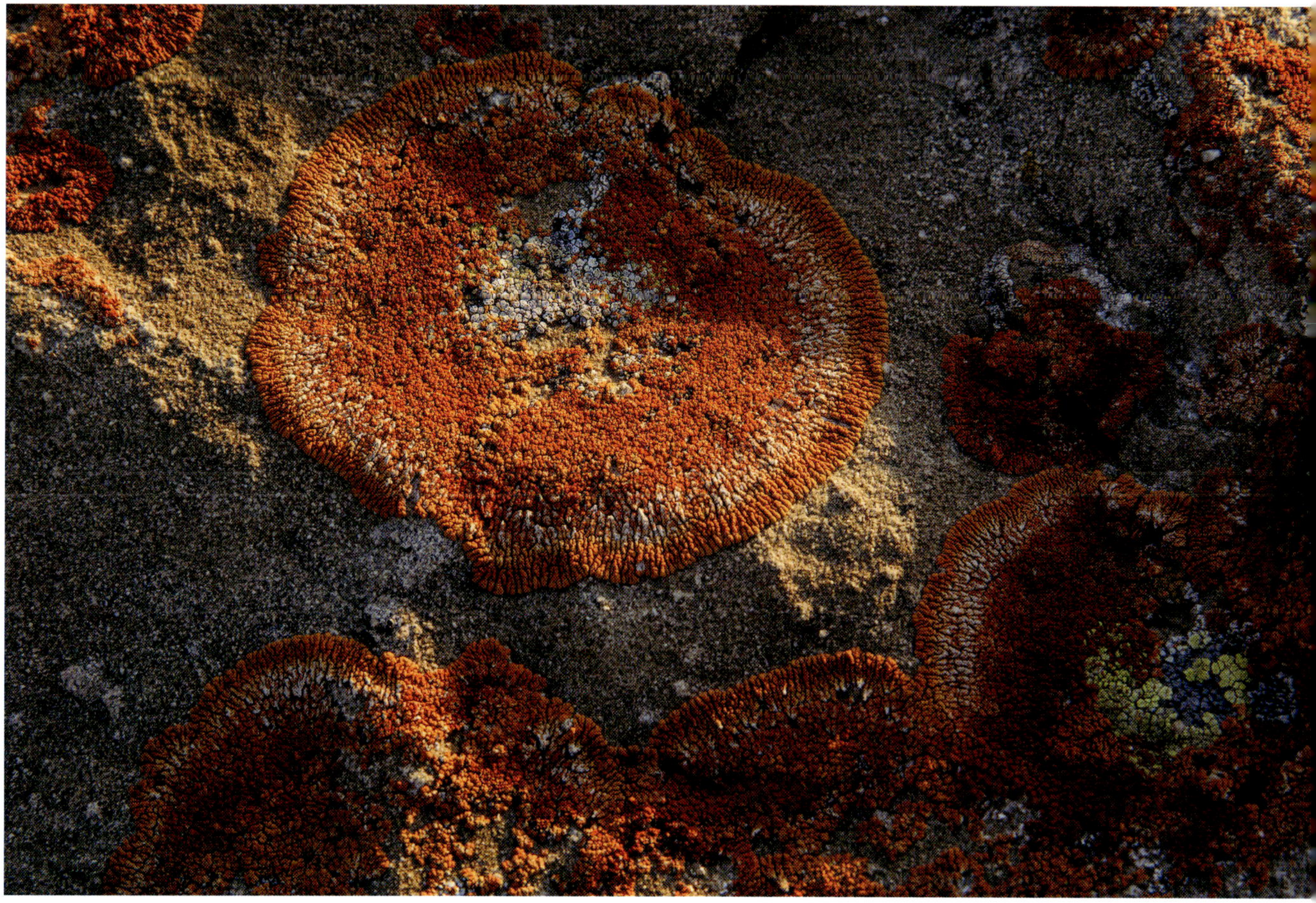

OPPOSITE Moonrise over Honeycomb Buttes Wilderness Study Area (Red Desert, Wyoming)
RIGHT Brilliant lichen patterns in the Jack Morrow Hills (Red Desert, Wyoming)

A western burrowing owl uses the tallest rabbitbrush for a hunting perch. Commotion beneath the brush suggested his family was in a burrow below (Red Desert, Sand Dunes Wilderness Study Area, Wyoming).

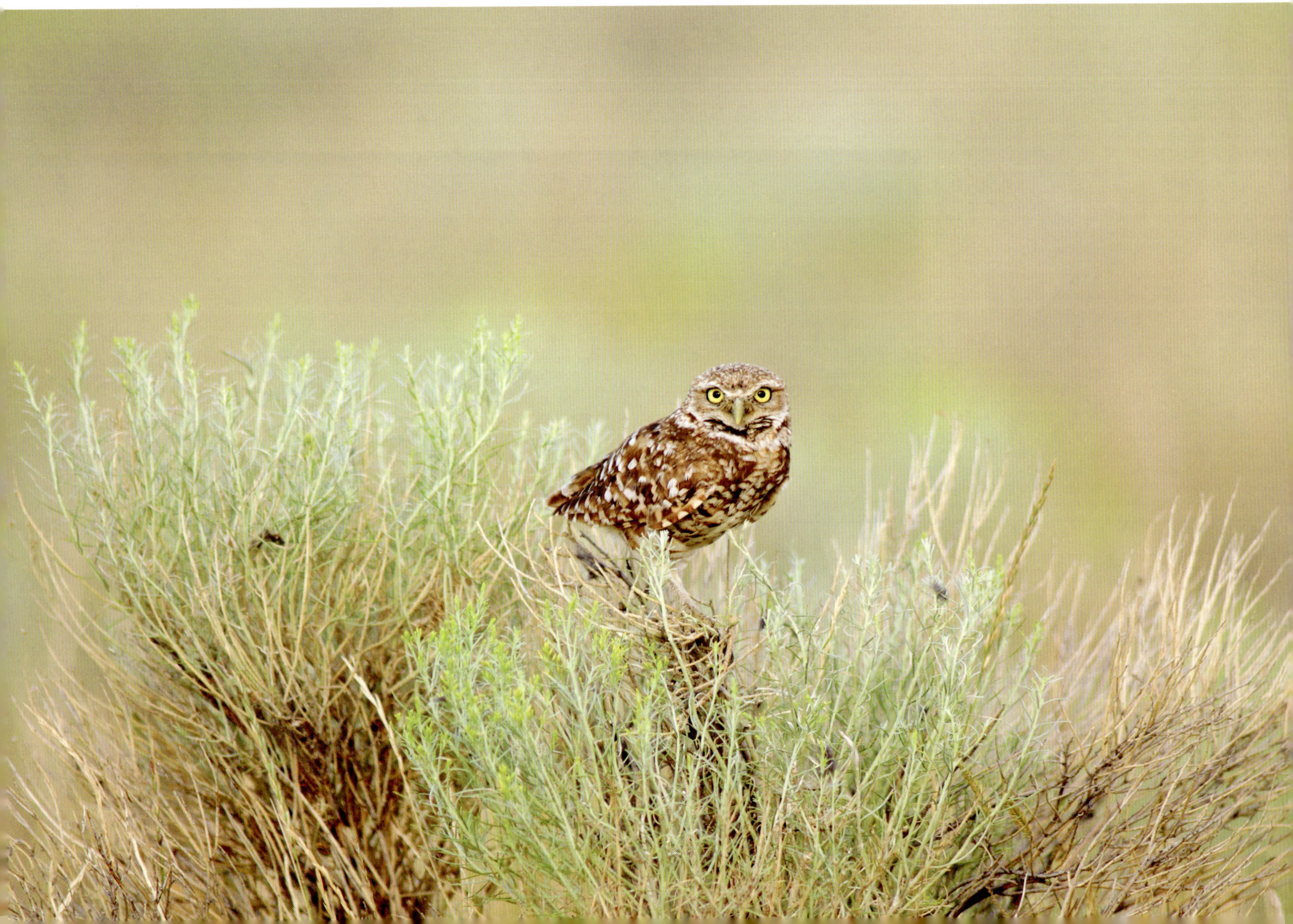

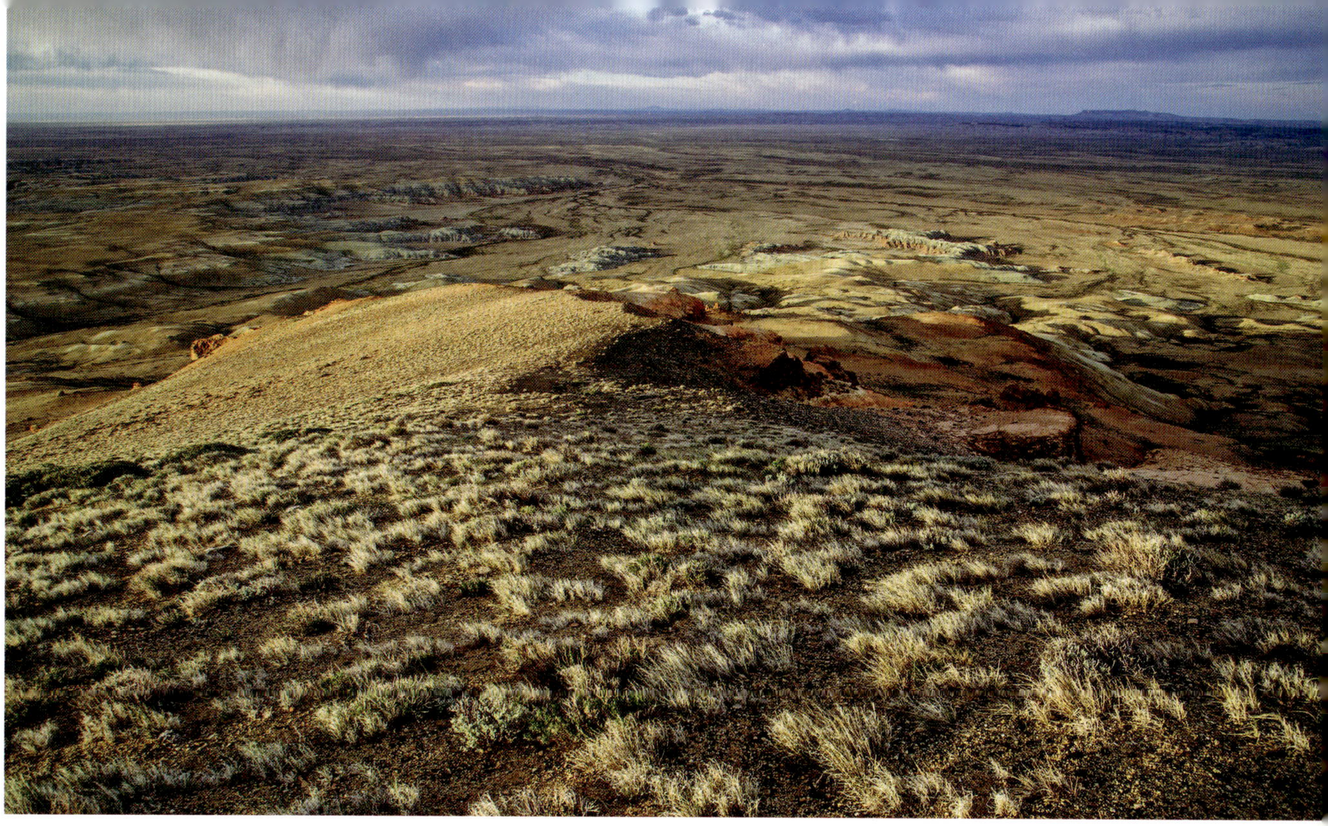

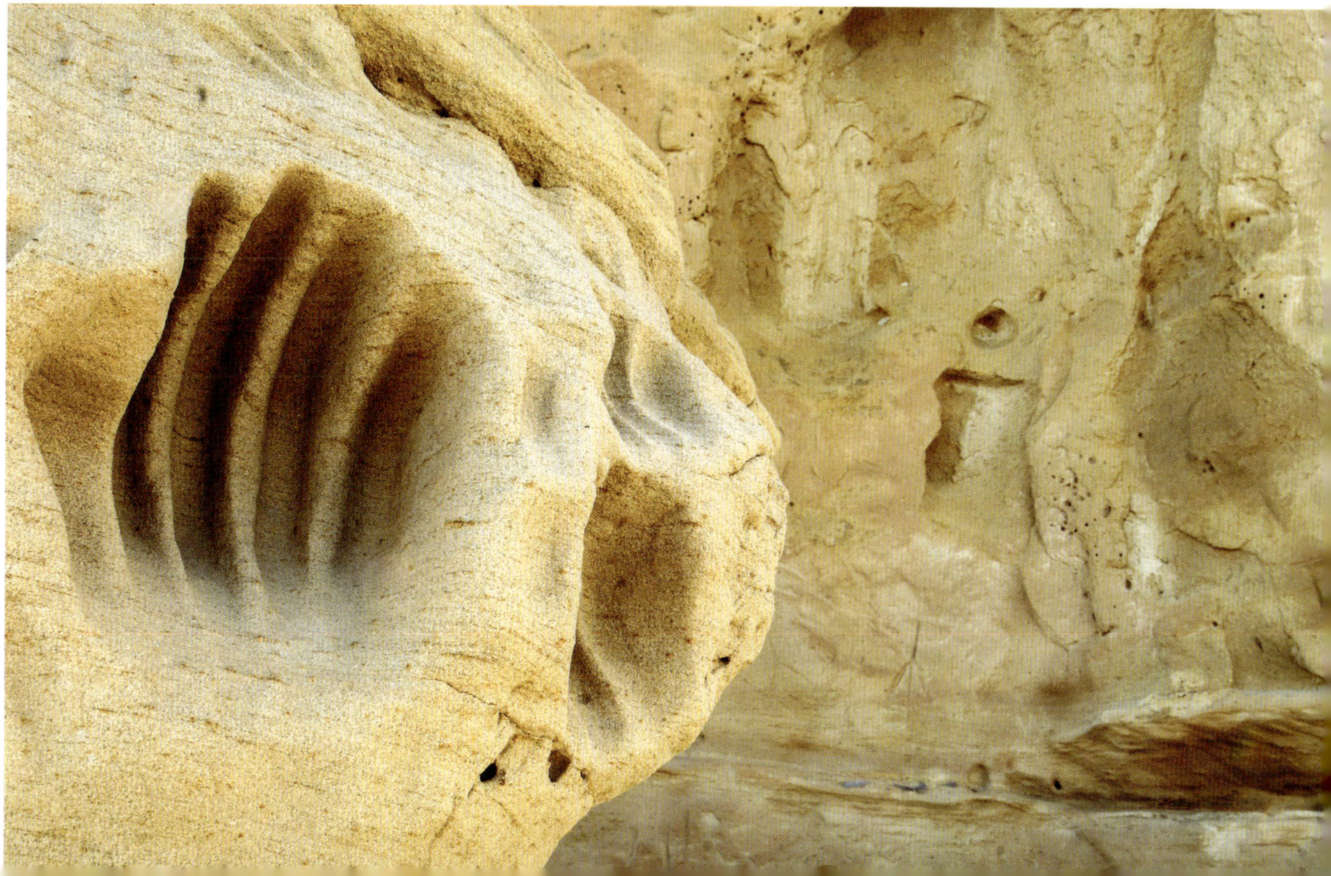

TOP RIGHT A southerly view of Great Divide Basin from the summit of 8431-foot (2570-meter) Continental Peak. This landscape-scale perspective includes much of the northern Red Desert, including distant Steamboat Mountain, across the largest unfenced area in the continental United States (Honeycomb Buttes Wilderness Study Area, Red Desert, Wyoming).

BOTTOM RIGHT This stone at the White Mountain Petroglyph Site is said to be a place where Native American women gripped the rock while giving birth (Red Desert, Sweetwater County, Wyoming).

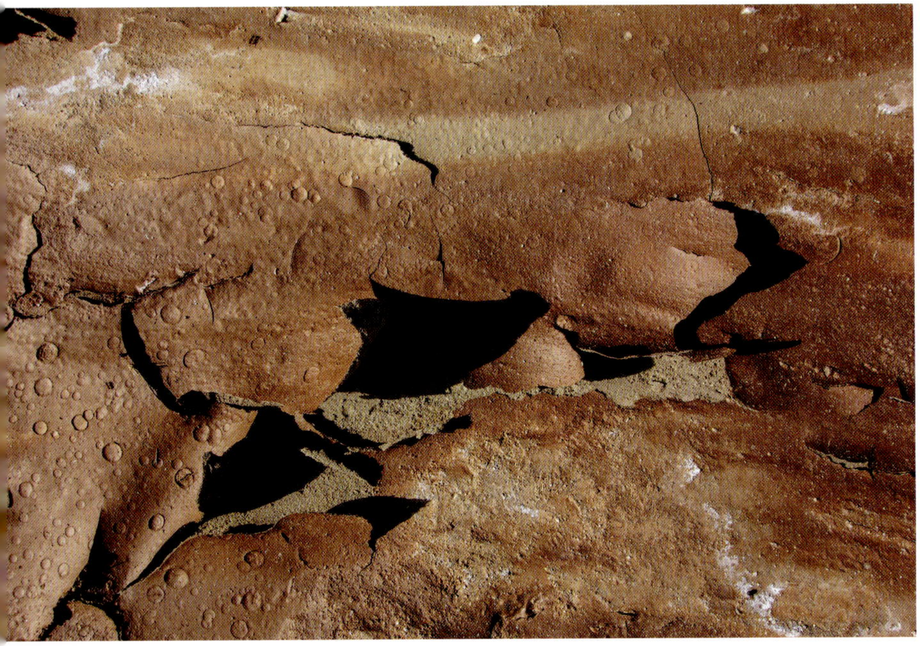

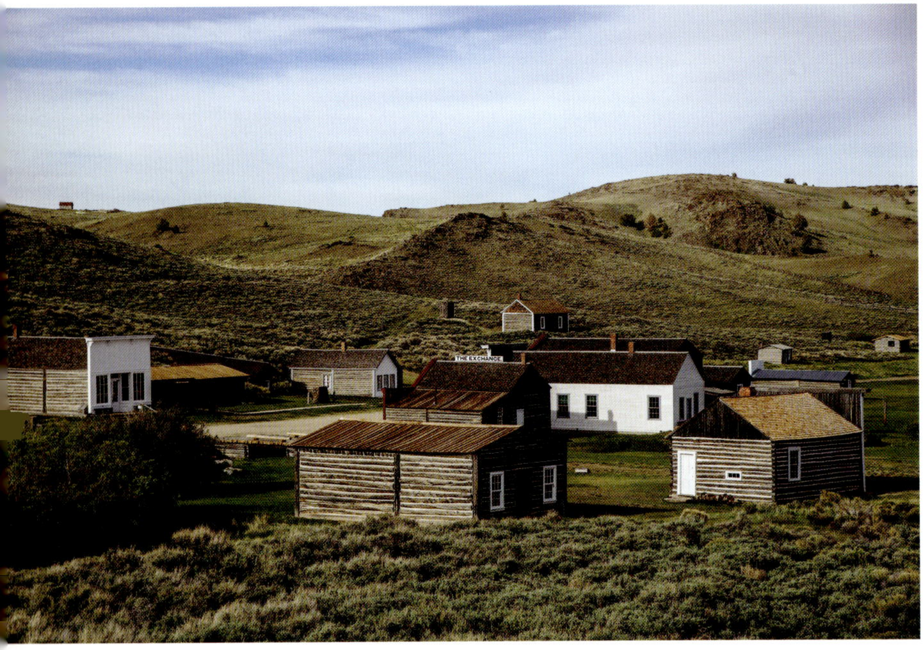

TOP LEFT Cracked mud in spring in the Honeycomb Buttes Wilderness Study Area of Wyoming's Red Desert

BOTTOM LEFT The restored South Pass townsite in Wyoming's Red Desert, which had a brief run as a booming gold town starting in 1866.

BELOW Soft morning light highlights the rugged landscape in Adobe Town Wilderness Study Area, located in Wyoming's southern Red Desert, Sweetwater County. A volcanic eruption in the Greater Yellowstone area 55 million years ago deposited ash that has been carved by wind and erosion over eons.

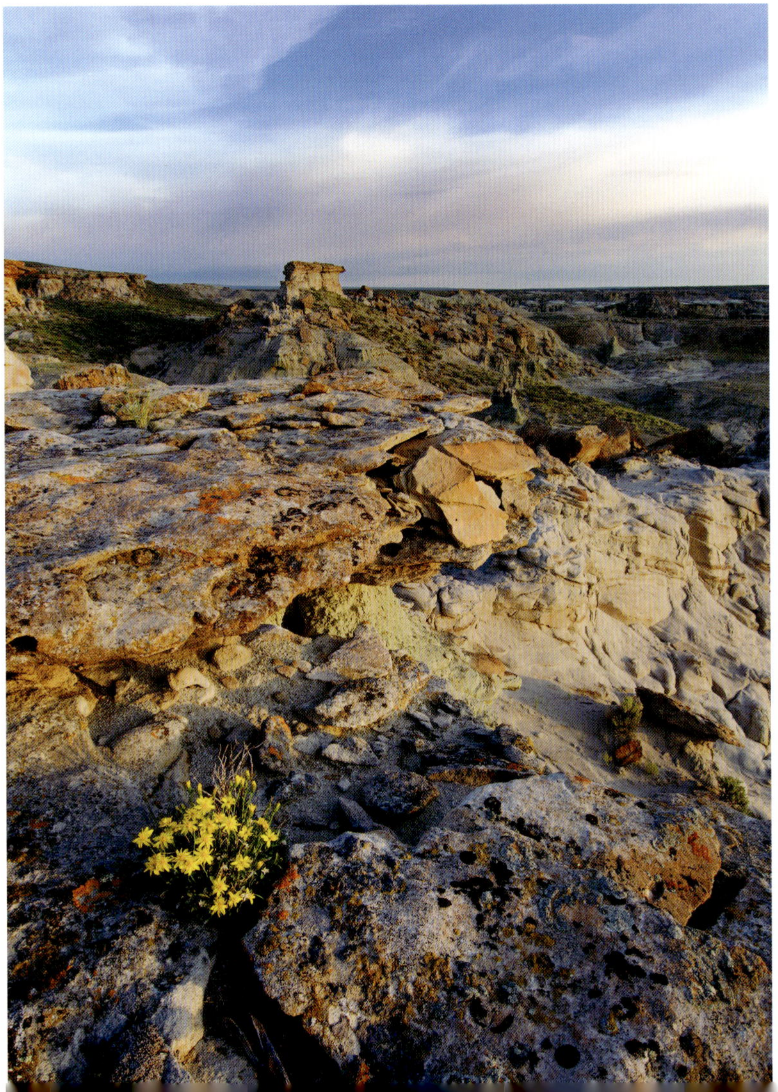

On a hike to the Killpecker Sand Dunes, tracks lead directly to the Boars Tusk, an iconic 400-foot-tall (120-meter-tall) volcanic monolith in the Red Desert (Sand Dunes Wilderness Study Area, Wyoming).

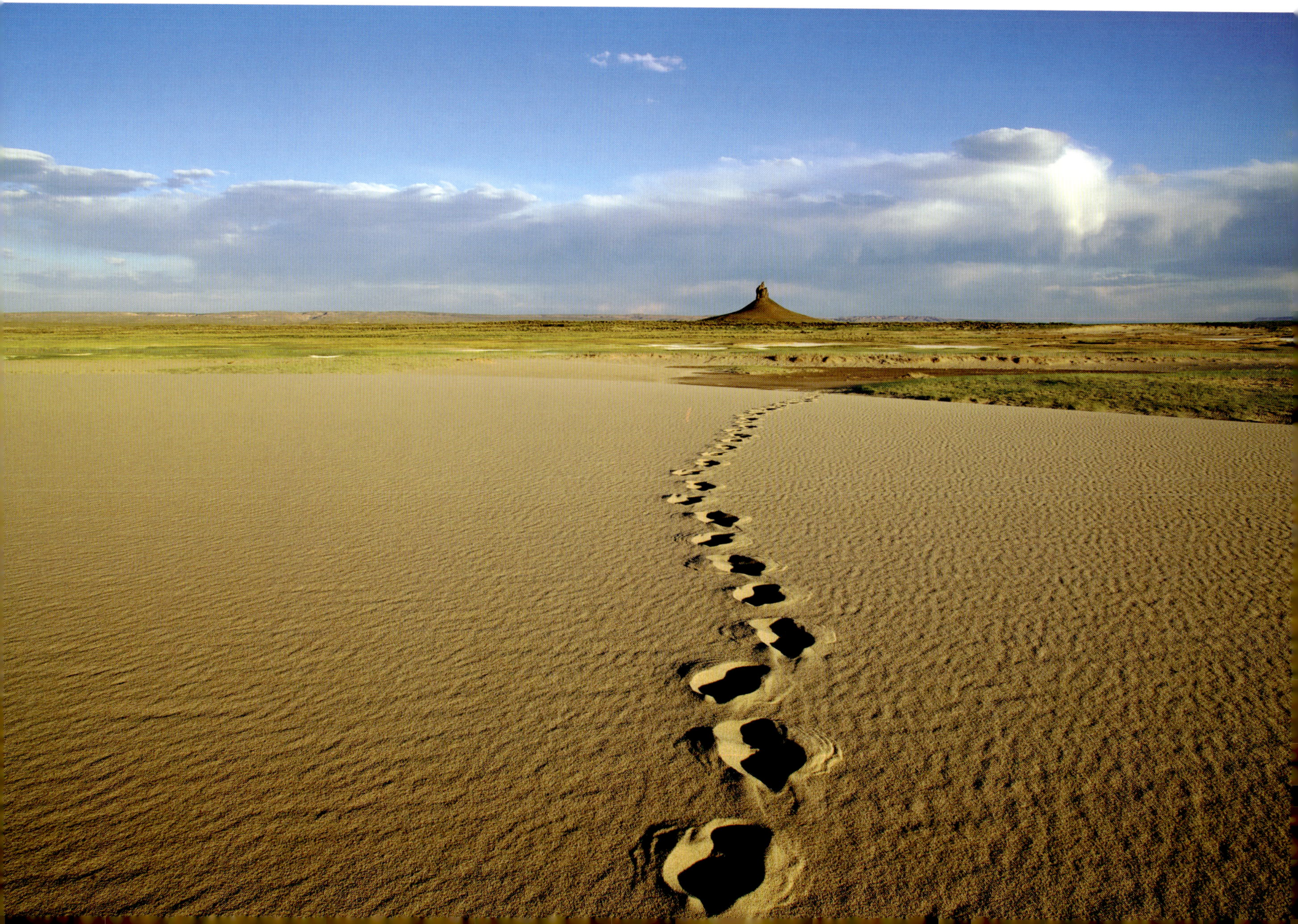

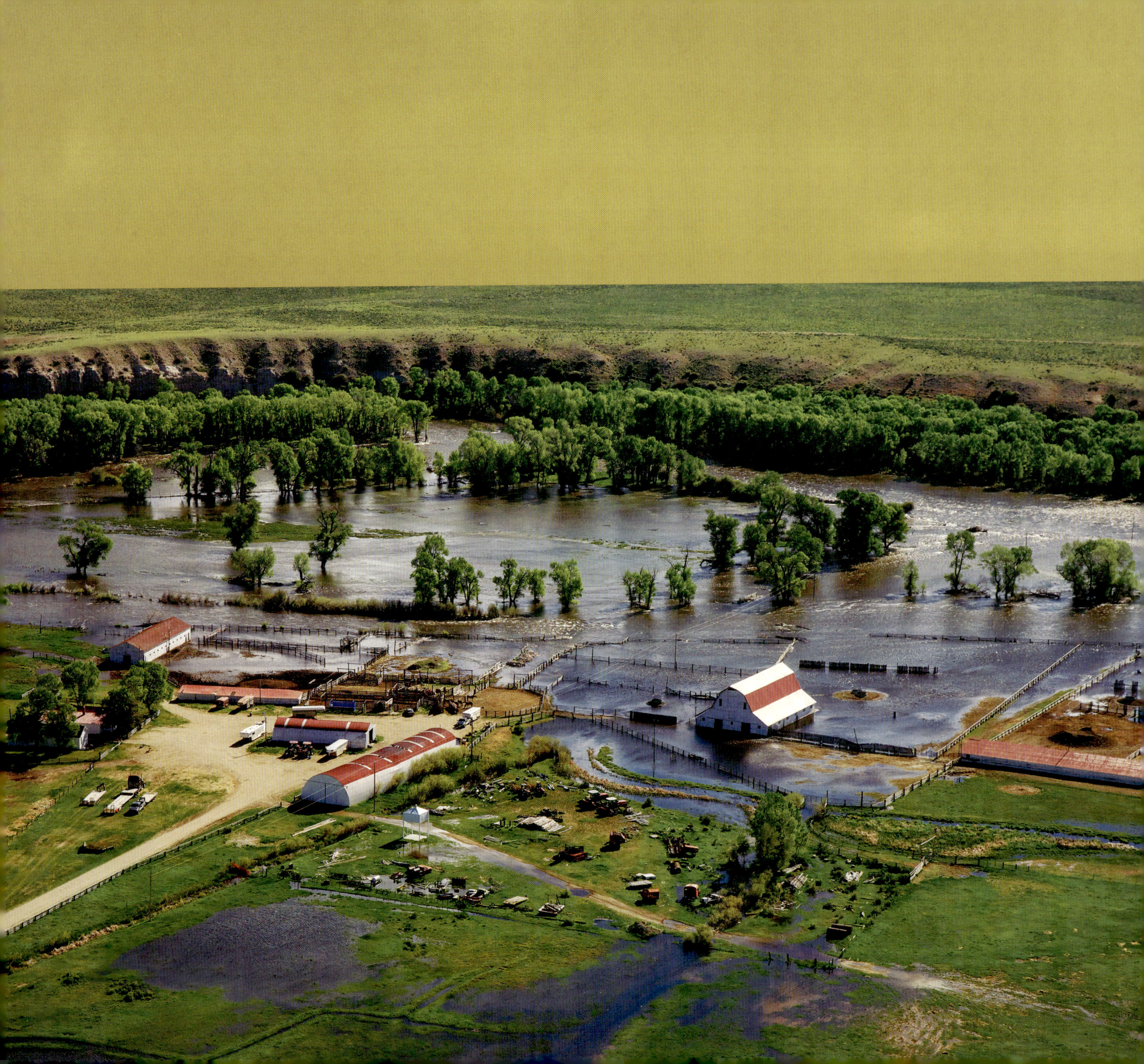

SOUTHERN WYOMING

OPPOSITE Heavy mountain snowpack and a cool, wet spring led to flooding in the West, such as here on the North Platte River near Saratoga, Wyoming, in the early summer of 2011.
RIGHT Greater short-horned lizards, iconic residents of the sagebrush-steppe, are unlike other lizards in both appearance and habits, resembling small, densely packed armored pancakes and feasting primarily on ants (Shirley Basin, Wyoming). Some horned-lizard species give birth to live young and may hibernate for eight full months. They escape predators by avoiding notice through camouflage, escaping quickly as they burrow into soil or leaf litter, looking uncomfortably spiny, and squirting blood from their eyes to surprise trespassers. Sagebrush provides shelter from sun, habitat for prey, a perch to reach a cooling breeze, and burrowing sites for nights and winter.

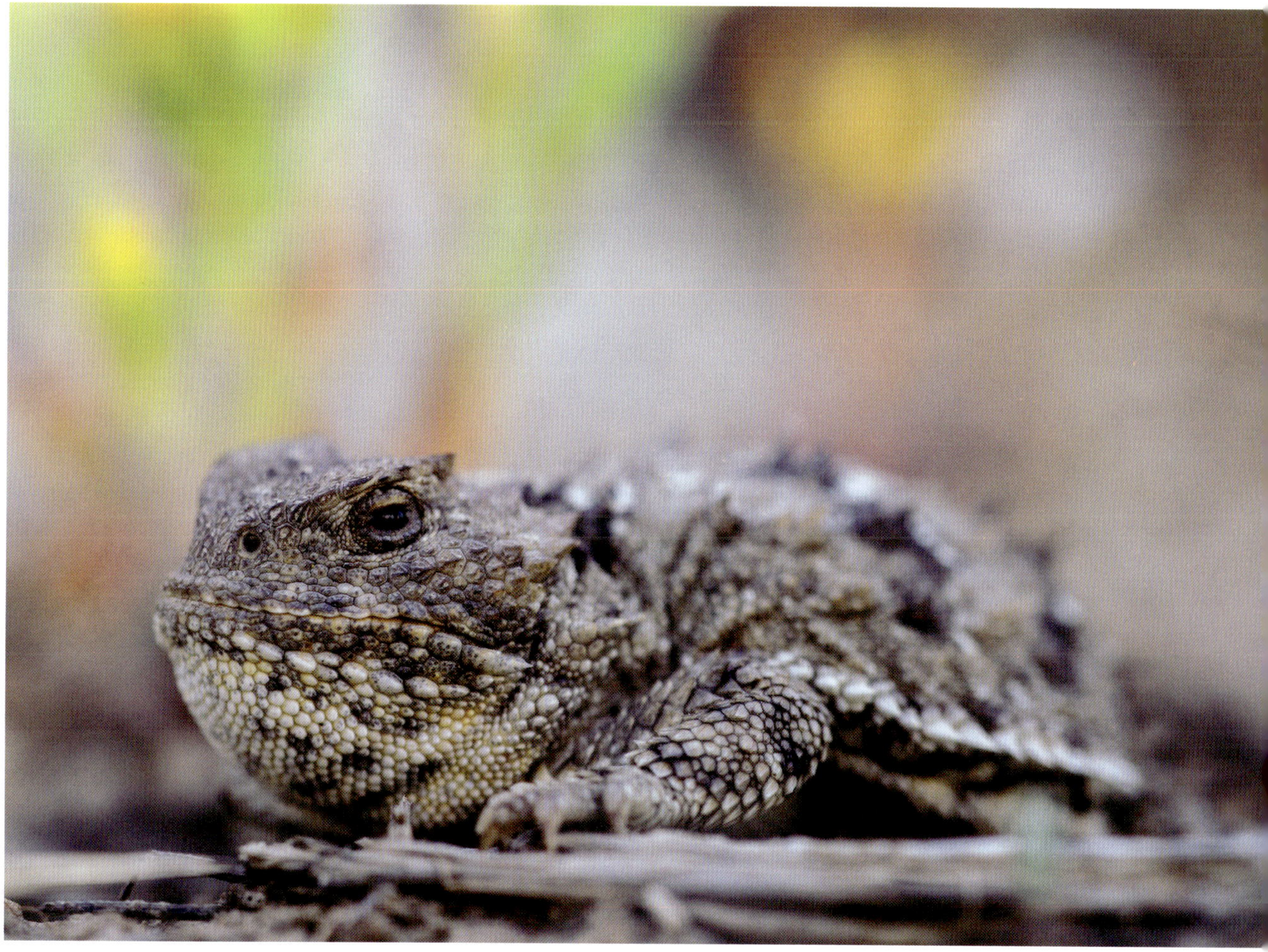

ALISON AND MATT HOLLORAN
MAKING SAGEBRUSH CONSERVATION A FAMILY AFFAIR

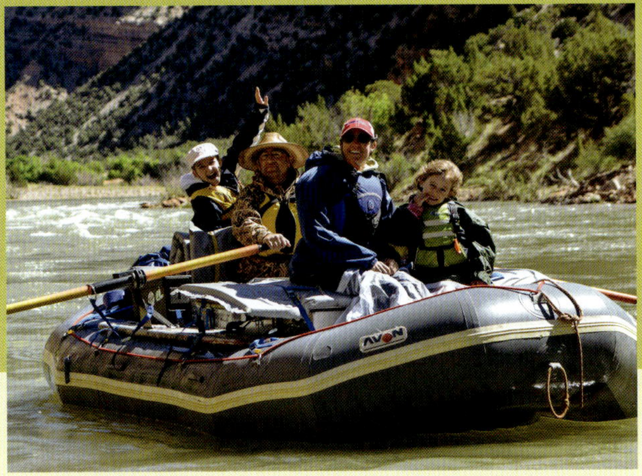

JUST WEST OF PINEDALE, Wyoming, is the Pinedale Mesa, also called the Pinedale Anticline—a rolling, treeless, sagebrush-carpeted ridge running north-south in the southern Greater Yellowstone Ecosystem. The Pinedale Anticline was an iconic sagebrush landscape, at least until new technology for hydraulic fracturing ("fracking") was tested in the nearby Jonah Field in the mid-1990s.

The Pinedale Anticline is also where Alison and Matt Holloran met in 1994, doing field studies while working on their master of science wildlife biology degrees at the University of Wyoming. Both were studying the impacts of natural-gas drilling on greater sage-grouse.

They recall how beautiful the area was "with very few wells on it. . . . When I shone a light, thousands of glowing orbs shone back," said Matt. Mule deer eyes. Alison said she fell in love with the sage the moment she stepped out there. As Matt puts it, "I really like the wide-open spaces where you can see as far as you can walk in a day. The sagebrush is one of the only places you can do that."

The Mesa, also critical winter range for the Pinedale Anticline mule deer herd, had been closed in winter to protect the deer migrating 60–100 miles (95–160 kilometers) between summer and winter ranges. But in 2004 the US Bureau of Land Management allowed year-round development on the Pinedale Anticline. Since that original plan to allow 700 wells over ten to fifteen years, another 4399 wells have been approved, including year-round drilling and disturbance of 12,000 acres (4855 hectares) in crucial mule deer winter range and sensitive sage-grouse habitats. The Pinedale mule deer herd has declined by 44 percent since 2001; greater sage-grouse have declined as well. Alison explained, "It has been completely and utterly transformed—several leks where I trapped (collared and released sage-grouse) for years during my master's degree work are now no longer active."

Matt was one of the first to investigate impacts of natural-gas drilling on greater sage-grouse, and his research throughout Wyoming led to his doctoral work on the greater sage-grouse's population-level response to natural-gas field development in western Wyoming. Alison's study of energy's impacts on greater sage-grouse connected her to Audubon Rockies, where she's advocated for sage-grouse and sagebrush habitat by developing a landscape-scale conservation plan with colleague Brian Rutledge,

"I really like the wide-open spaces where you can see as far as you can walk in a day."

Central Flyway Conservation Strategy and Policy Advisor. The science and vision of their Sagebrush Ecosystem Initiative began well before energy development's impacts on sage-grouse populations and grouse as indicator species were widely discussed. They have played a major role in shaping conservation strategies throughout the ecosystem.

In April 2009, during one of Matt's middle-of-the-night expeditions of catching and attaching telemetry collars to greater sage-grouse, I sat talking with him in his pickup on the Pinedale Anticline. Around 2:00 a.m. the conversation turned to how he felt about the transformation of a place where he fell in love with his wife, where he does so much important research, how it's hard to watch the steady decline. Matt said, "Two-thirds of this mesa is gone to development. That means I have one-third left here to protect and countless other energy developments where what we learned here can help us to do it better."

In their individual roles, Matt and Alison have developed a science-based vision of protecting greater sage-grouse by conserving their best remaining habitat—leks. "Doing things better" would require a 4-mile (6.4-kilometer) minimum buffer of undisturbed sagebrush habitat around the leks, a foundation of the core-area habitat strategy.

Matt is still trapping, collaring, and studying greater sage-grouse. His work has informed landscape-scale planning for sage-grouse recovery. Alison, executive director of Audubon Rockies, and her staff are propelling outreach and education with the Sagebrush Ecosystem Initiative, gradually changing how sage-grouse habitat is managed and valued.

Leaders in sagebrush and sage-grouse research, education, and outreach, Alison and Matt are also raising daughters Sage and Willow. Alison said, "I have a mom's perspective; I want these opportunities for my kids. No matter your position, we all have an obligation to this planet." Matt added, "If we (humanity) don't change the way we do things and deal with the potential consequences, the world won't be a place where we can live. We need clean water, clean air, and space not just for the wildlife and aesthetic beauty of it all, but for the continued existence of humans."

I asked Alison if it's possible to preserve the wild character of the West. "I think it is," she replied. "Otherwise, I would have given up." §

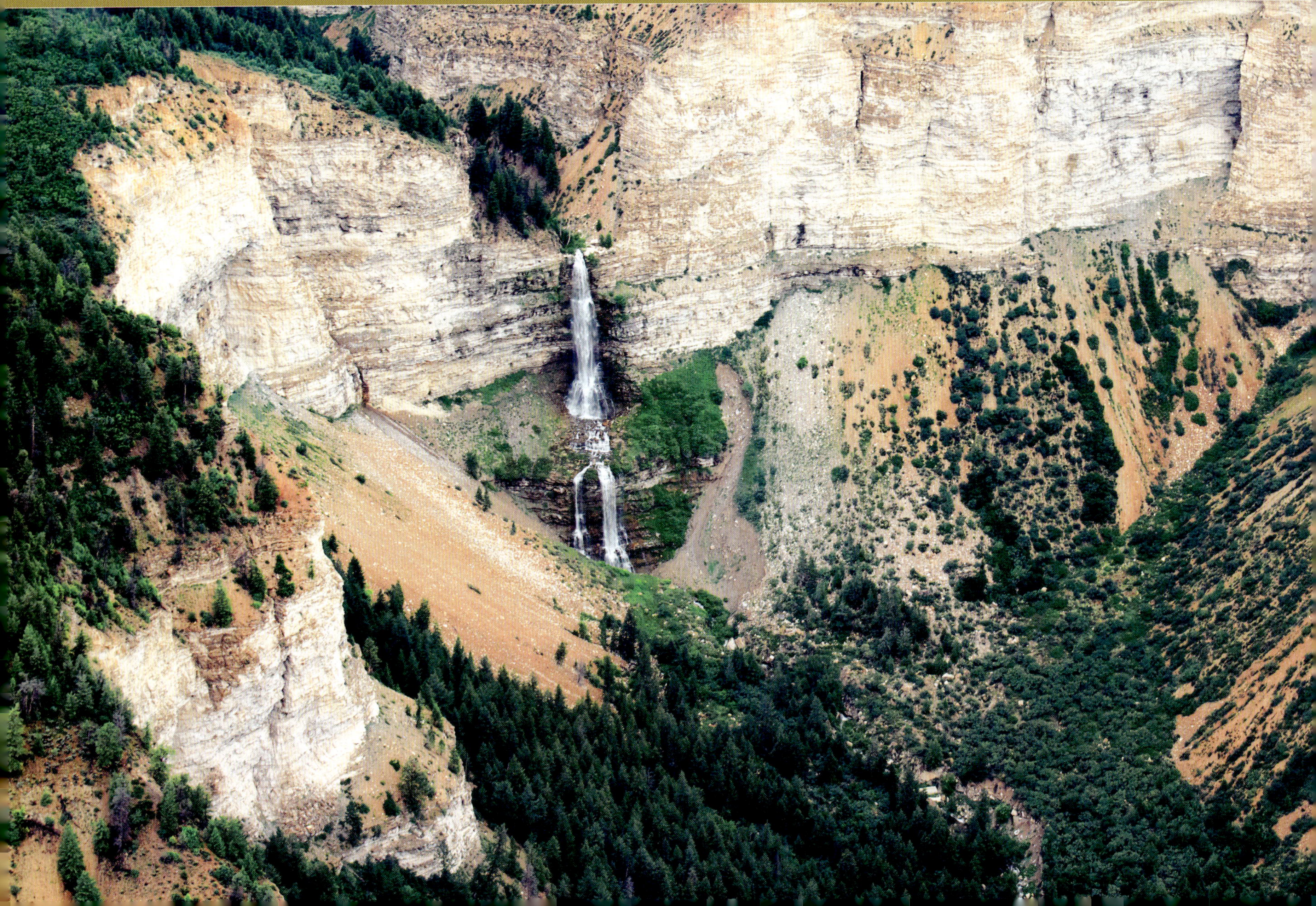

NORTHWEST COLORADO

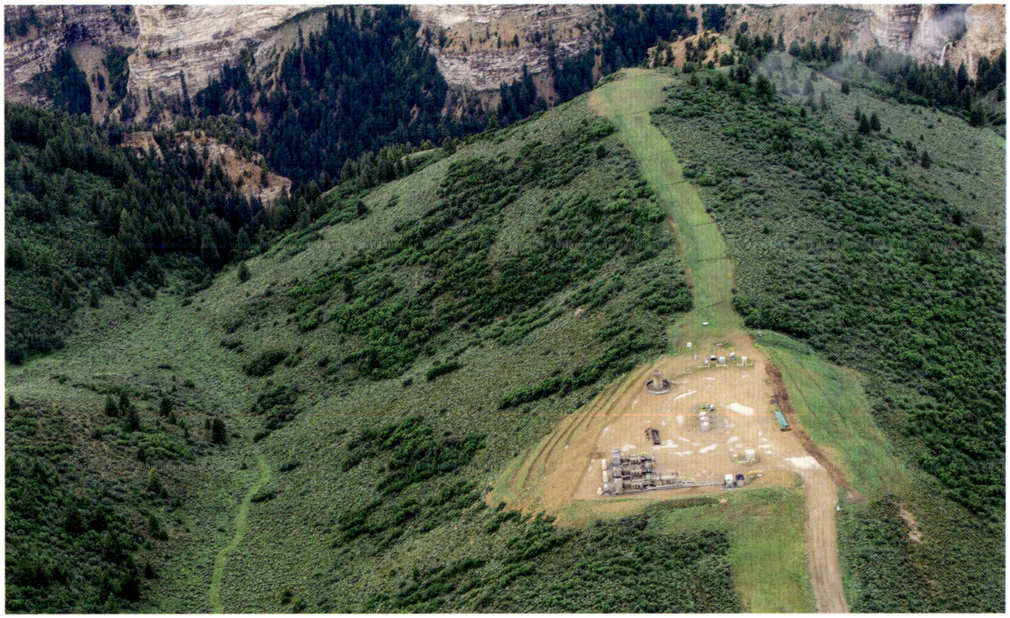

Two sides of the Roan Plateau.
OPPOSITE Parachute Creek spills over the Roan Plateau. Citing wilderness qualities and the need for protection of rare plants and a wide range of wildlife, conservation groups challenged natural-gas leases on the east "wild side" of the plateau and won an important victory in 2014, protecting most of the surface on the public land side and ensuring the Roan Plateau will remain wild.
TOP AND BOTTOM RIGHT Aerial views of natural-gas development on private lands on the west side of the Roan Plateau (Garfield County, Colorado)

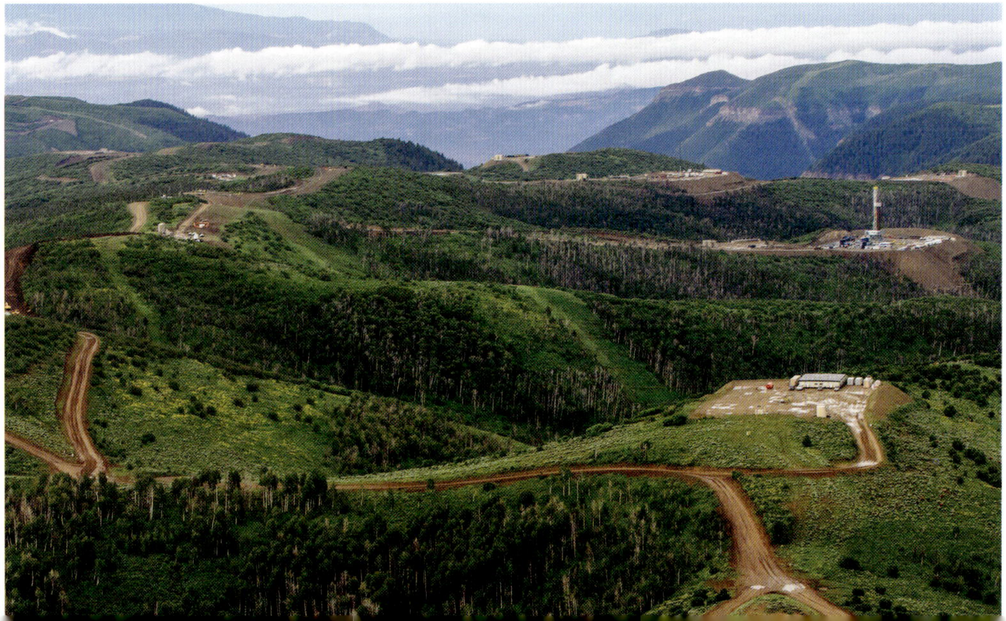

Lupine blooming in sagebrush in the spring (Moffatt County, Colorado)

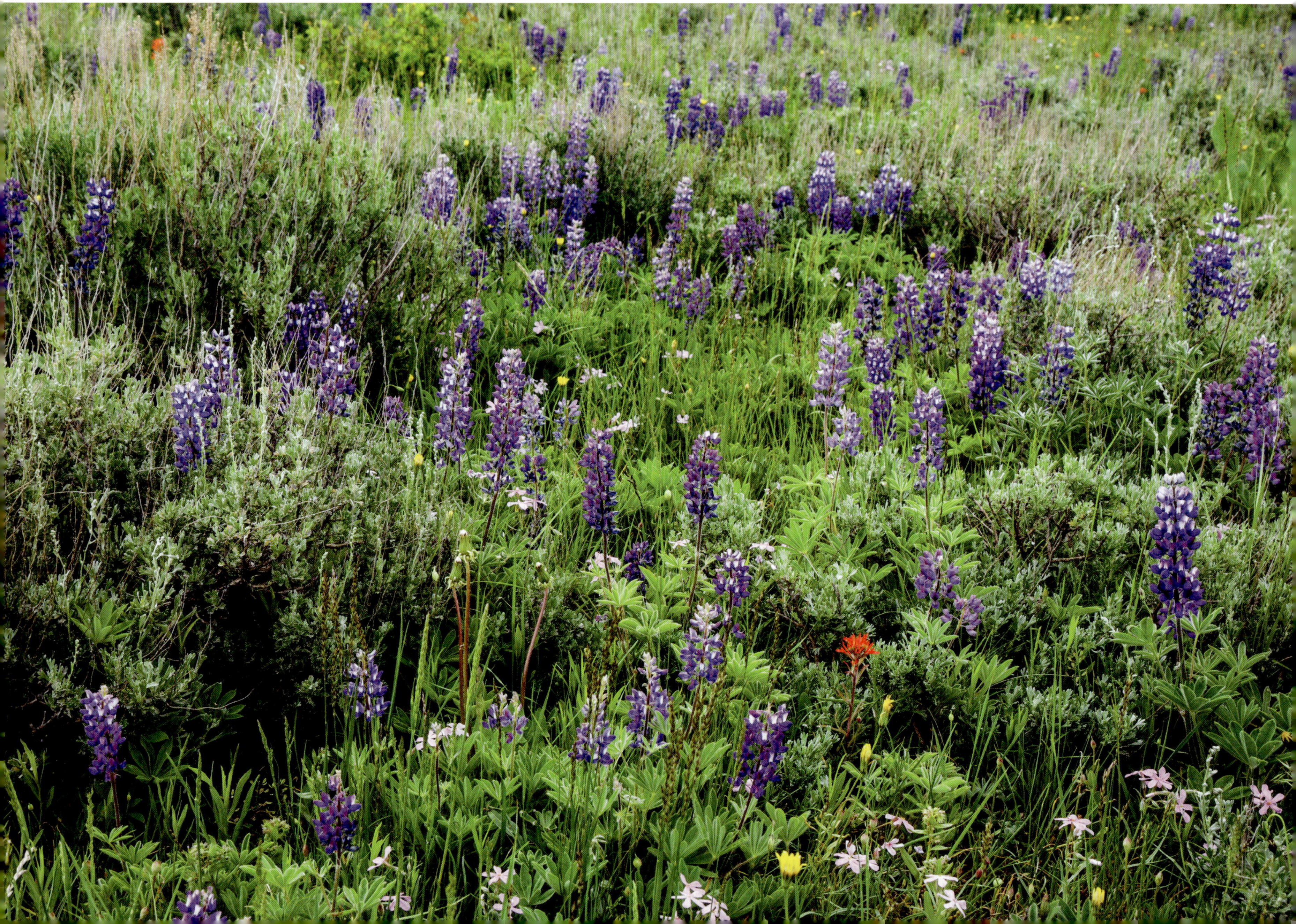

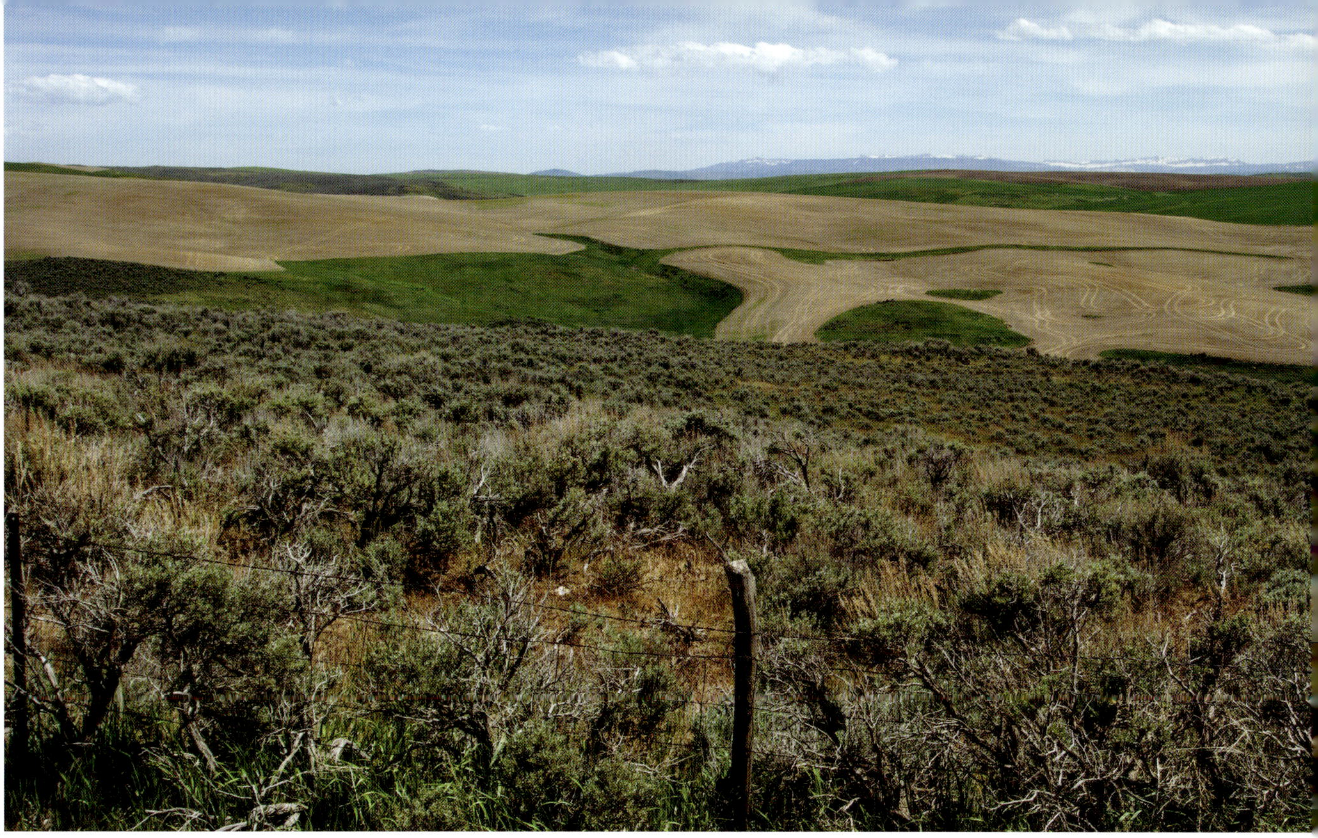

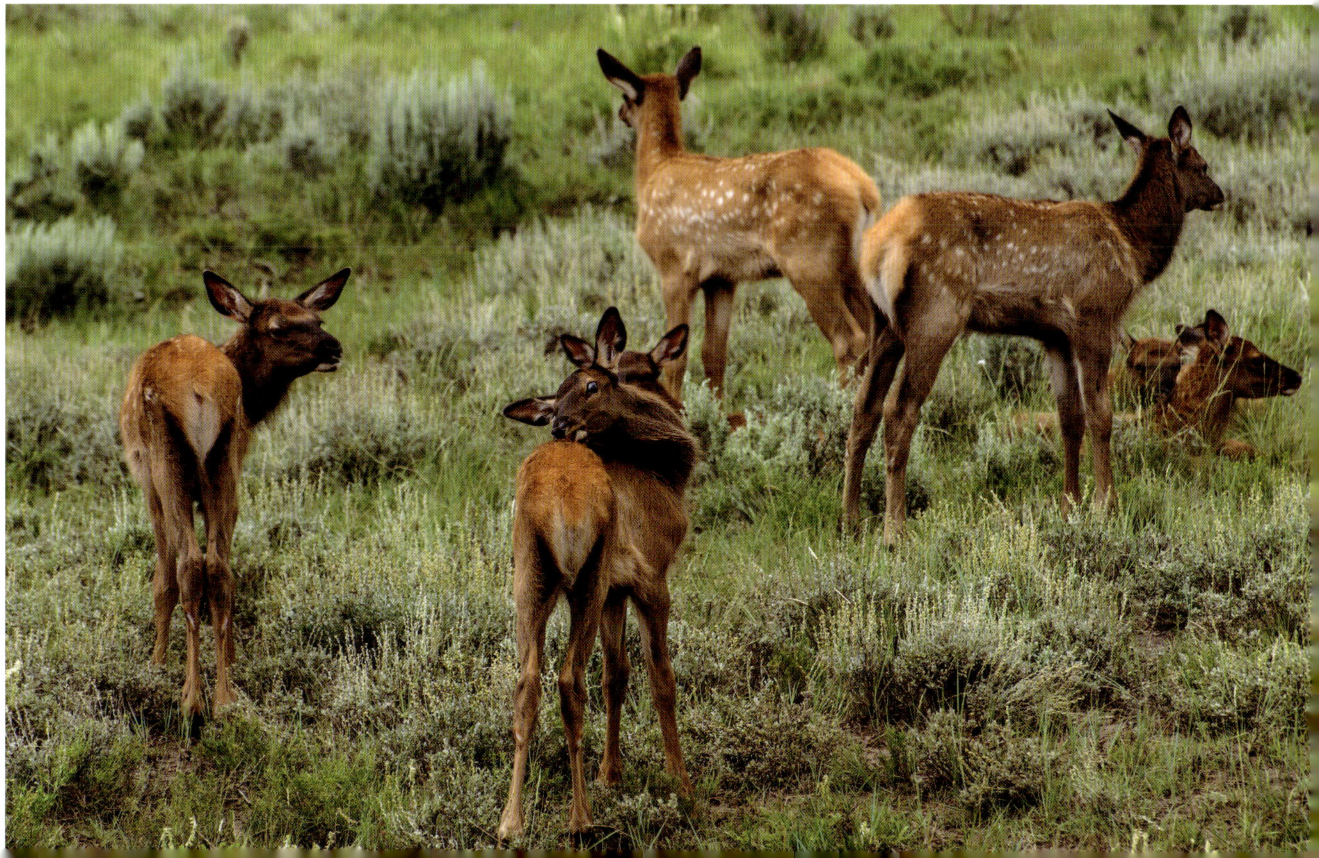

TOP RIGHT Crop conversion, where sagebrush has been removed, contrasts with native vegetation in northwest Colorado's sagebrush hills (Moffatt County).

BOTTOM RIGHT Elk calves in Piceance Basin sage (Rio Blanco County, Colorado)

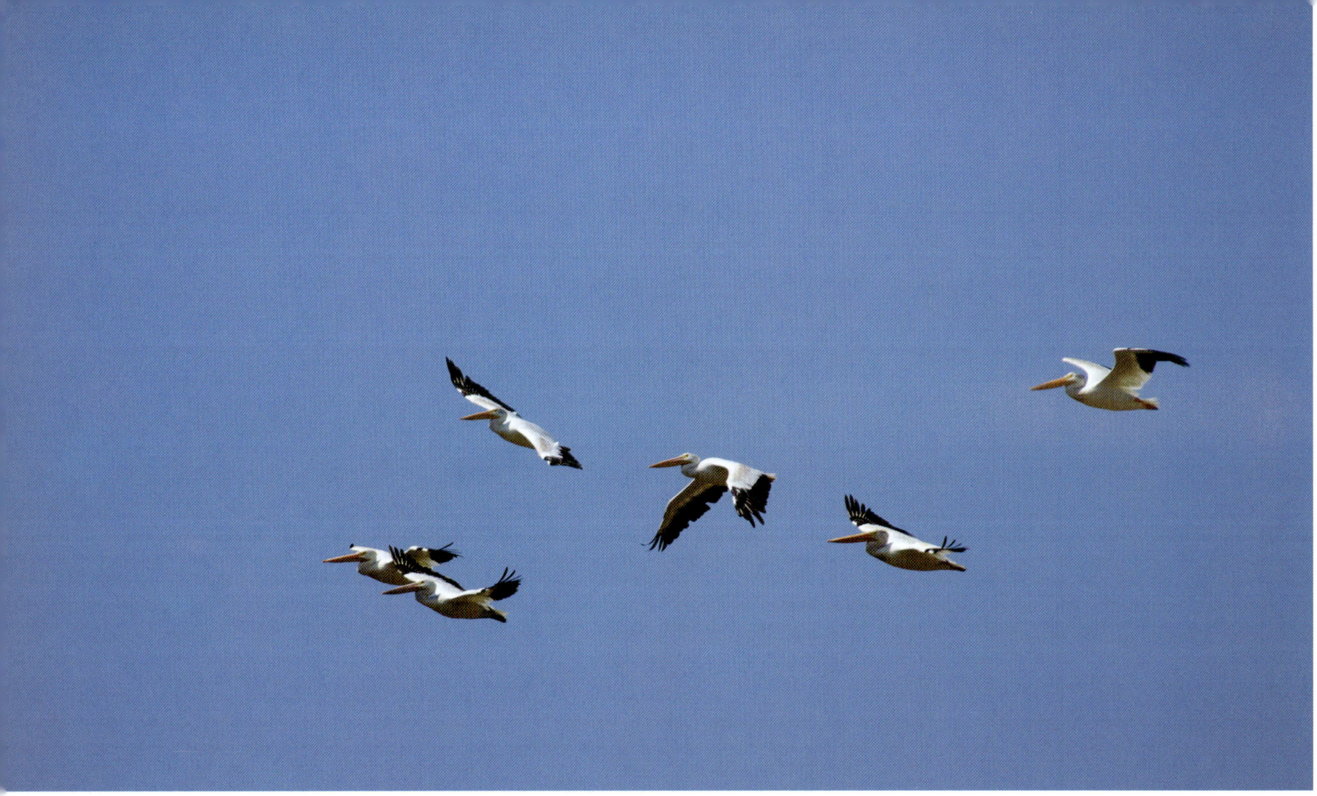

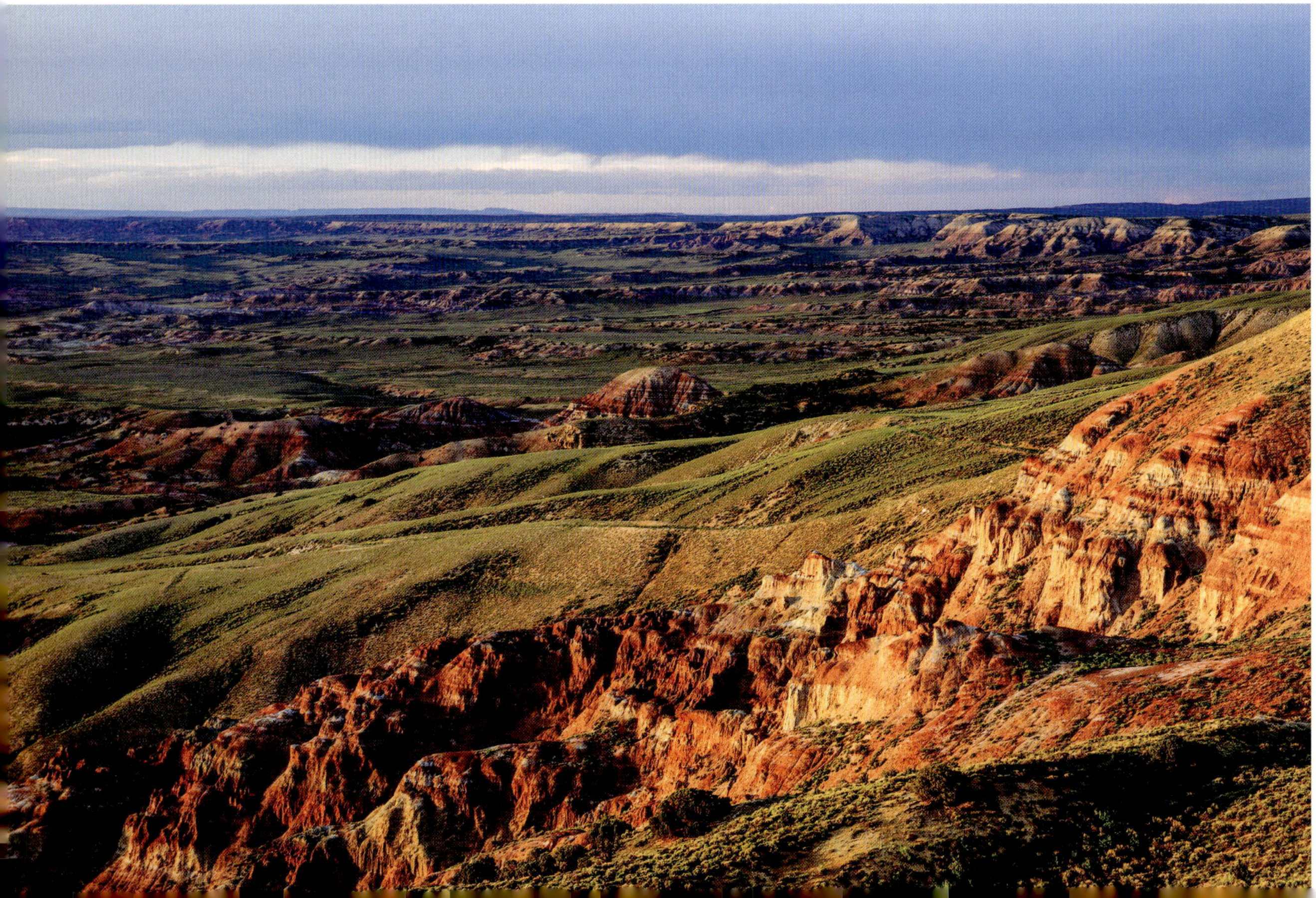

TOP LEFT American white pelicans in flight over Arapaho National Wildlfe Refuge, Colorado

LEFT Evening light settles on sagebrush badlands in Vermillion Basin, an area rich in scenic beauty, wildlife, and cultural resources in Moffatt County, Colorado. A collaborative grassroots conservation effort led to the Bureau of Land Management's 2010 decision to close Vermillion Basin to future oil and gas development.

OPPOSITE Confluence of the Green and Yampa rivers at Echo Park in Dinosaur National Monument, Colorado. Echo Park, named on the Powell Geographic Expedition of 1869, was a proposed dam site heavily contested by environmentalists in the early 1950s. The dam was never built, and the Colorado River Storage Project Act became law on April 11, 1956, stating that "no dam or reservoir constructed under the authorization of the Act shall be within any National Park or Monument."

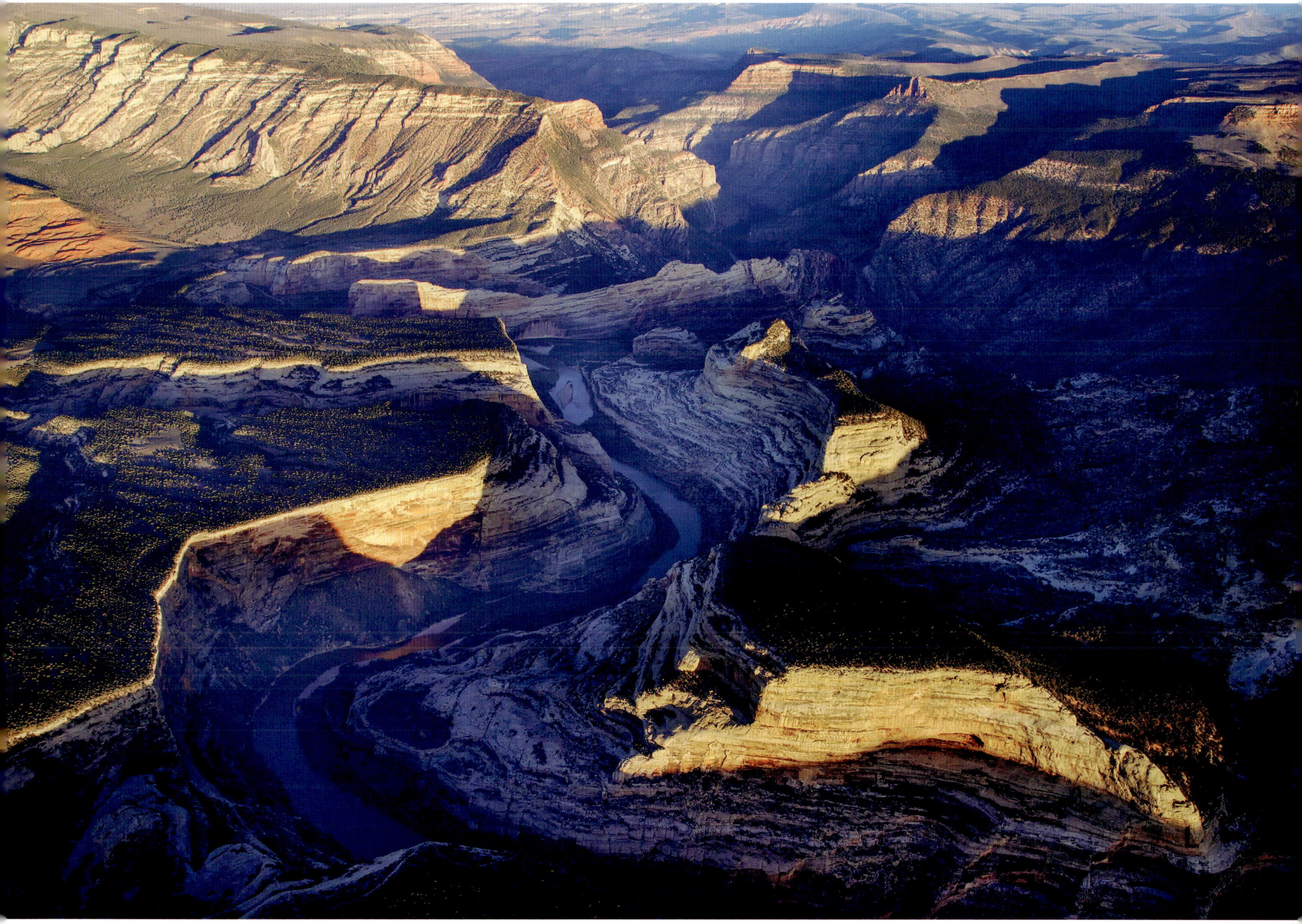

HEATHER STIRLING
PROTECTING COLORADO'S HISTORIC
ELK HEAD RANCH

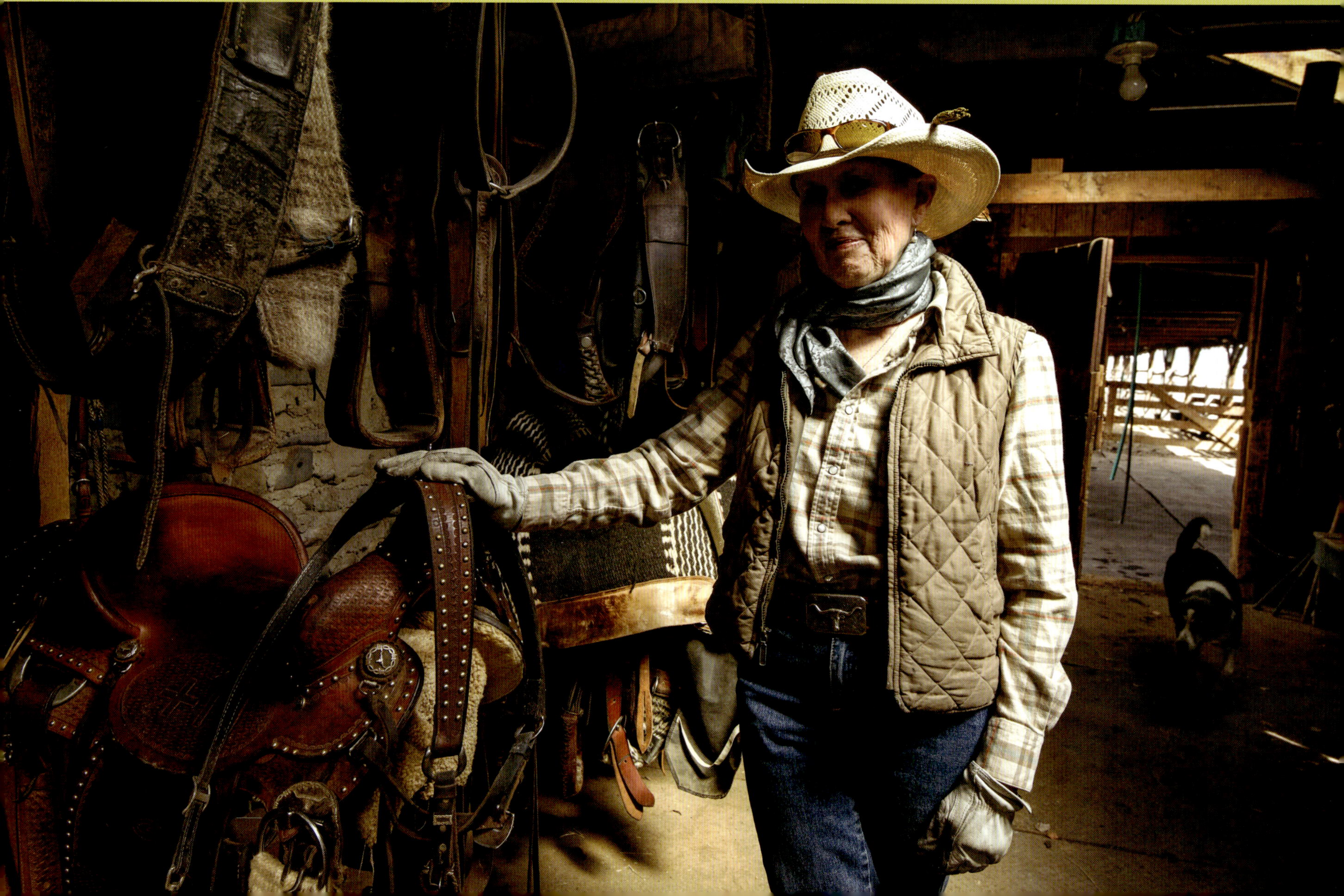

"This land, this place is very healing and spiritual."

COLORADO RANCHER Heather Stirling will leave the historic Elk Head Ranch better than she found it. Most of the 9500-acre (3845-hectare) working ranch is now under conservation easement, ensuring it will never be converted to a golf course or fancy housing development.

Stirling is one of more than one thousand ranching partners throughout the sagebrush ecosystem enrolled in the Sage Grouse Initiative (SGI) partnership, a Natural Resources Conservation Service program. The scale of protected lands is enormous, at 3.8 million acres (1.5 million hectares), or twice the size of Yellowstone National Park, with each SGI landowner helping to keep critical habitat whole. "Conservation easements, such as Stirling's, keep the vast sagebrush-steppe intact, where sage-grouse and working ranches thrive," said Tim Griffiths, the national coordinator for the SGI. More than one-third of sage-grouse habitat is on private lands that hold important brood-rearing habitat that are critical in the sage-grouse life cycle.

The Elk Head Ranch is situated at 6800 feet (2073 meters), 16 twisty miles (25 kilometers) above the town of Hayden, in northwest Colorado's Elkhead Creek Valley. The descent into the valley, backed by the rolling Elkhead Range and the appropriately named Bears Ears, a local mountain landmark, travels through dense cottonwood forest lining Elkhead Creek. When I first visited the ranch in April, winter white that lingered on Pilot Knob and Agner, Quaker, and Wolf mountains—which rim 180 degrees of skyline—made for a jaw-dropping vista.

Originally settled in 1883, the Elk Head became the foundation of Russian immigrant Isadore Bolton's sheep and registered Hereford cattle empire that stretched from northwest Colorado to Rawlins, Wyoming. The Elkhead homesteading community was made famous in Dorothy Wickenden's 2012 book *Nothing Daunted* about two society girls from Massachusetts who went west in 1910 to teach children in the new community. The legendary 1915 schoolhouse, made of blocky stone from hillsides above the ranch, and several other historic buildings grace the Elk Head Ranch.

Stirling and her (then) husband bought the Elk Head in 1974 and ran it together, raising three sons. When the marriage dissolved in 1991, Stirling assumed control of the ranch, with son Nathaniel now a working partner. The two "grow wild birds," offer elk hunting and bird-watching, and grow naturally raised beef cattle, wild mountain hay, and potatoes—a diverse working cattle ranch.

Thin and in full western dress topped with a cowboy hat, Stirling was raised in Michigan loving "guns, horses, and speed." She grew up "always on horses" and competed as a cowgirl mounted shooter. It was on a dude-ranch trip to Jackson, Wyoming, when she was eight that Stirling "fell in love with the West."

We hopped in a nimble ’80s ranch vehicle with no top and bounced around pastures, threading between Calf Creek, starting to swell with spring runoff, and wet meadows, where Nathaniel warned us about getting stuck. A dozen sandhill cranes, summer residents in the Yampa River Valley, flew from one wet pasture to another as Stirling proudly listed the range of wildlife that use the ranch.

“Big herds of elk are moving through now,” Stirling said, “and we have greater sage-grouse, sharp-tailed grouse, ruffed grouse, and dusky grouse. Did you see the turkeys on the way in?” She was worried about the beaver dams getting blown out on Elkhead Creek, now rising daily with the annual snowmelt. As we walked along the creek, Stirling picked up little bits of trash that arrived on the spring wind, as well as pieces of twine, saying, “I don’t know how these got here, but they’re bad for birds.”

With border collies weaving around our legs and retrieving sticks, Stirling spoke of hardship and well-being: “It’s really important for my peace of mind to have the harmony with people and animals. This land, this place is very healing and spiritual.”

In 2012, collaborating with the Colorado Cattlemen’s Agricultural Land Trust and the SGI, Stirling completed protection of the majority of Elk Head Ranch by placing an additional 3819 acres (155 hectares) in a conservation easement. SGI describes a conservation easement as a voluntary agreement, with built-in financial incentives that can include tax advantages, ensuring that a ranch will remain whole by limiting future development in perpetuity. It’s a way to keep ranchers on the land without subdividing, while stitching together protected lands with critical sage-grouse habitat that helps out a wide range of wildlife.

The easement offered some financial security for Stirling, who doesn’t think twice about making improvements on the ranch. Often quiet stewards, ranchers like Stirling won’t brag about what they’ve protected, but what’s good for rangelands is good for wildlife and the West. This conservation easement ensures the Elk Head Ranch’s rolling sagebrush hills, riparian areas, and priority habitat for the greater sage-grouse will remain intact forever. §

Indian paintbrush in spring at the Elk Head Ranch
in Routt County, Colorado

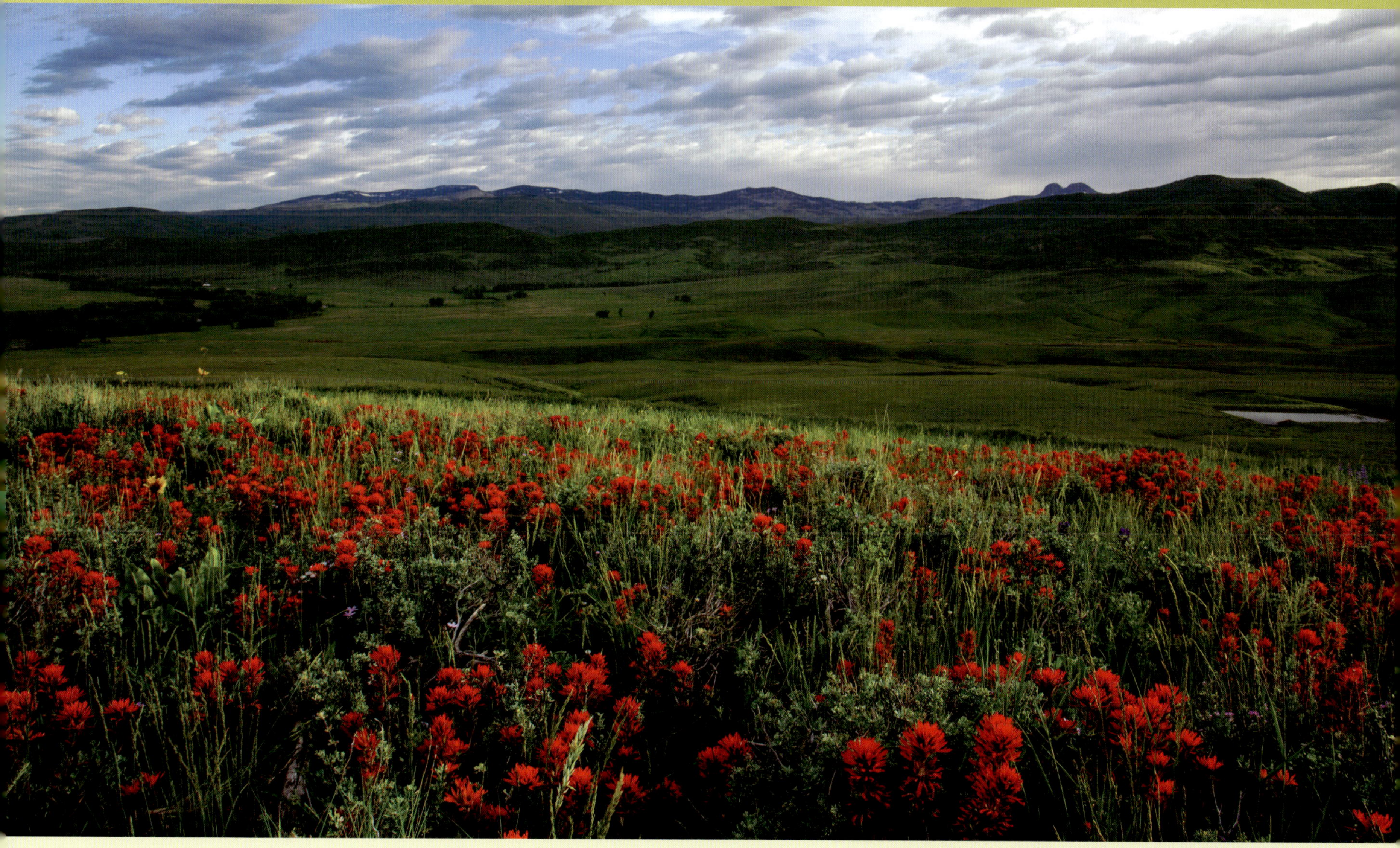

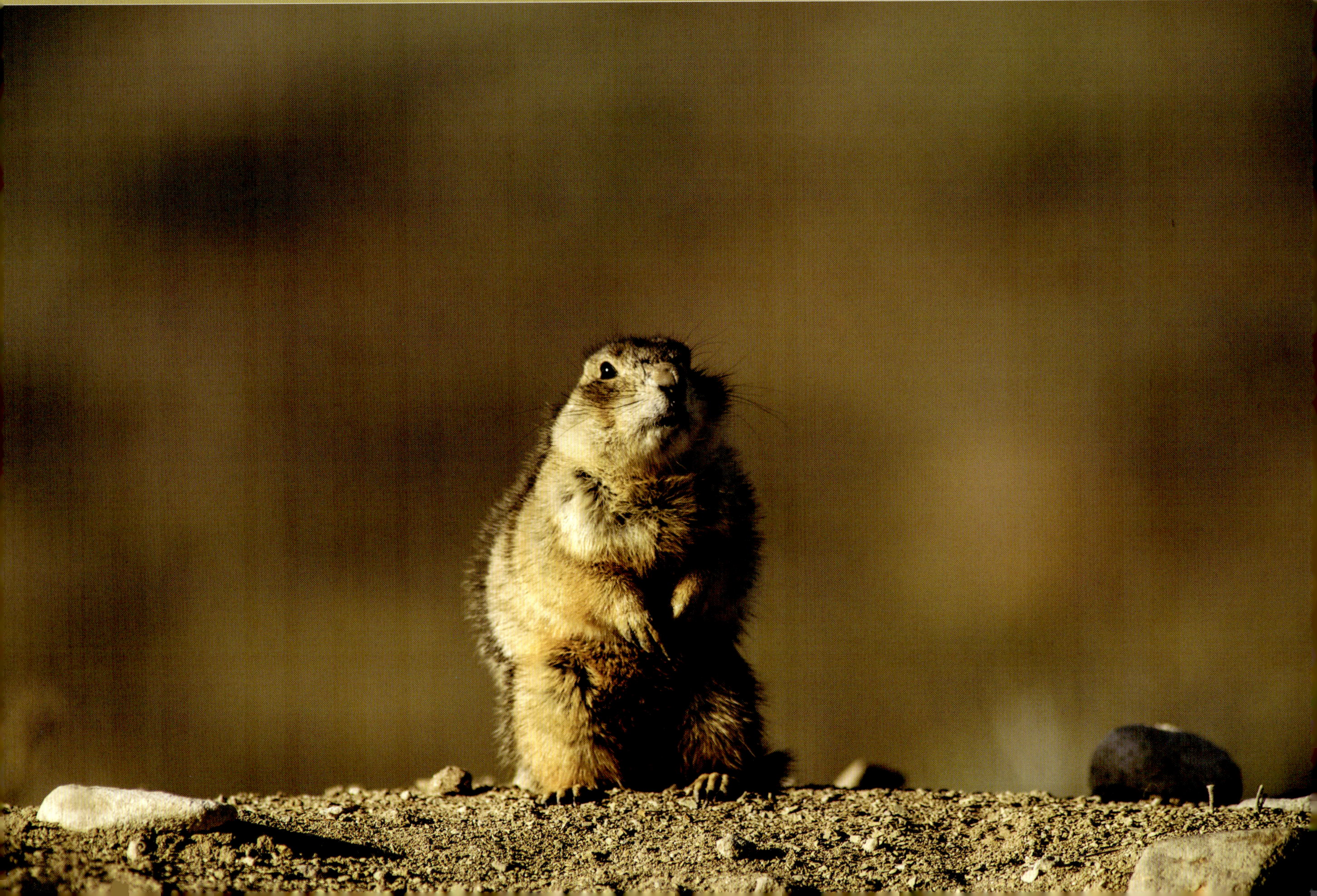

GUNNISON BASIN
COLORADO

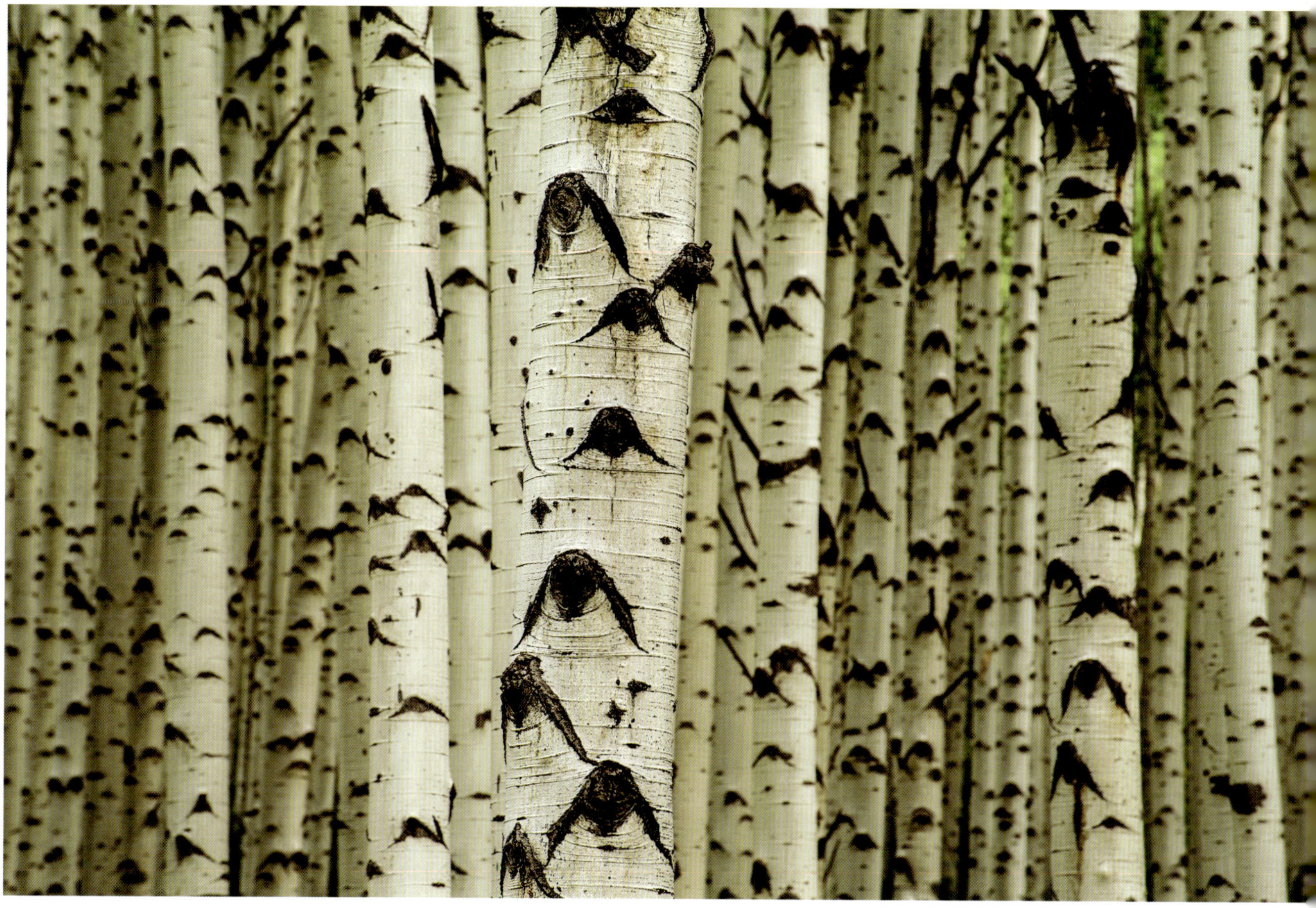

OPPOSITE Gunnison's prairie dogs are a colonial species that tunnel complex burrows for shelter at elevations ranging from 6000 to 12,000 feet (1830–3660 meters). The Gunnison's prairie dog is threatened by hunting, poisoning, and disease (Gunnison Basin, Colorado). RIGHT Aspen boles on McClure Pass, Gunnison County, Colorado

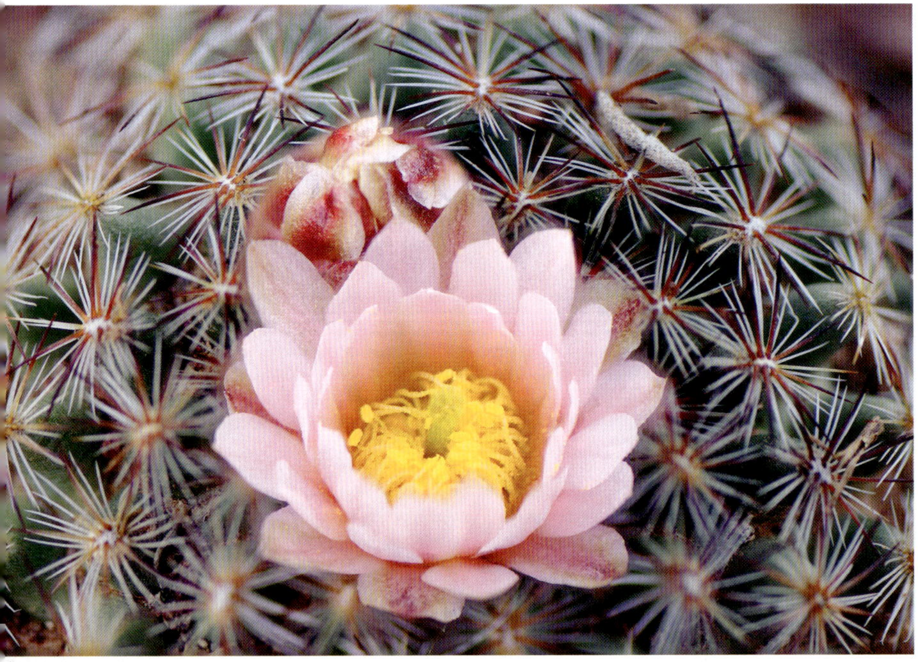

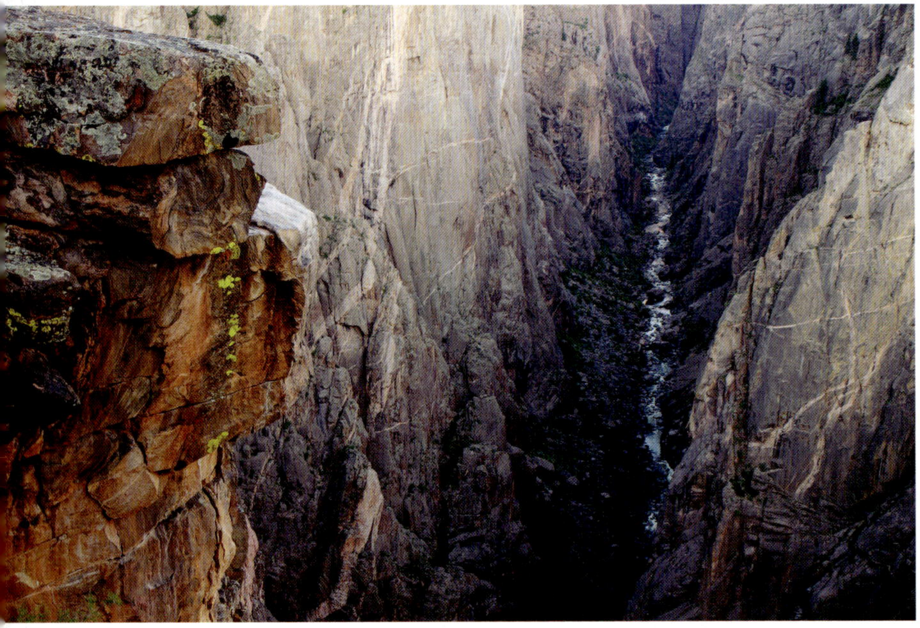

TOP LEFT A diminutive ball cactus blooms in mid-May with a flower smaller than a dime (Gunnison Basin, Colorado).

BOTTOM LEFT The Gunnison River flows 1700 feet (518 meters) below the north rim of the Black Canyon, where the river bottom receives only a sliver of sunlight daily (Black Canyon of the Gunnison National Park, Colorado).

BELOW A long-eared owl hunts in a small stand of aspen trees on a cold winter afternoon in Gunnison County, Colorado.

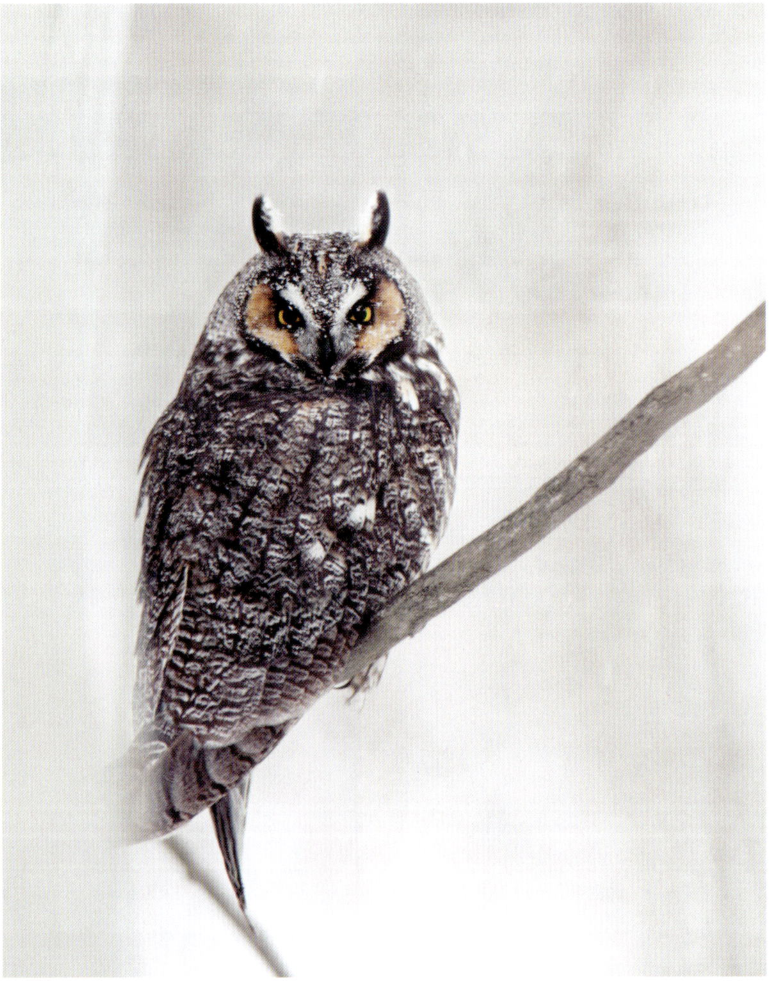

The frozen Gunnison River winds through Black Canyon, Colorado.

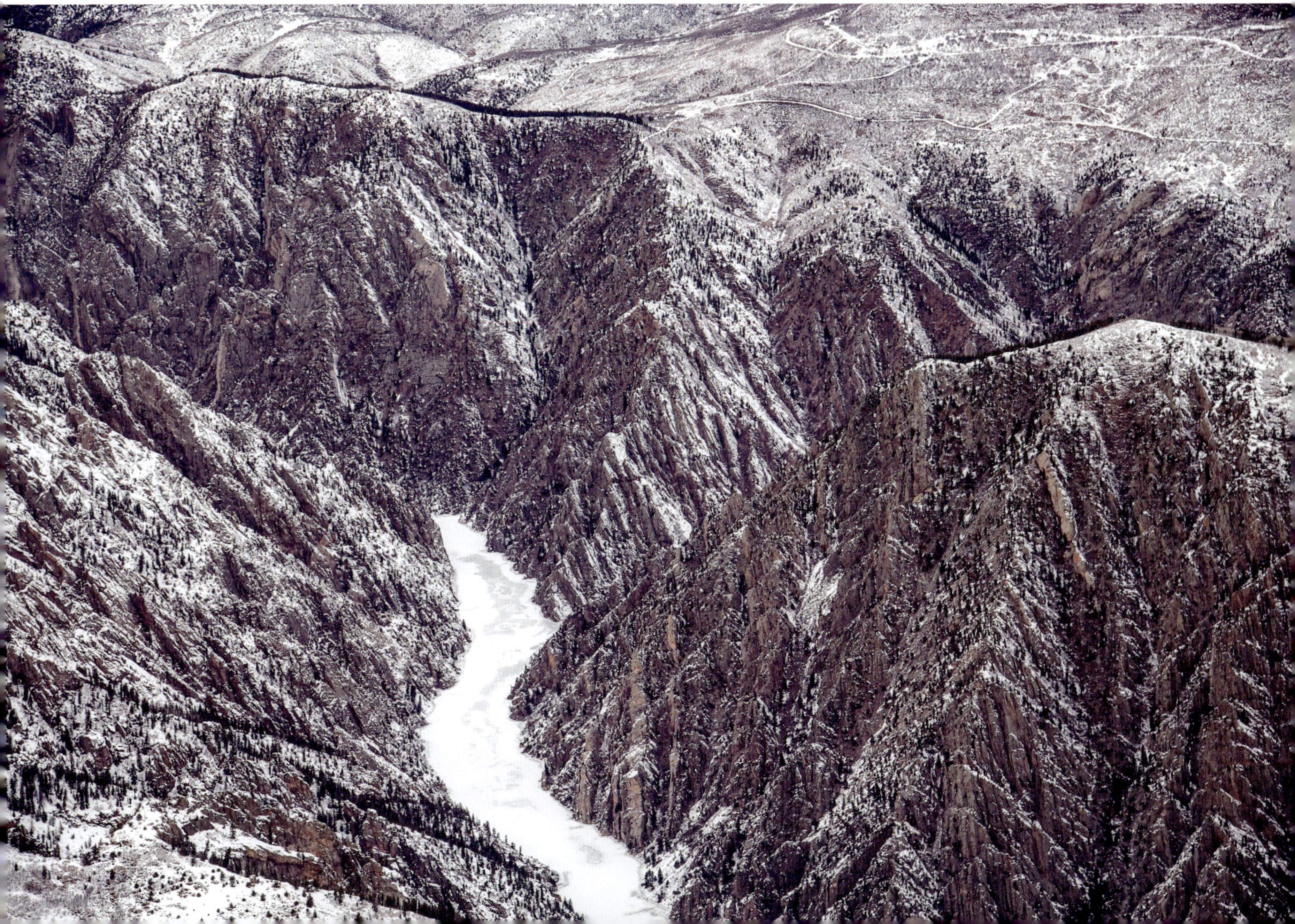

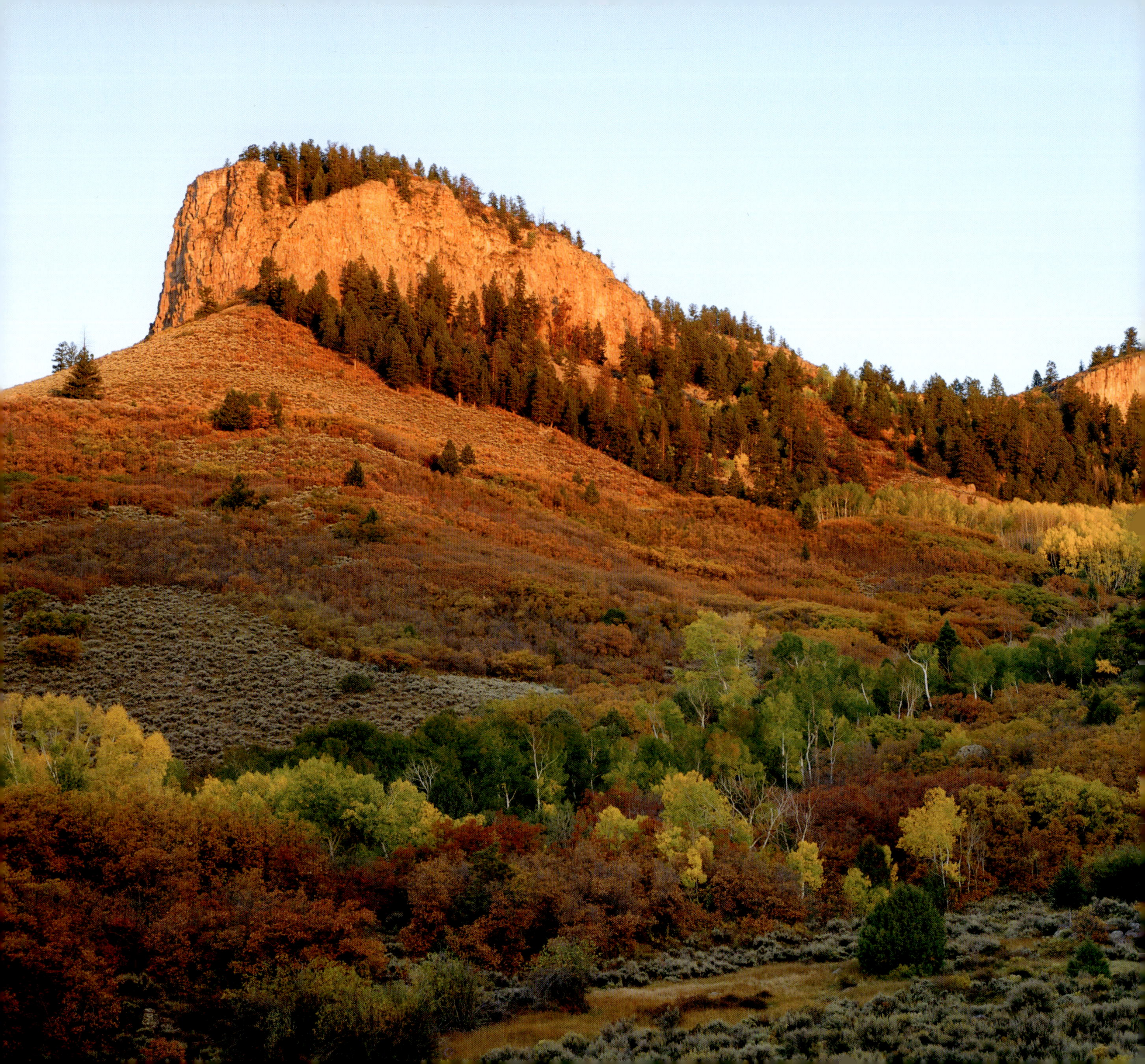

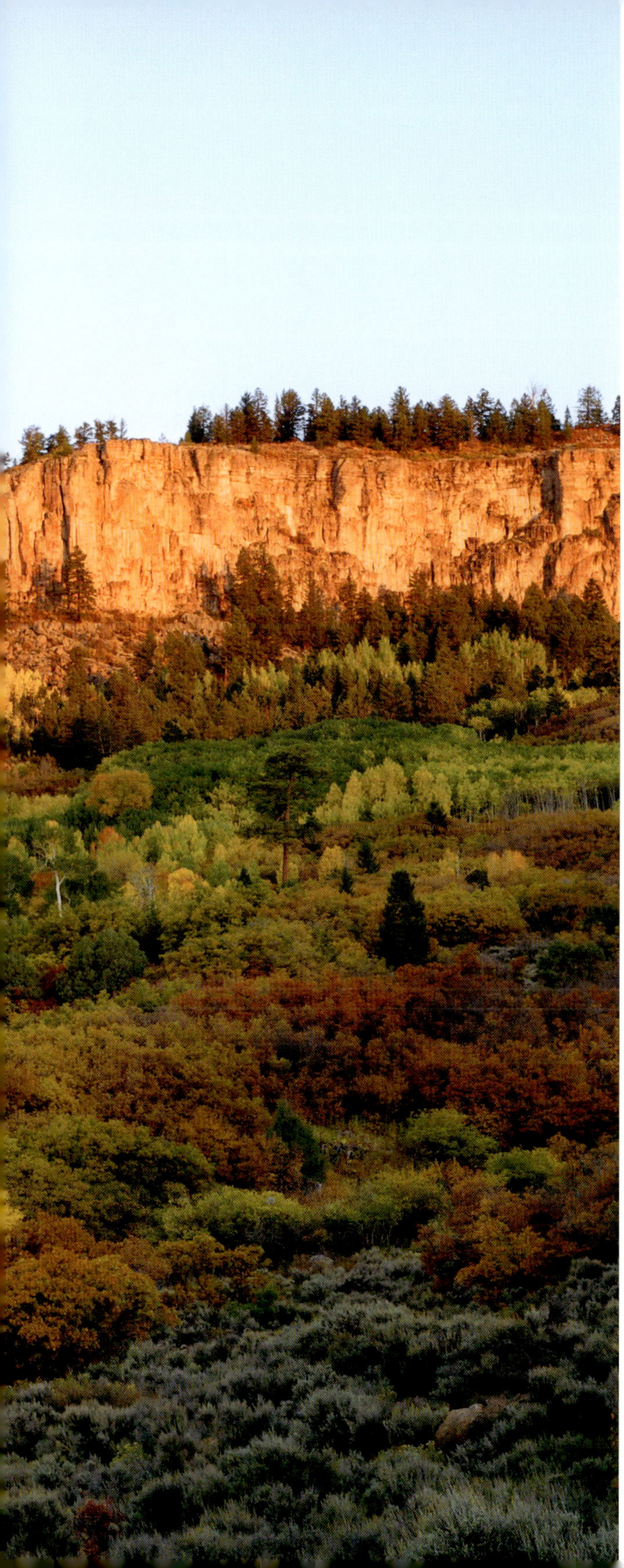

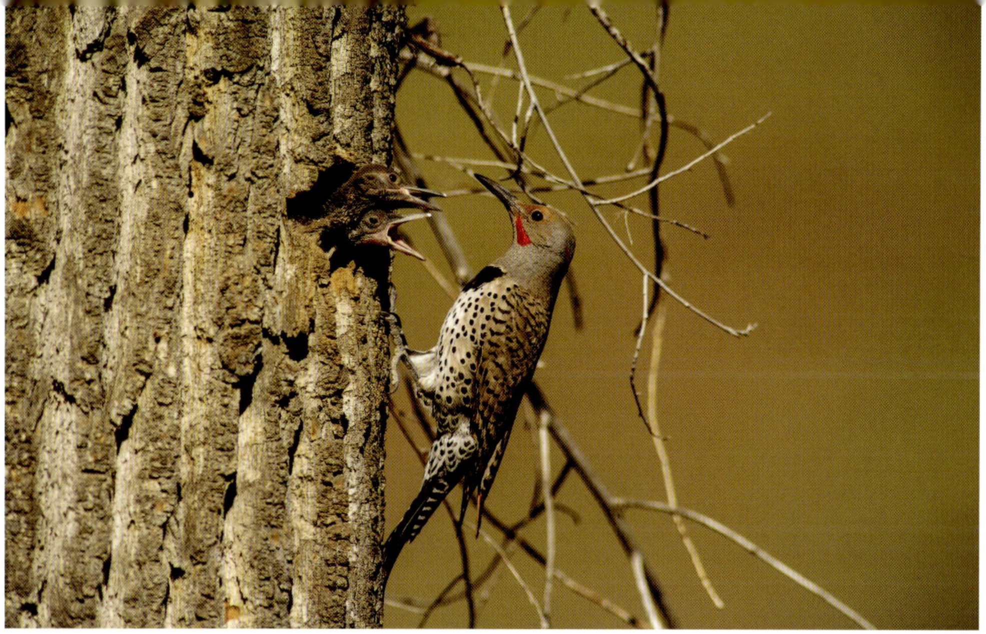

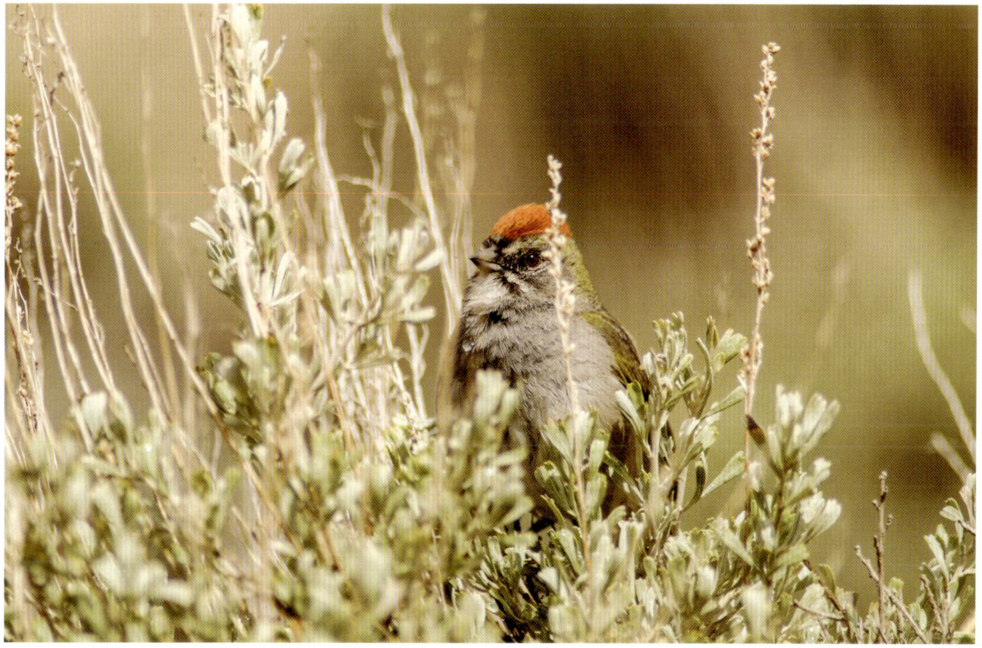

LEFT Black Butte is painted in warm hues by sunrise light, gold aspen, and red oak brush.
ABOVE, TOP A male red-shafted northern flicker feeds chicks at its nest cavity in a large cottonwood tree.
ABOVE, BOTTOM A green-tailed towhee male perches in Wyoming big sagebrush (Gunnison Basin, Colorado).

RIGHT White-tailed jackrabbit tracks in snow through sagebrush lead toward Cochetopa Dome, shrouded in clouds (Gunnison National Forest, Colorado).
BELOW White-tailed jackrabbits use winter camouflage to hide from predators in the snowy months.

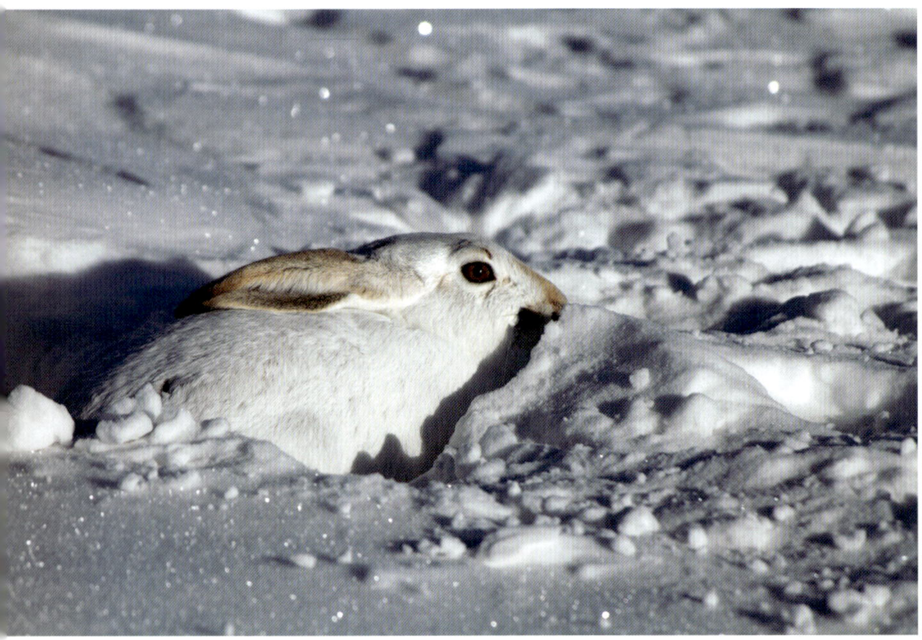

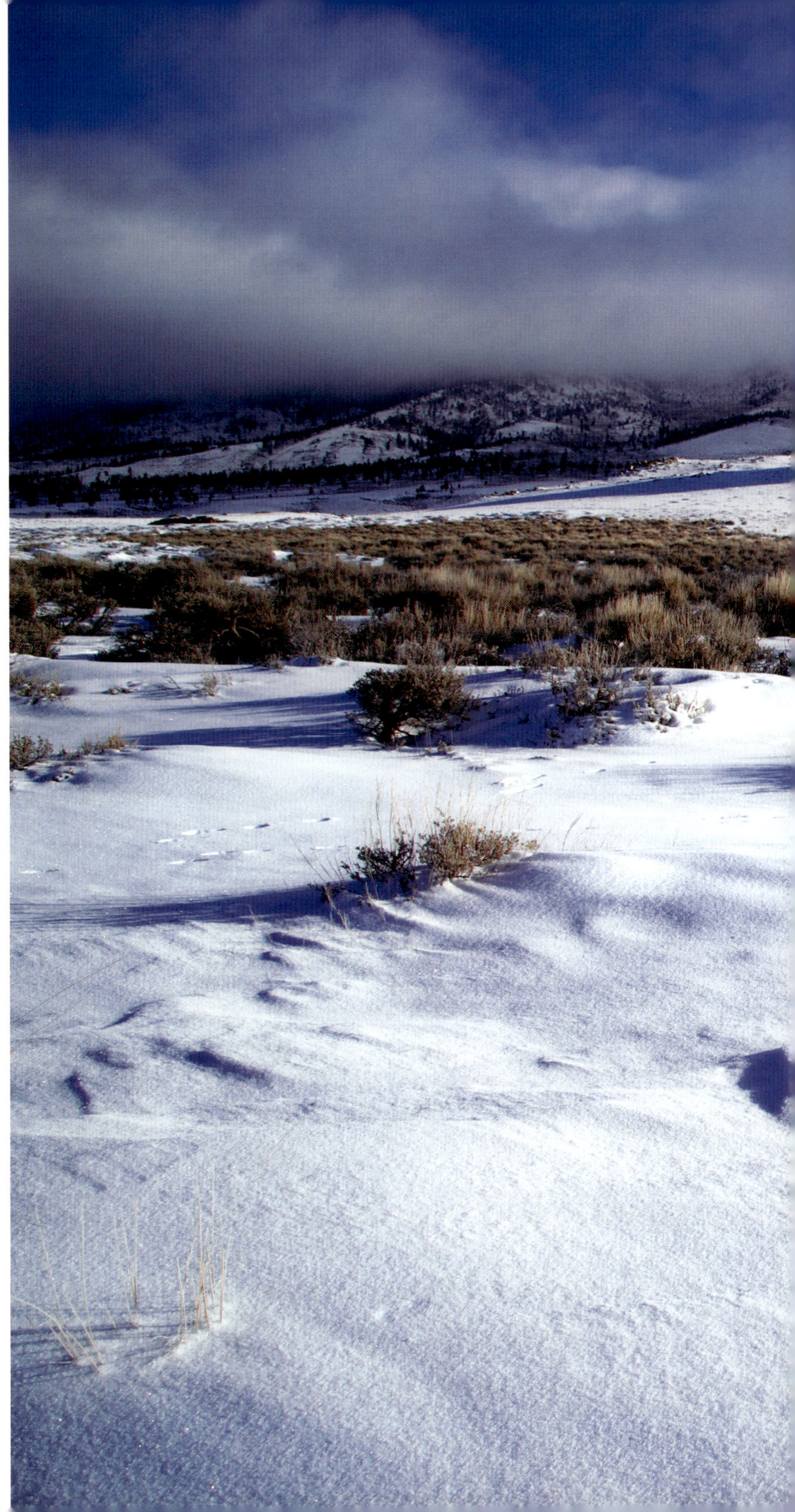

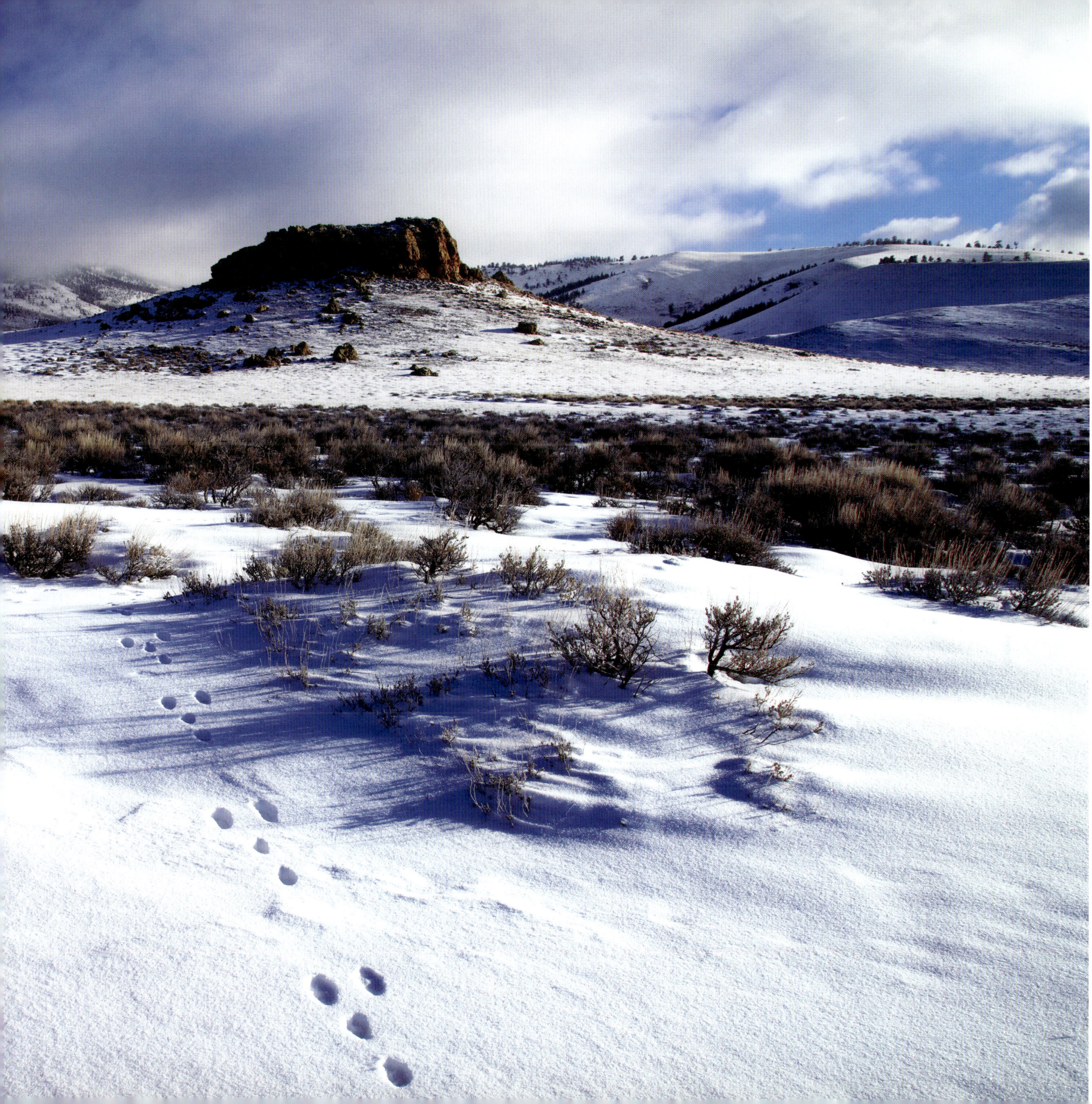

ROLAND McCOOK
PRESERVING THE UTE HERITAGE IN THE SAGEBRUSH SEA

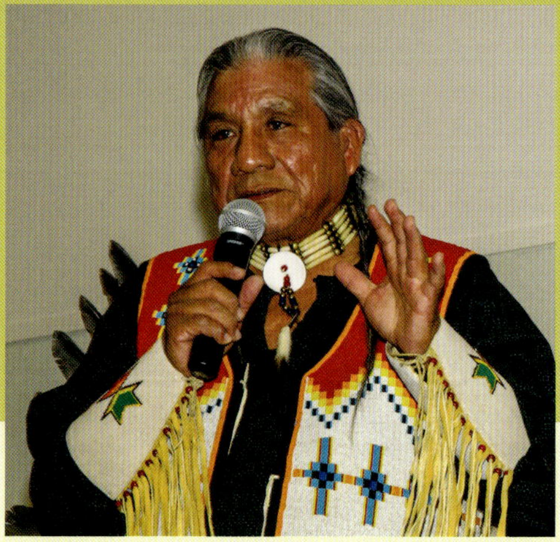

IN FULL NATIVE UTE Indian regalia, Roland McCook performs a sage-grouse dance, tapping feet, turning, tail feathers fanned. He wears the headdress passed down from his father, bright yellow leather leggings with dark rings, colorful beaded vest, moccasins, necklace, and armbands, his graying hair in braids. He chants in his native Ute language, unintelligible to this day's audience but powerful in its conviction. The Sage-Grouse Dance, performed by a number of western tribes, is inspired by the mating display of male sage-grouse, a species that the Utes hunted, mimicked, and revered for generations.

McCook then speaks of the Ute people, their reverence of the land and grouse, to an audience of academics, ranchers, conservationists, land managers, and folks interested in ongoing efforts to assist in the recovery of the sage-grouse. They listen to his impassioned message about keeping the Native American traditions alive.

Indigenous to the northern Colorado Plateau, primarily in western Colorado and eastern Utah, numerous bands of Ute people were hunter-gatherers, roaming the sage, mountains, and desert. Men fished and hunted elk, deer, antelope, bison, rabbits, and other small mammals and birds. Women gathered grasses, nuts, and berries, using a digging stick to collect roots of yampa, sego lily, and yellow pond lily. Wickiups, made of willow covered in brush, were portable housing. Bear root and sage leaves prepared as tea were among many plants with healing properties. The land gave all they needed; the Ute were *of* the sagebrush landscape.

The Brunot Treaty, ratified by Congress in 1874, led to the explulsion of the Ute Tribe that occupied all of western Colorado; seven bands in all were pushed from traditional lands. As the Ute People walked to a new reservation in Utah, white settlers staked claims behind them. Today the Northern Ute Tribe is located in Fort Duchesne, in northeastern Utah, the Southern Ute Tribe is in the town of Ignacio, in southwestern Colorado, and the Ute Mountain Utes are in Towaoc, Colorado.

McCook remembers sage tea being used for stomachaches, headaches, and other ailments, but many of the old Ute ways and language have been lost. His family participated in traditional Bear dances that celebrate the coming of spring, the midsummer Sun Dance—the most important Ute spiritual ceremony—and powwows.

When McCook was eight or nine, his father took him on his first deer hunt. They dressed the deer and took it directly to McCook's

"My work is a mission to preserve and teach the Native American spirit and to keep our image alive."

elderly uncle. Roland's father gave the deer to the uncle, telling McCook, "This is what you do with your first kill, so you're not selfish. Share what you have."

McCook is the great-great-grandson of Chief Ouray and his wife, Chipeta. Ouray, recognized as one of the great nineteenth-century Indian chiefs—statesman, peacekeeper, linguist—spoke Ute, English, Spanish, and Apache. Ouray sought peaceful settlement between the Utes and encroaching white people, made numerous trips to Washington, DC, to negotiate for his people, and met US presidents Lincoln, Johnson, and Grant.

In his family's tradition, Roland's father told him when he was seventeen, "You should follow the footsteps of the white man. When you learn to talk and walk, then come back to teach your people."

And so McCook attended college in Utah, his colorful career spanning different worlds: civil engineer for the Bureau of Indian Affairs (BIA), fire manager for the Bureau of Land Management, realty manager in the BIA, director of housing and development for the Ute Nation. In 1995 he was elected to the tribal government, serving nine years, including two years as chairman. Afterward, he spent eight years at the Smithsonian Institution returning Native American human remains and sacred objects to their tribes. While at the Smithsonian, McCook "took full advantage of the National Archive" by studying Ute history.

Roland McCook never forgot his father's words, and in 2011 he became a full-time lecturer and motivational speaker. Today he lives in Montrose, Colorado, in his people's ancestral lands. He teaches audiences of all ages and races about the Ute people and Native American history, not willing to "let the Ute people fade away into the past and be forgotten." He says, "My work is a mission to preserve and teach the Native American spirit and to keep our image alive." Having experienced racism at a cultural program at the University of California–Berkeley in 1961, he also teaches Native American youths by example to "never use your color as an excuse."

When we parted, McCook was headed to Aspen, Colorado, to give a Ute blessing at the "peace pole" presented to John Denver in the late '80s. McCook made white sage smudge sticks for the blessing and, in the tradition of Chief Ouray and Chipeta, he would use sagebrush smoke to carry his message of peace and tolerance, love of the land, and honoring heritage to the people who live today in the beautiful Roaring Fork Valley of his homeland. §

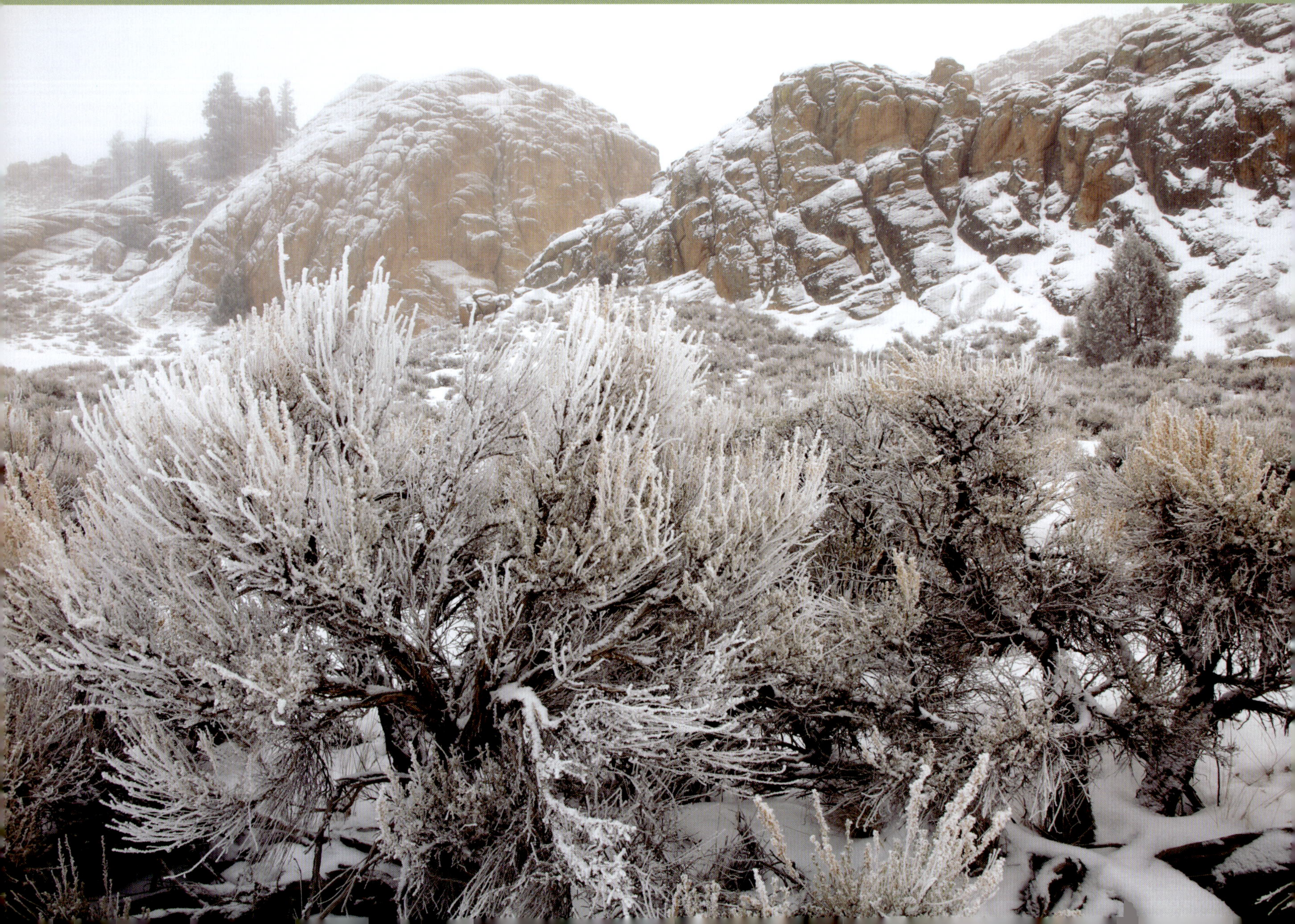

EPILOGUE

Rick Bass

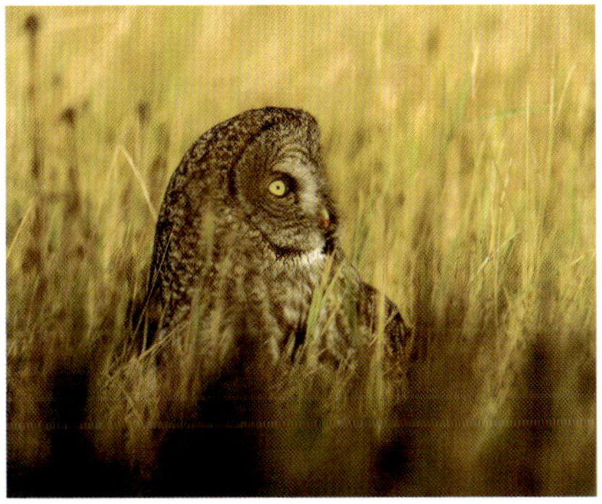

I LIVE IN MONTANA, over on the west side of the Continental Divide—the wet side, where in my green valley it's been raining steadily, even as just a couple of valleys east, a quarter-million-acre wildfire howls, consuming houses and people, a world in tumult.

Two days ago a solitary sandhill crane flew croaking upriver. They're rare, up here in the rain forest where I hide out, writing, reading, hiking. They come and go, from one stronghold to another.

I know where they come from. Like so many supremely crafted species, they are part of the landscape: stalking, probing, piercing the land with their long bills, sometimes inches deep, hunting, redistributing nutrients in their travels. That which they consume, they expend—nothing more, nothing less—and in their wake, they leave beauty and music.

There is nothing in nature that is not connected to everything else. But in our reaching ever deeper, we are changing the equation of the world for ourselves and all else. We are tipping.

The sandhills, improbably, remind me of the sea: not any ocean of today, but the old inland sea of several hundred million years ago. It's still out there, just dry now—petrified, though

ABOVE Great gray owl on prey (Jackson Hole, Wyoming)
OPPOSITE Fresh snow and fog cover sage and stone at Hartman Rocks, Gunnison Basin, Colorado.

not lifeless. Quite the contrary: covering eleven states and nearly 166 million acres (67 million hectares), the inland sagebrush sea is home to a hundred species of birds and all manners of other animals, which this book explores. Before I came to the center of this wet green place, I lived in Utah, in the center of the inland sagebrush. It's where I first fell in love with the West—the most wonderful kind of awakening. Back then, I perceived myself to be in the center of something—wilderness—rather than at the edge, which seems a natural enough response when something as tiny as a human being finds him- or herself in any new environment, but particularly one as spacious as those of the American West. If you can't see the edges, where else can you be but in the center?

After a while, however, a pilgrim becomes more acclimated to his or her surroundings: takes in the scent of things, hears and sees more, gets out and walks around, and comes to understand that on the blue planet, there are borders and boundaries to all things, and that as wild country becomes fragmented and ever smaller, there are ever more edges and ever fewer centers. Wild country becomes smaller. We've all witnessed it: another road, another parking lot, another pipeline, another pit, another hydraulic fracturing site, another ten thousand fracking sites—and the garden that we began with is defiled further.

Nonetheless, the garden's inhabitants—ourselves included—still wander these smaller pockets, these places where the edges collapse inward upon what was once a greater—even vast—expanse of western space.

What stitching, then, might hold together these smaller pieces? The paths of the travelers themselves, perhaps—the weavings and careful wanderings of cranes, mule deer, humans, pronghorn, short-horned lizards, and the curious and specific lives of sage-grouse, masterful umbrellas for all the natural history that lies beneath their wings in a land where rain is scarce and where, even underground, there is a finite amount of water. If the country is still healthy enough to hold a sage-grouse, it is likely healthy enough to hold much else, maybe even everything else. The canary in the coal mine is the grouse in the sage.

What a beautiful paradox, that this eleven-state basin of sagebrush country possesses so little water, vital water, where once there was an entire sea's worth. The water is long gone, but still remaining are the most pleasing shades of blue: blue of frost, feathery blue-green of sage, blue of nightfall, blue of turquoise, blue of hope.

The sweet, dry sea over which the sage-grouse presides—one of the most uniquely adapted creatures in this landscape of the petrified and desiccated old ocean—still teems, though what is fully at stake may not yet be entirely visible to us. As the margins of the big wild country needed by sage-grouse and so many other species contract, however, we are starting to get a clearer sense. There is still much to lose; there is still much to hold on to. It's as if sometimes we can't see what's right in front of us or we assume that it's our birthright and will be here forever, with no need on our part to tend to or caretake that treasure.

Is there time? I don't know. I hope so. And I think this is an informed hope, not a false one. The photos and essays in this book comprise a spirit that is borne of the landscape Wallace Stegner called the geography of hope. Certainly, it is the landscape I am most familiar with, though this is not to say hope does not also have in it elements of pain and fear. In these warming days, so many vital landscapes exist at the precipice of loss or, beyond loss, extinction. It's easy to become numb, fatigued. It's easy to give up. It's important not to. The condition of life demands our full effort in order to best honor it.

Why does this pale-blue dry sea matter? In my mind, there is no end to the reasons, but the language of defense can itself be dry: *ecosystem services, ground water storage and filtration, species diversity and distribution, carbon sequestration,* whatever.

What I find myself thinking about most often are two things: *Does beauty still matter?* And, *are we truly a moral animal or only when it suits us to be so?* It's easy to give lip service to the rights of other beings, so long as that existence does not compromise the pace of our pell-mell rush to get to wherever it is we think we want or need to be, with our own hungry paths rushing across the crafted trails of ancient others.

We can see the edges of sagebrush loss—so many roads criss-crossing the desert, so much noise, so much dewatering now—like never before.

We need to hold on to whatever centers—large unbroken areas of sage—there still are.

I look at these photographs, I read these essays, and I remember the way the sage country smells and looks in first light. In many ways, the sharpness of senses felt there—the scent of the inland sagebrush sea, and particularly the way sounds such as crane grate, lek boom, raven call carry there, with all else silence: no rig clatter, no highway groan, no sonic engine-roar, only the pulse of the old blue sea awakening each morning—gives me hope. There is work to be done, but where there is love, there is also always its inseparable companion, hope.

Things are happening too fast, these days, out in the dry old blue sea. We need a time-out. We need to think things over. We need to buy back some of the rarest thing, *time,* for every living thing out there, ourselves included. §

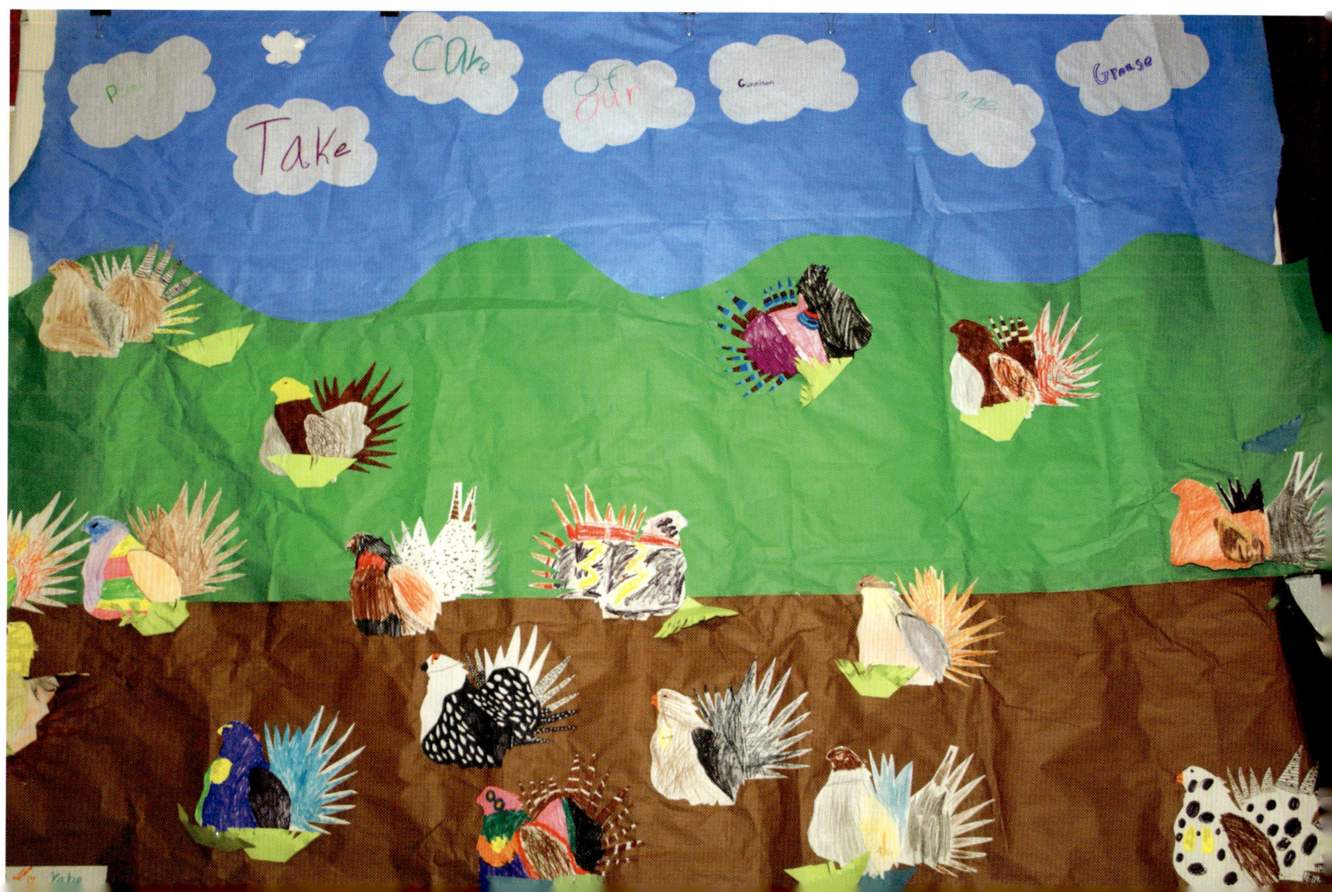

"Please Take Care of Our Gunnison Sage Grouse" reads a mural by Brownie Troop No. 10512 at the 2011 Gunnison Sage Grouse Summit. Saving the namesake grouse is a community-wide effort throughout the Gunnison Basin, Colorado.

PHOTOGRAPHER'S NOTES

HUDDLED AND HIDDEN in a portable camouflaged hunting blind, I'm completely surrounded by Gunnison sage-grouse, one of North America's most threatened birds, loudly displaying on fresh April snow. At the invitation of the Gunnison Sage Grouse Strategic Committee, I've been allowed two mornings to photograph these imperiled birds on their lek. I can feel my heart pounding with anticipation when I finally see the birds strutting in blue predawn light. I guide my lens slowly to make the first photograph, and suddenly more than a hundred birds explode from the lek as a golden eagle glides a few feet off the ground, then right over my blind, its wings covering the width of the flimsy structure.

The interplay of golden eagle and grouse is all part of the natural process, the top avian predator seeking an easy meal under the cover of near-darkness. While there's nothing more thrilling than seeing a soaring golden eagle at close range, watching the rare grouse disappear before I could make an image was immensely frustrating. My day was over before first light—and the same thing happened the next morning. In my first field season of this project, I left Gunnison, Colorado, with no images. Gunnison sage-grouse became my grail bird, and each spring I made a pilgrimage to Gunnison and to a satellite lek near Telluride, Colorado, where just a few birds were hanging on.

At the beginning of this project, I had only an idea and a partly developed project plan, so I reached out to Brian Rutledge and Alison Holloran at Audubon Rockies and then a number of western conservation groups doing great work in the sagebrush ecosystem. Pointed in the right direction, and inspired by the western landscape, I set out to make images.

Telling a photographic story in a dynamic landscape naturally requires a variety of approaches and tools. I begin by visualizing a photograph or series of images. That mental picture then sets a whole process in motion: a study of animal behavior, habitat, seasons, and logistics because everything happens on a seasonal clock. Then I develop a field plan, often to locate a small creature in a massive area. Portable blinds are an essential tool to make images without disturbing the behavior of some creatures like sage-grouse and songbirds. I photograph from my vehicle too, using the same principle of blending in, where wildlife are accustomed to seeing cars.

My primary photo tools are Nikon 35mm cameras, with Nikon lenses ranging from 14mm to 600mm, as well as a couple of smaller point-and-shoot cameras. Camera bags, tripods, tent, camping gear, duffels, and all the bits and pieces of gear were stuffed in my truck, which started with 28,000 miles (45,000 kilometers); it's still running strong at 172,000 miles (277,000 kilometers) today. I often shoved everything to one side and slept in a narrow space in the back of the truck, which offered some comfort in grizzly bear country; occasionally I set up a two-person tent, rarely camping in the same spot two nights in a row.

At the beginning of this project, I was a neophyte aerial photographer and ultimately flew eight missions with five LightHawk pilots, each committed to making sure I made the images I was after. The perspective of flight changes the way we perceive the land. With the grand-scale aerial view, you can see how the landscape is sewn together, where roadless areas connect, and how fragmentation chops up the land. I also set camera traps that are designed to be triggered by wildlife crossing an infrared beam; these traps haven't worked out so well for me—yet. This land and its wildlife teach humility, patience, and persistence.

Most of my work is straightforward photography: carrying a photo backpack and tripod into the sagebrush to make images, usually with a plan, other times while chasing light. Traveling on foot, skis, snowshoes, horse, and mountain bike, I've seen this ecosystem reveal itself in ways I never could have imagined. This sagebrush study has taught me to just be, in one place, a speck in a giant landscape waiting for a defining moment.

A point must be made about wildlife photography. I don't photograph captive animals raised in game farms, and I personally believe that raising animals in game-farm cages for the pleasure of photography is unethical and without conservation value. The wildlife represented here is but a snapshot of a functioning ecosystem. Each species is a role player. It's my hope that as we better understand complex wildlife relationships, we'll value animals equally and stop blaming predators for human-created problems.

Building the story over six years has been about establishing alliances, photographing on assignment, addressing urgent issues, and understanding local views about our federal lands—all while keeping an eye on the big picture. The project has to be grounded in good science, and middle-of-the-night sage-grouse trap-and-release excursions established a solid foundation for fieldwork with grouse, songbird, reptile, aquatic, and ungulate biologists. Colorado rancher Duke Phillips once told me, "All these lands are stitched together by people." I've gained great inspiration from folks who dedicate their careers to studying species, ranch sustainably, and protect and write about the sagebrush ecosystem; a number of them are featured in this book.

In the last field season before completing this book, I finally made the Gunnison sage-grouse images you see here: photographs that I had pictured in my mind's eye throughout the project. I've learned that time, patience, persistence, failure, and—perhaps most important—collaboration can carry a project through seasons, even years. In the field, I have developed a mantra of "stay until you learn something" to temper my impatient tendency to move on when things aren't working out as planned. The mantra has served me well photographically and as a metaphor to listen, learn, and let this project unfold naturally.

I sometimes wonder what I'll say when someone asks me why I do this work, why I chose the sagebrush ecosystem for a long-term project. I'm a westerner first, and part of the answer is wrapped up in how we choose to see the West: as more than lines on a map, and on a grand scale—managing and valuing the land accordingly, sustainably. Culturally, we made a value judgment long ago that wildlife should be protected from extinction and allowed to roam freely across the landscape, that extraordinary places will be free for all to enjoy. All Americans own the same deed to much of this land. Those values matter, as does love of the land, and I simply want my grandkids and your grandkids to have the opportunities we've had. It's no more complicated than that. §

For more information about Braided River's Sage Spirit project, visit www.sagespirit.org.

ABOUT THE CONTRIBUTORS

Rick Bass is the author of thirty-one books of fiction and nonfiction, including most recently a novel, *All the Land to Hold Us*. His work has been anthologized in *Best American Short Stories*, *Best American Travel Writing*, *Best Spiritual Writing*, and *Best American Nature Writing*. For more than twenty years he has been active in helping to protect the last roadless lands of Montana's Yaak Valley—the lowest-elevation, wettest habitat, and narrowest bottleneck of the United States' section of a critical migration corridor that stretches to the Yukon. He lives in the Yaak Valley, where he is a board member of the Yaak Valley Forest Council (www.yaakvalley.org).

Kris Fulsaas grew up in central Washington, roaming the sagebrush steppes of the Columbia Plateau on foot, horseback, and bicycle. She has worked on several Braided River books, including *The Last Polar Bear*, *Planet Ice*, and *The Wild Edge*, as an editor, project editor, and writer. She has edited many books for Mountaineers Books, as well as other book publishers, magazines, and companies large and small. An award-winning editing instructor for the University of Washington and the Editorial Freelancers Association, she resides in Seattle.

Patrick Magee is a wildlife biologist at Western State Colorado University in Gunnison, where for eighteen years he has taught courses in ecology, ornithology, mammalogy, wildlife ecology and management, and Colorado ecoregions. He is director of the Thornton Biology Undergraduate Research Program and develops collaborations with state and federal conservation agencies to study wildlife ecology. He has been a member of the Gunnison Sage-Grouse Working Group and serves on three subcommittees of the Gunnison County Sage-Grouse Strategic Committee. In 2000 he formed Sisk-a-dee (www.siskadee.org), a nonprofit organization that focuses on conservation of the Gunnison sage-grouse and the sagebrush ecosystem. He lives in Gunnison with his son, Rory.

David Allen Sibley, son of ornithologist Fred Sibley, began watching and drawing birds at age seven. Since 1980 David has traveled throughout North America in search of birds, both on his own and as a leader of bird-watching tours. His comprehensive guide to bird identification—*The Sibley Guide to Birds*—was published in the fall of 2000, the updated second edition in 2014. Other books include *The Sibley Guide to Bird Life and Behavior*, *Sibley's Birding Basics*, *The Sibley Field Guide to Birds of Eastern North America*, *The Sibley Field Guide to Birds of Western North America*, and *The Sibley Guide to Trees*. He is the recipient of the Roger Tory Peterson Award for lifetime achievement from the American Birding Association and the Linnaean Society of New York's Eisenmann Medal. David lives in Concord, Massachusetts. You can learn more about David's work at www.sibleyguides.com.

Todd Wilkinson has been writing about the environment, politics, culture, and the arts for most of the last thirty years. Most recently, he authored the critically acclaimed biography *Last Stand: Ted Turner's Quest to Save a Troubled Planet*, which was distributed to every member of Congress and to United Nations ambassadors from every country on Earth. Over the years, Todd's work has appeared in two dozen magazines and newspapers, ranging from the *Wall Street Journal* and *US News and World Report* to the *Christian Science Monitor* and *Audubon*. He is the author of several books, including *Science Under Siege: The Politicians' War on Nature and Truth*. A noted expert on wildlife art, he is also editor and founder of the online magazine *Wildlife Art + Nature Journal*. He lives in Bozeman, Montana.

ACKNOWLEDGMENTS

I often think of the West as a small, extended community with folks pitching in to do their part. It would be impossible to express my gratitude to all of the people who've informed and supported this project, and I'm humbled by the generosity of so many who have openly shared what matters most to them. Thank you to my wife and best friend, Marla, whose love, support, banana bread, and little notes and photos in my duffel smoothed out many bumps in the road on this journey.

Extended fieldwork requires resources, and I offer my sincere thanks to friends, family, and organizations who supported this work, especially "Effie" Norris, Patagonia, Sisk-a-dee, and Laramie Audubon Society.

Thanks to friend, colleague, and mentor Michael Forsberg, for generously sharing your hard-earned knowledge. Alison Holloran and Brian Rutledge of Audubon Rockies were there from the beginning, as was LightHawk, whose outstanding pilots generously donated all of the aerial missions. Thanks to Audubon, The Wilderness Society, Wyoming Outdoor Council, Conservation Colorado, Greater Yellowstone Coalition, Sierra Club, Trust For Public Land, and the Gunnison Basin Sage-grouse Strategic Committee.

Researchers dedicate their careers to quietly studying the smallest of details. Thanks to Dr. Matt Holloran, Dr. Anna Chalfoun, Dr. Tracey Johnson, Dr. Hall Sawyer, and Reilly Renshaw Dibner.

Dave Burke, Roland McCook, Heather Stirling, Dan Smitherman, and the Hollorans allowed me to tell their personal stories here and I'm grateful. Thanks to David Allen Sibley, Todd Wilkinson, Dr. Pat Magee, and Rick Bass for essays that inspire and speak for the West.

Special thanks to visionary Braided River Publisher Helen Cherullo for patiently watching this project unfold, then bringing it to life with the amazing Braided River staff: Deb Easter, Janet Kimball, Kris Fulsaas, Lace Thornberg, Jill Eikenhorst, Kate Rogers, Emily White, and Doug Canfield.

Finally, special thanks to my mom, LaVerne Showalter, and dad, Guy Showalter, who first showed me the West.

SAGE SPIRIT PARTNERS

Audubon ROCKIES

At Audubon, we work with a hundred-year vision. We ask "what if?" What if we all thought about what we want our world to look like in one hundred years? When we take a longer view, we begin to see it is not about one bird or plant or species; rather, it's about the condition of ecosystems, landscapes, the entire world, and it's about how our actions are helping or hindering the sustained conservation of all species—including ourselves. We may look to a single species, such as the greater sage-grouse, to give us insight into the health of an ecosystem, but our focus is on using the information gleaned from certain populations to inform conservation goals for the ecosystem as a whole. In the case of the sagebrush steppe, in one hundred years we want a world with thriving greater sage-grouse populations as well as thriving populations of the myriad species reliant on the same habitats on which greater sage-grouse depend.

That is one of the reasons Audubon is a supporter of Braided River and Dave Showalter's *Sage Spirit* effort. Dave's beautiful photos tell a story that no words can express. Thus, *Sage Spirit* has the ability to reach out to audiences and help garner the interest and find the common ground that sometimes seems so elusive. It reflects our holistic approach of working with diverse partners to ensure that science informs policy decisions and that our communities are knowledgeable about the importance of such a unique ecosystem.

Ultimately, Audubon wants to be able to look back and point to concrete, effective, and worthwhile conservation efforts that say to future generations: We had the foresight to consider the world you inherited. We asked "what if?" and we answered with action.

For more information about Audubon's efforts on the sagebrush ecosystem, visit http://rockies.audubon.org/sagebrush-ecosystem-initiative.

SAGE SPIRIT PARTNERS

For 123 years, the Sierra Club has worked to safeguard wild places and the wildlife they sustain. Our commitment continues today in our work to preserve natural landscapes of the Intermountain West. Sagebrush landscapes, stretching as far as the eye can see, support centuries-old pronghorn, mule deer, and elk migration routes, many of which we only recently have recognized and hardly understand. These high-elevation steppes are known as habitat for the sage-grouse, but other birds, large and small mammals, amphibians, and reptiles also depend on the unique characteristics of the sagebrush ecosystem, and on our protection of that ecosystem. We are proud to join with Dave Showalter and Braided River on *Sage Spirit* in working to preserve the sagebrush landscapes that are the essence of the West.

For more information about what we can accomplish together, please visit www.sierraclub.org/wyoming.

The Sierra Club Wyoming Chapter dedicates our support of *Sage Spirit* to the memory of our friend and colleague, Steve Jones. Steve loved these landscapes heart and soul, and he spent his life exploring, enjoying, and working to protect these irreplaceable parts of Wyoming and the Intermountain West.

Wildlife are the icons of America's great landscapes. Like bison for the Great Plains, grizzly bears for the Northern Rockies, and caribou for Alaska, the sage-grouse has emerged as a symbol of the West—with sagebrush habitat being some of the wildest public lands in the American West.

The Wilderness Society works to protect large connected landscapes that are key habitat for wildlife like the sage-grouse in the face of expanding energy development and a changing climate. By focusing on the sage-grouse, we find that we protect habitat for a broad range of species, including prized big game like mule deer and elk. The sage-grouse and other wildlife found in the sagebrush habitat are interwoven in the fabric of the West. Unfortunately, habitat for the sage-grouse is rapidly declining, threatening the bird itself, in addition to all of the wildlife that call the area home.

The threats to sage-grouse are emerging in the center of the debate over how these wild sagebrush ecosystems will be managed and how development and conservation will be balanced in the future to best protect land and wildlife.

As the debate heats up, Dave Showalter's book is as timely as it is important. *Sage Spirit* brilliantly illustrates the incredible value of these wild places—an important reminder as we debate the future of our communities and wildlands. We know that once people see Dave's stunning photos and understand the wildlife and ecological richness of these lands, they will be compelled to help protect these places for future generations.

BRAIDED RIVER

BRAIDED RIVER®, the conservation imprint of Mountaineers Books, combines photography and writing to bring a fresh perspective to key environmental issues facing western North America's wildest places. Our books reach beyond the printed page as we take these distinctive voices and vision to a wider audience through lectures, exhibits, and multimedia events. Our goal is to build public support for wilderness preservation campaigns and inspire public action. This work is made possible through the book sales and contributions made to Braided River, a 501(c)(3) nonprofit organization. Please visit BraidedRiver.org for more information on events, exhibits, speakers, and how to contribute to this work.

Braided River books may be purchased for corporate, educational, or other promotional sales. For special discounts and information, contact our sales department at 800.553.4453 or mbooks@mountaineersbooks.org.

THE MOUNTAINEERS, founded in 1906, is a nonprofit outdoor activity and conservation organization, whose mission is "to explore, study, preserve, and enjoy the natural beauty of the outdoors . . ." Mountaineers Books supports this mission by publishing travel and natural history guides, instructional texts, and works on conservation and history.

Send or call for our catalog of more than 600 outdoor titles:
Mountaineers Books
1001 SW Klickitat Way, Suite 201
Seattle, WA 98134
800.553.4453
www.mountaineersbooks.org
Manufactured in China on FSC®-certified paper, using soy-based ink.

FSC www.fsc.org MIX Paper from responsible sources FSC® C008047

Library of Congress Cataloging-in-Publication Data
Showalter, Dave (Photographer)
Sage spirit : the American west at a crossroads / Dave Showalter ; foreword by David Allen Sibley. — First edition.
pages cm
Includes bibliographical references.
ISBN 978-1-59485-916-8 (paperback)
1. Sagebrush—West (U.S.) 2. Sagebrush—Ecology—West (U.S.) 3. Biotic communities—West (U.S.) 4. Corridors (Ecology)—West (U.S.) I. Title.
QK495.C74S44 2015
577.4'40978—dc23
2015002264

First edition, 2015

Publisher: Helen Cherullo
Acquisitions and Developmental Editor: Deb Easter
Project Manager and Developmental Editor: Janet Kimball
Developmental Editor: Linda Gunnarson
Content and Copy Editor: Kris Fulsaas
Cover and Book Designer: Jane Jeszeck/Jigsaw, www.jigsawseattle.com
Braided River Development and Communications: Lace Thornberg and Jill Eikenhorst
Cartographer: R. Jay Flaming, Big Water Consulting
Scientific Advisor: Patrick Magee
Photo credits: Page 39, top, courtesy of the Sweetwater County Historical Museum, Wyoming; Page 39, bottom, courtesy of the Denver Public Library, Western History Collection, call number X-30698; Page 144 by Mark McKinstry; Page 103 by Chris Boyer
Map credits: Page 6: US Geological Survey and Colorado Cartographics, Ltd.; Page 7: US Information and Energy Administration; Doherty, Kevin E., Jason D. Tack, Jeffrey S. Evans, and David E. Naugle. "Mapping breeding densities of greater sage-grouse." BLM report no. L10PG00911 (2010). Greater sage-grouse critical habitat information provided by James Lindstrom, Cartographer, US Fish & Wildlife Service; Gunnison sage-grouse data from USF&W/Colorado Ecological Services Field Office; Page 35: SageMap, Dr. Michael A. Schroeder, Washington Department of Fish and Wildlife, February, 2002. http://sagemap.wr.usgs.gov/; Page 103: US Geological Survey and Colorado Cartographics, Ltd.
Front cover: *A greater sage-grouse male displays on a lek (Sublette County, Wyoming).*
Back cover: *American bison graze on Antelope Flats, Grand Teton National Park, Wyoming.*
Frontispiece: *Scattered feathers are commonly seen on sage-grouse leks, signs of an attack by a golden eagle or coyote (San Miguel County, Colorado).*
Title page: *Clark's Fork of the Yellowstone River, Wyoming*
Contents page: *Greater sage-grouse arrive at the lek well before dawn to display for females, undaunted by a spring snowstorm (Sublette County, Wyoming).*
Page 176: *A Gunnison sage-grouse male displays broad white bands on tail feathers and thick filo plumes on top of the head, often called a ponytail—two primary features that distinguish the species from greater sage-grouse (Gunnison Basin, Colorado).*

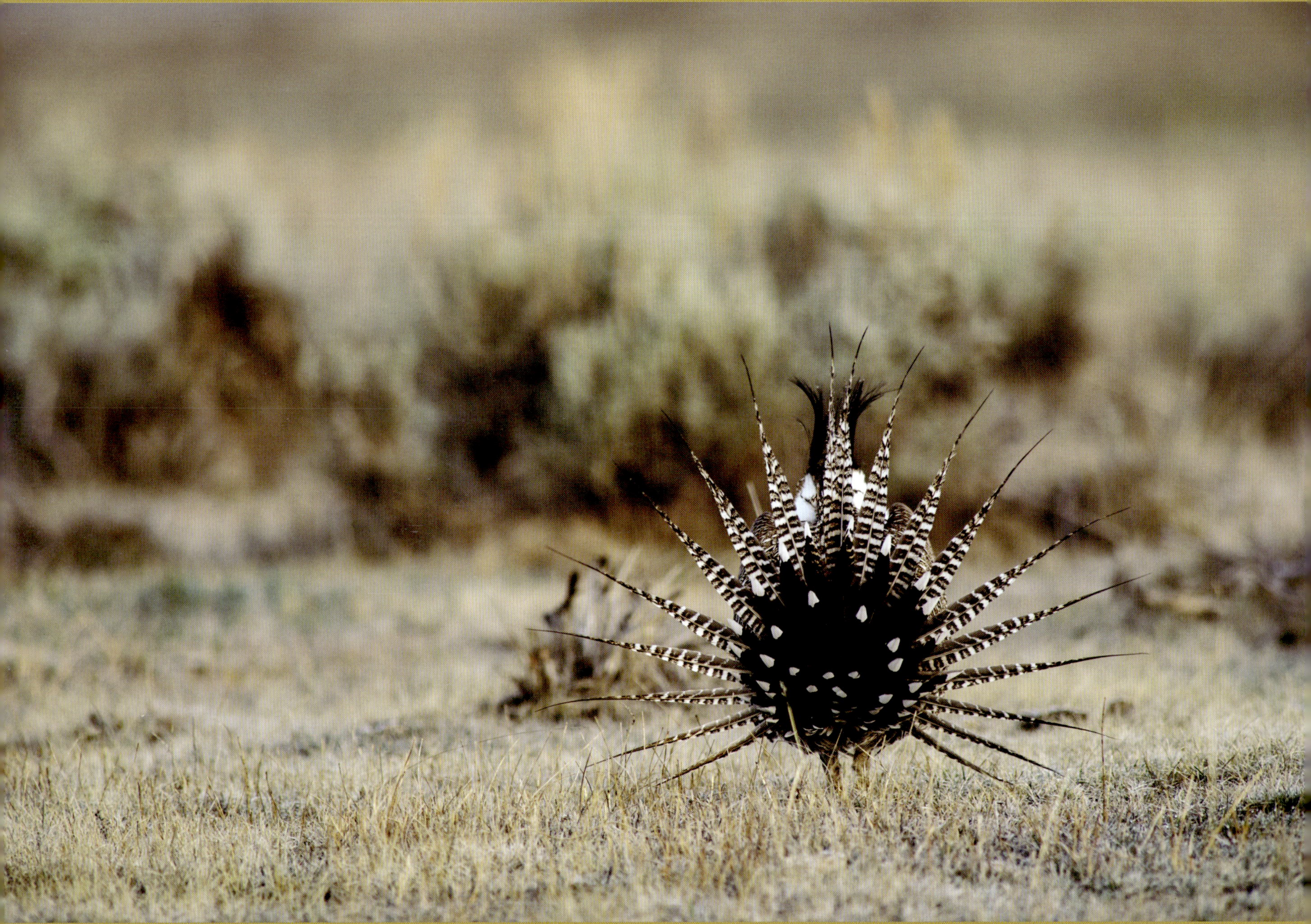